Sir Charles Bent Ball

The Rectum and Anus, their Diseas and Treatment

Sir Charles Bent Ball

The Rectum and Anus, their Diseas and Treatment

ISBN/EAN: 9783744718530

Printed in Europe, USA, Canada, Australia, Japan

Cover: Foto ©berggeist007 / pixelio.de

More available books at **www.hansebooks.com**

THE
RECTUM AND ANUS

Their Diseases and Treatment

BY

CHARLES B. BALL

M.Ch. Univ. Dub. F.R.C.S.I.

SURGEON TO SIR PATRICK DUN'S AND SIMPSON'S HOSPITALS; UNIVERSITY
EXAMINER IN SURGERY; LATE MEMBER OF SURGICAL
COURT OF EXAMINERS, ROYAL COLLEGE OF SURGEONS, IRELAND

WITH 61 ILLUSTRATIONS AND 4 COLOURED PLATES

SECOND EDITION

CASSELL AND COMPANY LIMITED
LONDON PARIS & MELBOURNE
1894

ALL RIGHTS RESERVED

PREFACE.

THE voluminous literature which has, from time to time, appeared on the diseases of the lower bowel, renders it necessary that some explanation should be given why an addition is made to the already long list.

Improvements in wound treatment within the past few years have produced such important changes in the practice of surgery generally, that in almost all branches of the subject it has become necessary to modify hitherto expressed views, and recast our teaching in several essential particulars. To this general rule the surgery of the rectum forms no exception; and if in the following pages the busy practitioner succeeds in finding a fair account of our knowledge of this subject, as it exists at the present date, the aim of the author will have been accomplished.

One of the most pleasing duties in connection with a work of this kind, is the expression of indebtedness to professional friends for valuable assistance. To the medical officers of the Richmond Hospital, Dublin, I am much indebted for allowing me to make use of their invaluable collection of original drawings :

from this source several of the chromo-lithographs and wood-cuts have been obtained.

I would also acknowledge with thanks the permission accorded to me to make use of specimens in the Hunterian Museum; the Museums of Trinity College, Dublin; the Royal College of Surgeons in Ireland, and of King's College, University College, Middlesex, and Steeven's Hospitals.

From Professor Purser and Dr. P. S. Abraham, I have received much assistance in the pathological part of the work.

To my colleagues, Dr. T. E. Little, and Dr. W. G. Smith, I am greatly indebted for the interest which they have taken in the work, and the care with which they supervised the pages as they issued from the Press.

C. B. BALL.

September, 1887.

PREFACE TO SECOND EDITION.

During the six years that have elapsed since the publication of the first edition of this work, the progress which has been made in Rectal Surgery has been at least equal to that which has characterised the modern surgery of other regions of the body. It has, therefore, been found necessary to revise thoroughly and rewrite largely the present edition.

Amongst the more important alterations will be found what I believe to be a more accurate pathology of painful fissure, with a simplified treatment based thereon; excision of piles has largely taken the place of the older and less radical procedures, so that the subject has been more amply discussed; while fuller information will also be found concerning the treatment of simple stricture, and prolapse, by means of excision. In the treatment of cancer situated high up by trans-sacral incision much real progress has also been made.

Several of the illustrations have been replaced, while many new ones have been introduced.

To my colleague, Professor W. G. Smith, my best thanks are again due for revising the proof sheets.

I can only express the hope that the efforts I have made to bring the work up to date will render it deserving of as cordial a reception as that which the first edition received.

C. B. BALL.

24, *Merrion Square, Dublin.*
January, 1894.

CONTENTS.

CHAPTER	PAGE
I.—The Diagnosis of Rectal Disease	1
II.—The Congenital Malformations of the Rectum and Anus	20
III.—Proctitis	49
IV.—Periproctitis	57
V.—Rectal Fistulæ	69
VI.—The Relations of Pulmonary Phthisis to Rectal Fistula	108
VII.—Ulceration of the Rectum	113
VIII.—Irritable Ulcer, or Fissure of the Anus	132
IX.—Non-Malignant Stricture of the Rectum and Anus	148
X.—Symptoms of Non-malignant Stricture of the Rectum	165
XI.—Treatment of Non-malignant Stricture of Rectum	180
XII.—Syphilis of the Rectum and Anus	189
XIII.—Prolapsus Recti	198
XIV.—The Ætiology and Pathology of Piles	228
XV.—External Piles	239
XVI.—Symptoms of Internal Piles	247
XVII.—The Clinical Relations and Complications of Internal Piles	251
XVIII.—Palliative Treatment of Internal Piles	261
XIX.—Operative Treatment of Internal Piles	268

CHAPTER	PAGE
XX.—Benign Neoplasms of the Rectum and Anus	296
XXI.—Malignant Neoplasms of the Rectum and Anus.	321
XXII.—The Pathology of Malignant Neoplasms of the Rectum and Anus	326
XXIII.—Symptoms of Rectal Cancer	350
XXIV.—Treatment of Rectal Cancer	358
XXV.—Colotomy	380
XXVI.—Pruritus Ani	402
XXVII.—Atony of the Rectum	409
XXVIII.—Irritable Rectum and Neuralgia	415
XXIX.—Injuries of Rectum and Anus	417
XXX.—Foreign Bodies in the Rectum	421
XXXI.—Diverticula of the Rectum	427
Index	433

LIST OF ILLUSTRATIONS.

	PAGE
Longitudinal Section of Male Pelvis . . . . . *Facing page*	1
Hegar's Retractor	9
Allingham's Four-bladed Speculum	10
Clover's Crutch	11
Olive-headed Flexible Bougie	15
Second Variety of Congenital Malformation	26
Third Variety of Congenital Malformation	27
Fourth Variety of Congenital Malformation	33
Congenital Malformation (Amussat's case)	34
Congenital Malformation (Cruveilhier)	38
Congenital Malformation (*atresia ani vesicalis*)	42
Congenital Malformation (*atresia ani urethralis*)	44
Congenital Malformation (*atresia ani vaginalis*)	45
Vulvar Anus in Child aged 19 months	46
Brodie's Probe-pointed Fistula Director.	87
Allingham's Instrument for the Introduction of Elastic Ligature	98
Allingham's Screw Ointment-Introducer	117
Ulcers originating in the Sinuses of Morgagni . . .	120
Follicular Ulceration of Rectum	122
Tuberculosis of the large Intestine	124

	PAGE
Section of the Colon from a Case of Dysentery	128
Nervous Supply of Anus	133
Diagram of the Nervous Relations of Irritable Ulcer of the Anus	135
Pathogeny of Anal Fissure	136
Non-malignant Stricture of the Rectum.	161
Diagram illustrating Dieffenbach's Procto-plastic Operation	188
Partial Prolapsus Recti	201
Complete Prolapsus Recti	202
Complete Prolapsus in a Child	204
Pessary for Prolapsus Recti	212
Paquelin's Thermo-cautery	215
Section of Intero-external Pile	237
Benham's Pile-crusher	270
Coates' Clamp for Excision of Piles	281
Lee's Pile Clamp	287
Adenomatous Rectal Polypus	297
Double Rectal Polypi producing Prolapsus Recti	298
Multiple Polypi of Rectum	299
Multiple Polypi of Rectum	300
Adenoma of Rectum	302
Section of Adenomatous Polypus	303
Section of Adenomatous Polypus	305
Fibrous Polypus of Rectum	309
Papilloma of Rectum	311
Anal Papilloma	312
Papilloma of Anus	313
Angioma of Rectum	318
Section of Angioma of Rectum, showing cavernous Structure	319
Lymphoma of Rectum	320

LIST OF ILLUSTRATIONS.

	PAGE
Margin of Cancerous Nodule	328
Cylinder-celled Epithelioma of Rectum	334
Adenoma of Rectum, from Case of Rectal Cancer	336
Colloid of Rectum, showing perforation of Muscularis Mucosæ by new growth	339
Case of Colloid Cancer of Rectum, natural size, removed by trans-sacral Incision	340
Large Sarcomatous Tumour of Anus and lower part of Rectum, with secondary Tumours on the inside of the Thigh	341
Sarcomatous Infiltration of Rectum, producing long tubular Stricture	342
Melanotic Sarcoma of Rectum	347
Diagram showing the method of passing Sutures	360
Author's method of performing Laparo-colotomy in the left Linea semilunaris with fixation of the gut by leaded Button Suture through the Meso-colon	395
Villous Condition of Mucous Membrane	400
False Diverticula of Rectum	430

LIST OF COLOURED PLATES.

PLATE I. . . . . *Frontispiece.*
    Epithelioma Ani with complete Rectal Fistula

PLATE II. . . . . *To face page* 123.
    Fig. 1.—Lupoid Ulceration of Rectum.
    Fig. 2.—Hæmorrhoidal Ulceration of Rectum.

PLATE III. . . . . *To face page* 192.
    Fig. 1.—Eczema of Anus.
    Fig. 2.—Condylomata of Anus.

PLATE IV. . . . . *To face page* 256
    Fig. 1.—Prolapsed and Gangrenous Internal Piles.
    Fig. 2.—Prolapsed Internal Piles.

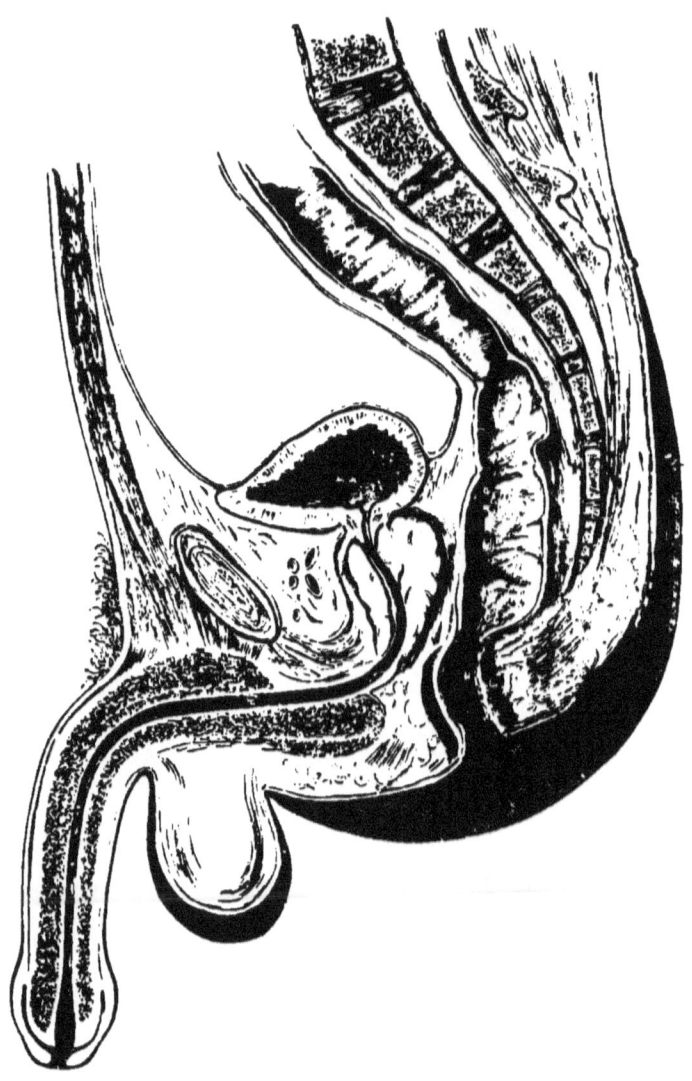

LONGITUDINAL SECTION OF MALE PELVIS.
(Drawn to scale from the average of three frozen sections.)
The relations of the anal canal and rectum to the other pelvic structures are well shown

THE RECTUM AND ANUS:
THEIR DISEASES AND TREATMENT.

CHAPTER I.

THE DIAGNOSIS OF RECTAL DISEASE.

THE early diagnosis of rectal disease is attended in some instances with difficulty, partly from the fact that patients suffering from these diseases do not seek the advice of the surgeon until they have suffered for some time, when the severity of the pain, considerable loss of blood, or great difficulty in defæcation, overcomes the repugnance to local examination. In other cases, more especially in commencing malignant disease, where early diagnosis is of such paramount importance, the subjective phenomena may not in the first instance point to the rectum as the seat of lesion..

It is essential, therefore, that the surgeon should pay particular attention to those symptoms which, although not directly pointing to disease of the lower bowel, are frequently caused by such condition; and when the statements of the patient render it probable that rectal disease exists, he should insist on a complete local examination. Amongst the more vague symptoms which should arouse the suspicions of the medical attendant we may enumerate the following: Slight morning diarrhœa, which continues for a long time, alternating with attacks of constipation; flatulent dyspepsia; sense of weight about the pelvis; dull pain about the sacrum, with pain or œdema of the left leg. This last symptom is one to which Hilton has drawn particular attention.* Progressive anæmia may be the result of abundant and frequently repeated hæmorrhage from the rectum, which sometimes continues for a considerable time before it is noticed by the patient, the loss occurring painlessly at time of defæcation. And, again, owing to the close sympathy between the uterus and adjacent mucous tract, disease of the latter may lead to the erroneous impression that the cause of the patient's suffering is to be found in the organs of generation. I have lately had under my care a woman who had been for months under treatment for uterine disease without benefit, and whose symptoms were at once and permanently removed by the division of the base of a small rectal ulcer. Similarly we sometimes find that irritability of the bladder is the symptom to which the patient directs special attention.

In order to elucidate the symptomatology as fully as possible, it is well, in the first place, to let the patient describe his case in his own words, then, by a few well-directed but not leading questions, we may be able to complete the history. The questions to be asked may be usefully directed in reference to the following principal points.

Pain.—Inquire if the pain is severe or not? Whether it is situated in the rectum, or is complained of elsewhere? What is its relation to defæcation? Is it worse during evacuation, shortly afterwards, or is it independent of the act? Is itching, sense of fulness, throbbing, or heat, complained of?

Protrusion.—Is there any swelling or protrusion at the anus? Does this occur only at defæcation, or does it appear at irregular times? Does it

* "Rest and Pain," p. 283. Third edition.

disappear spontaneously, can the patient return it, or is it constantly present?

Discharge.—Bloody, mucous, or purulent? Fœtid? Mixed with fæces? Occurring before or immediately after defæcation, or independent of the act?

Fæces.—Diarrhœa, or constipation? Consistence? Of normal size? Tape-like, or lumpy?

Having now clearly obtained the subjective phenomena, we should, if there is a suspicion even of rectal disease, insist on a complete local examination. Under no circumstances is the attendant justified in prescribing without careful examination, although often he will be asked by his patient to prescribe something for "the piles." I have seen a case of extensive malignant disease of the rectum, which had passed all hope of useful treatment under the care of a medical man, who had rested satisfied with the statement of the sufferer that she was suffering from piles, and who never had made an examination.

Of some diseases the symptomatology is tolerably diagnostic; as, for instance, if the patient has severe pain continuing for some hours after defæcation, if the motions are small and tape-like, and if they are streaked on one side with bloody mucus, there is a very strong probability that there is a painful fissure present; but without examination we cannot possibly say that the fissure is the *only* rectal disease; indeed, in a large proportion of cases we find more than one pathological condition present.

Examination.—If possible the patient should have the bowel emptied by an enema immediately beforehand. In some cases this is absolutely essential; as, for instance, where it is necessary to use a speculum for the exploration of the higher portions of the mucous membrane.

Position.—By far the most convenient position for ordinary examination is the semi-prone of Marion

Sims. The patient lies on a rather high couch on the left side, the right shoulder turned away from the surgeon, the left arm brought backwards from under the body, and the right thigh well flexed upon the abdomen. By separating the buttocks we obtain a good view of the anus and surrounding skin; pathological changes in which are easily noticed. By gently feeling the anal margin, we may be enabled to detect deep-seated hardness, which may be due to the presence of a fistula; tenderness at any part of the circumference will serve to direct our attention to that portion more particularly. By drawing open the anus and directing the patient to bear down, a good view will be obtained of the muco-cutaneous junction; the condition of sphincter as to laxity can be observed, and any discharge noted both as to quality and quantity: a full view of external and sometimes of internal piles being also thus obtained. By separating the radiating folds of the anus, we may be enabled to see the commencement of fissure, a small external pile frequently serving to direct our attention to this lesion. The presence of fistulous openings and the more obvious anal diseases, together with the existence of oxyurides, may also be determined. A digital examination should now be instituted, and it is by this means that the most important information is to be obtained, the educated finger being able to recognise with certainty almost all varieties of rectal disease. In order, however, that the fullest information may be obtained, it is evident that familiarity with the normal parts is essential. In a digital examination of the normal rectum the following points may readily be made out, and the student will do well to appreciate thoroughly all the details in every examination he makes, as it is by these means that the familiarity with parts necessary to rapid and accurate diagnosis can alone be acquired.

1. In the male (*see* Plate facing page 1). As the finger enters from the anal canal into the rectum proper a ridge can be felt; immediately above this some of the little anal valves, if well developed, and behind them the rectal pouches; passing the finger round posteriorly, first the coccyx and lower part of the sacrum can be made out, with the sacro-sciatic ligaments spreading out laterally; the spine of the ischium on each side can also be felt, while lymphatic glands, if enlarged, can be recognised in the hollow of the sacrum. Laterally the condition of the ischio-rectal fossæ and the bony wall of the pelvis terminating in the tuberosity of the ischium on each side can be determined. In front, the prostate gland is easily appreciated, and below and in front is a pouch of the rectum, from which the urethra as it passes through the triangular ligament can be felt. A finger in this pouch is often of use to the surgeon in guiding the point of a catheter through the prostatic urethra, and also in recognising whether the instrument has entered a false passage. The posterior surface of the prostate can be fully explored, and above and behind the prostate the bulging wall of the bladder can be felt, with possibly the vesiculæ seminales on each side.

2. In the female. On the posterior and lateral aspect the relations of the bowel are similar to those in the male, except in so far as they are modified by the greater capacity of the female pelvis. In front the cervix and os uteri can easily be felt through the recto-vaginal septum, while higher up the body and fundus of the uterus can be explored; while occasionally, if displaced, one or both ovaries can be felt. It is astonishing how frequently mistakes are made by practitioners who have failed to render themselves familiar with the normal arrangement of the pelvic viscera. I have frequently known the fundus uteri diagnosed as a pathological tumour pressing on the bowel, and a rugged

and indurated os uteri felt *through* the recto-vaginal septum mistaken for the lower opening of a strictured rectum.

In directing his investigation the surgeon should bear distinctly in mind what he is about to look for, and prosecute his examination in a definite and systematic manner. Having filled the nail with soap, he should cover the finger well with some stiff lubricant: ordinary oil is not satisfactory. This may appear a small matter, but it makes the difference between little or considerable pain to the patient, and, as a result, passiveness or resistance to the examination. The thymol, or eucalyptol jelly, sold in collapsible metal tubes for obstetric purposes, answers the purpose admirably; and, moreover, has the advantages of ready portability and asepticism.

The finger should now be introduced by a gradual boring motion, with a direction, at first, slightly forwards. This should be carried out slowly, so as to give the sphincters time to relax; if attempted suddenly, spasm will to a certainty be induced. As the finger enters, the condition of the sphincter may be noted; *i.e.* whether it is relaxed, normal, or spasmodically contracted. A firm, long-continued resistance is very characteristic of spasm, the result of chronic irritation. The finger should now be steadily passed up to the fullest extent, and by telling the patient to bear down forcibly, the rectum can be explored for a considerable distance. Malignant infiltration, or stricture, can be detected, if situated within reach. By sweeping the finger round the mucous membrane its condition can be noted; a general smoothness, and absence of the normal folds indicating atony. Ulceration may be recognised; and the attachment of polypi can be felt. In examination for this disease it is of importance that the investigation should be conducted as directed, from above downwards; as, otherwise, the tumour may be pushed up out of reach, the pedicle, in these cases, often being of considerable length. Fæcal accumulation in the rectal pouch is, of course, recognised without difficulty; and the condition of the surrounding contents of the pelvis should also be noted. The finger is now partly withdrawn, passing the pulp round the entire circumference of the mucous membrane, as this is done the internal openings of fistulæ and ulceration being carefully felt for. As the margin is approached internal piles may be perceived, but the fact should always be remembered that internal hæmorrhoids, unless they have been previously thickened by inflammation, are extremely hard to recognise by the touch ; indeed, the surgeon is more likely to be deceived by the sensation conveyed to the finger by internal piles than by any other rectal disease. I have frequently seen cases in which these growths, although scarcely appreciable to the touch, were found, upon ocular examination after dilatation of the sphincter, to be of considerable size. This is, no doubt, due to the fact that they are so soft and movable that they resemble closely in feel the normal columns and folds of the mucous membrane. Where, however, they have been thickened, either by present inflammation or by the growth of connective tissue in them, no difficulty can be experienced in their detection.

Immediately inside the anus we may be able to feel the upper portion of a painful ulcer ; and can sometimes recognise the fact that the extreme sensitiveness is confined to one portion of the surface. In some cases, complete examination is facilitated by having the rectum distended with warm water, by means of an enema ; while in other cases, when disease is situated high up, additional information can be gained by making the patient stand up and directing him to make an expulsive effort.

Having now completed our digital examination we can usually come to a fair conclusion as to the nature of the case. Should an examination by the speculum be deemed necessary, I am decidedly of opinion that it should be prosecuted under the influence of an anæsthetic, as without the assistance afforded to us by this means but little additional useful information can be obtained. The great number of specula which have been introduced is, I think, a strong indication of the difficulty that has been experienced in their use. If we attempt to use one without anæsthesia we find that the patient suffers considerable pain, he strains violently, the rectum contracts round the speculum, the mucous membrane prolapses into the orifice, and but little can be seen.

If, therefore, further examination is necessary; or if, as a result of our digital examination, we come to the conclusion that operative measures may be required, it is advisable to give an anæsthetic, preferably æther; having previously got the consent of the patient to complete at the same time any necessary operation. Anæsthesia having been produced, and the sphincter completely dilated, we can, without difficulty, institute an examination with a speculum; by far the best is some modification of the duckbill of Marion Sims, such as recommended by Van Buren, which has a notch to receive the sphincter, and a handle by which it can be held well out of the operator's way.* Or the retractors of Hegar, which are used by obstetricians, answer the purpose admirably (Fig. 1). For a complete view of the entire circumference of the mucous membrane of the bowel for a distance of a few inches from the anus, Allingham's four-bladed speculum will be found useful (Fig. 2). It is to be introduced closed, and gradually expanded to the desired extent, and then fixed by turning the

* "Diseases of Rectum," p. 392.

thumb-screw. It has the great advantage of being self-retaining, thus leaving both hands free for any necessary manipulations.

Artificial light may sometimes be required in making these examinations, and for this purpose the little electric lights, now commonly sold by instrument makers, worked by four specially-constructed Leclanché

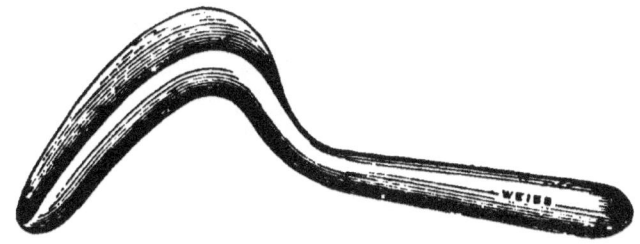

Fig. 1.—Hegar's Retractor (scale ½).

elements, answer admirably. By this means the bowel, as far as it is distended by the speculum, can be thoroughly illuminated, and the surgeon enabled to diagnose any morbid conditions of the mucous membrane that may be present. For extensive examinations by means of specula it is best to retain the patient in the lithotomy position by means of Clover's crutch (Fig. 3). In using this instrument it is convenient to pass the strap round the end of the couch instead of round the patient's neck. Interference with respiration is thus avoided.

The tubular vaginal specula I have not found of much use in rectal surgery.

If, however, dilatation of the sphincter has been thoroughly accomplished during anæsthesia, the last two inches of the rectum can be brought sufficiently well into view without any instrument beyond the fingers or simple retractors. No single step in the practice of this department of surgery has done so much as the introducing of forcible dilatation of the sphincter as an aid to the diagnosis and treatment of rectal disease. And the credit of the establishment of this practice, as a means of diagnosis, and as a preliminary step in rectal operations, is undoubtedly due to Van Buren of New York.*

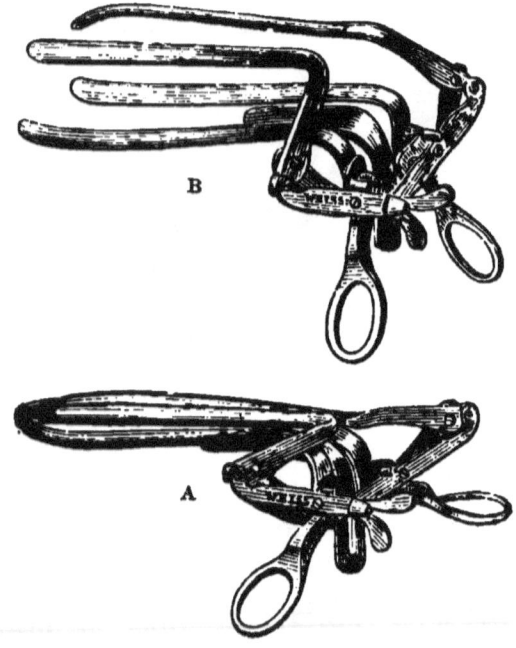

Fig. 2.—Allingham's Four-bladed Speculum (scale ½). A closed. B open.

Dilatation of the sphincter, for curative purposes, was first recommended by Recamier for cases of fissure,† but it differed in some respects from the way in which the operation is now conducted; it was more a kneading of the spasmodically contracted sphincter. This was carried out first with one finger, afterwards with two fingers, in the rectum, and the thumb outside. This

* "Diseases of Rectum," p. 207.
† Academy of Medicine of Paris, 1829.

pinching of the muscle was done at regular intervals and frequently repeated, to which he applied the term *massage cadencé*. Although it would appear that his results were at the time good, the method fell into disuse; and the objections to its adoption were further increased by the modification of the operation

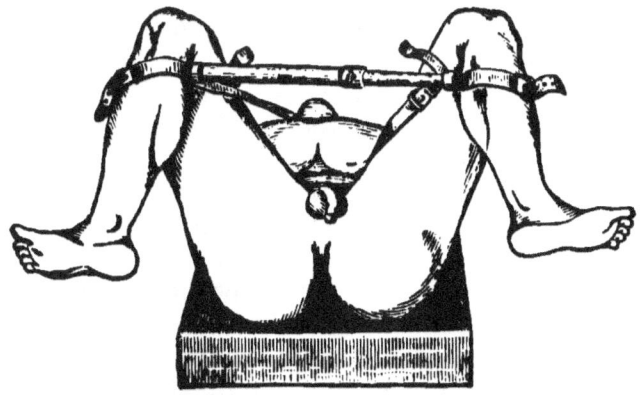

Fig. 3.—Clover's Crutch, for retaining a patient in the lithotomy position.

proposed twenty years later by Maisonneuve. He first introduced the index finger, then one by one the others, till finally the whole hand passed the sphincters, then by closing the fist and suddenly withdrawing the hand the dilatation was effected. This would appear to justify the appellation of barbarous, so frequently since lavished on the operation of dilatation; but as at present practised it is essentially a scientific and safe procedure, unattended with the extensive bruising and laceration which must of necessity have attended the method of Maisonneuve. The only argument which can be urged against it is, that it may be followed by temporary incontinence. I have, however, only seen it upon one occasion give rise to any serious trouble, and then it only continued for a few weeks. Most authors who have had large experience do not even allude to this unpleasant sequela.

The way in which this operation should be practised is as follows: First one and then the other thumb should be insinuated into the anus, the other fingers being laid flat over the buttock on each side; care should be taken to introduce the thumbs high enough to stretch the entire external sphincter; then by firmly grasping the tuber ischii on each side, dilatation of the anus should be gradually but firmly proceeded with, several minutes being occupied in the process; the muscle will be found by degrees to give way, and after a time the sense of resistance will disappear. In order to effect this a very considerable amount of force is necessary. It is well also to adopt the plan advocated by Allingham,* of dilating as well in the antero-posterior direction. The resistance offered by the muscle is soon at an end, and we can examine the interior of the bowel without further trouble, and perform whatever operation is required with ease and certainty. The mucous membrane is occasionally slightly torn, and ecchymosis exists about the anus for a few days; but no trouble is to be apprehended from either. The partial paresis of the sphincter, if the operation has been fully carried out, usually lasts for about a week. This is a point of very great importance in the treatment of operation cases, as the physiological rest so afforded to the part tends, in a very marked manner, to favour healing. And, moreover, the after-pain is minimised, the chief suffering being caused by the pinching of the wounded and tender parts by the sphincter muscle; this, of course, is absent where hyper-distension has been efficiently performed. Retention of urine, which formerly was a frequent source of annoyance after these operations, is now less frequently seen.

* "Diseases of Rectum," p. 128. Fourth edition.

The explanation of the atony of muscle produced by stretching is a point of considerable interest; we find many examples of this condition in human pathology, the most familiar being the atony induced in the bladder by a hyper-distension with urine; and we know that in this case, after a catheter has been used to relieve the accumulation, the muscular fibres of the vesical wall do not at once recover their tone. Similarly we have further examples in the intestinal tube caused by flatulence and fæcal accumulation, and in the gravid uterus from hyper-secretion of liquor amnii.

What is the immediate cause of this temporary paresis? is it due to alterations in the muscular fibres themselves, or is it due to changes in the nerves supplying them? Probably the result is attributable to changes in both of these structures. That altered innervation is present we know by the fact that then we have diminished sensibility, and in cases where the trunks of large nerves are stretched forcibly for therapeutic purposes we frequently find a more or less complete paralysis of the muscles supplied. Here it is, of course, evident that no injury has been inflicted on the muscular fibres themselves.

In the female, if a suspicion exists of disease in the vagina or uterus, the surgeon should institute such examination as will satisfy him of its nature. Dr. Horatio Storer, of Boston, has recommended the eversion of the rectal mucous membrane through the anus by means of the fingers passed into the vagina, and pressing on the recto-vaginal septum; this method is, however, comparatively useless, as only a small portion of the anterior wall can by this means be brought into view. Possibly in some cases the small perinæal pile, which is tolerably common in women, can be everted and its attachment made out by this means.

The diagnosis of disease which is situated beyond the reach of the finger is attended with very considerable difficulty; but fortunately by far the greater number of cases of disease which present themselves to us are situated within three or four inches of the anal margin, and are consequently well within reach.

So difficult is it to diagnose disease which is situated high up, that Syme wrote "there is good reason to suspect the honesty of a man who pretends to enter a stricture which is beyond the reach of the finger."

By directing a patient to stand up, and at the same time to make a violent expulsive effort, at least an inch more of the canal can be felt than by the ordinary examination; and sometimes, with one finger in the rectum and the other hand making deep pressure above the brim of the pelvis, as in the bi-manual examination adopted by obstetricians, additional information may be gained.

Bougies are instruments to which recourse is had when we wish to estimate the calibre of the upper portion of the rectum or end of the sigmoid flexure, and a great variety have been introduced; the one in common use, made of gum-elastic and rather stiff, is undoubtedly the worst form, and its use is attended with considerable danger, perforation of the gut having frequently followed its introduction, the low degree of sensibility of the upper portions of mucous membrane favouring this accident; indeed in many recorded cases the perforation has been occasioned by the patient passing a bougie himself.

The best instrument of the kind is the *bougie à boule*, in which the top is made of ivory or hard rubber, and mounted on a very flexible thin whalebone stem (Fig. 4); by means of the finger it can be guided up the rectum and directed through a stricture. Allingham recommends pewter stems. Or the bougie recommended by Kelsey* may be used, which is made of soft red rubber, similar to the flexible catheter. This instrument is so pliable that injury of the rectal wall would be with it almost an impossibility; and, as it is a perforated tube, it is possible to inject tepid water through it should the point become engaged in any of the folds of mucous membrane. By this means the obstructing portion is lifted away from the end of the bougie, and progress can again be made. Goodsall† has recommended the use of conical soft rubber tubes filled with small shot. They make excellent bougies, and they are so soft and flexible that injury to the rectal wall is almost an impossibility with them; they can, moreover, be kept in place for many hours without much discomfort. The introduction of any of these instruments is one of considerable difficulty, and it requires quite as much practice as the passage of an urethral catheter; the danger of perforation is as great, and the result of such an accident very much more serious. In the one case a fatal peritonitis will in all probability be started, while in the other a false passage will probably be the result. At best, however, their instruments as means of diagnosis are of no very extended value, and it is only in occasional cases that we can get absolute evidence of a stricture high up by this means—namely, when we feel the point of the instrument grasped and held by the constricted intestine. This of course is

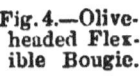

Fig. 4.—Olive-headed Flexible Bougie.

* "Diseases of the Rectum and Anus," p. 56.
† *Lancet*, June 23, 1888.

only possible where the anus has been fully distended. With the olive-headed bougie previously recommended, we may in rare cases be enabled to get important evidence of the length of a stricture and to define its limits. When the bulb at the end has passed through the stricture, its end can be felt free in the intestine above; and as it is withdrawn it comes suddenly down on the upper termination of the constriction, and so indicates the distance from the anus.

The *colonoscope* of Bodenhamer, or the *endoscope*, is of no practical utility, owing to the very limited portion of mucous membrane which is brought into view.

The only other means of exploring the higher portions of the rectum is by the introduction of the whole hand, and this is a proceeding which within the last few years has been more frequently adopted, owing chiefly to the writings of the late Professor Simon, of Heidelberg, to whom undoubtedly is due the credit of first having introduced the method as a means of diagnosis; although several cases are previously on record of the hand having been introduced into the rectum for the purpose of removing foreign bodies.

This operation, which obviously is one of considerable severity, must not lightly be undertaken; but in properly selected cases is a most valuable aid to diagnosis and treatment. The conditions which render this proceeding advisable may be classed under two heads: first, those in which disease is thought to be present in the upper part of the rectum and sigmoid flexure; and secondly, those in which it is adopted as a means of diagnosing diseases of the other pelvic and abdominal organs, and it is probably in the latter cases that the best results have hitherto been obtained. The diagnosis of stricture at the upper portion of the rectum or termination of the sigmoid flexure is not easily made, even with the hand in the rectal pouch, the muscular intestinal wall contracting so strongly on the hand, and the folds into which the normal mucous membrane is thrown serving to obscure the diagnosis. Walsham* records cases in which the diagnosis was not made by this means, where post-mortem examination proved the presence or stricture of the sigmoid flexure. The directions given by Simon are as follows.†

As a preparatory measure the entire large intestine should be thoroughly cleared out. This is best done by the hydrostatic method which is so strongly recommended by Hegar, and is accomplished as follows: The patient is placed in the knee-elbow position, so that gravity may assist the operation, and a long flexible tube is introduced into the rectum. To this is attached one end of a flexible rubber tube two feet long, and a funnel is connected with the other end. The funnel is now held well above the patient's body, and warm water slowly poured into it. In this way, several quarts of water may be introduced, and the entire colon thoroughly distended. When this is subsequently evacuated, it washes out all the contained fæces. An anæsthetic should now be administered, preferably æther, to profound narcosis; the patient placed in lithotomy position, and the sphincter stretched. Two fingers should now be fully introduced, and, by degrees, the third and fourth, and finally the thumb, in form of a cone. By a gradual boring motion the knuckles can now be made to pass the sphincter, steady pressure being made with the left hand above the pubes; several minutes being occupied in this procedure. In order to facilitate the introduction, incision is sometimes necessary. When the hand

---

* St Bartholomew's Hospital Reports, vol. xii. p. 223; 1876: "Some remarks on the introduction of the whole hand into the rectum."

† Langenbeck, *Archiv f. klin. Chir.*, vol. xv. p. 1; 1872.

has fully entered the rectum it should be held quiet for a short time, till spasm subsides, and then slowly pressed onwards, the fingers being alternately opened and closed, so as to dilate the intestine. Simon limits the distance to which the knuckles should be introduced as 17 to 19 cm. (7 to $7\frac{1}{2}$ inches), but the fingers may be extended well up into the sigmoid flexure. Other surgeons, however, have passed the hand higher up, Walsham, who has made a number of experiments on the dead subject, having shown that the hand may be passed into the descending colon without producing injury; but the prudent surgeon will never allow the hand to pass the rectal pouch, as the peritonæum is a comparatively inelastic membrane, and if the hand is passed entirely within that portion of the intestine which is surrounded by this membrane, grave danger will be encountered. Indeed, the cases in which this operation has been followed by fatal rupture are tolerably numerous. Weir[*] records two, Dittel[†] one, and Heslop[‡] two. In the latter cases the rupture took place through an ulcer in the neighbourhood of a stricture, and would suggest greater caution in the manual exploration for disease of the intestine itself than for the purpose of investigating the other pelvic and abdominal viscera. Of the structures felt through the intestinal wall the most important is the bladder in the male. In the female the uterus and uterine appendages can be more thoroughly explored by this than by any other method. The internal iliac vessels with their branches can also be traced. Walsham has pointed out the importance of examining cases of gluteal aneurism in this way, so that exact knowledge may be obtained as to whether the disease is confined to the exterior of

---

[*] *New York Medical Record*, March 20th, 1875.
[†] *London Medical Record*, July 15th, 1875.
[‡] *Lancet*, May 11th, 1872.

the pelvis, a plan which would manifestly determine the nature of operative proceeding. The bony wall of the pelvis can also be investigated; and with the finger in the sigmoid flexure, the lower end of the left kidney and the last rib on the same side can be explored. Some American surgeons speak of feeling the spleen and sweeping the hand over the under surface of the liver. Such procedures, if possible, must be attended with such extreme danger that I cannot conceive any circumstances that would justify a prudent surgeon in resorting to an exploration of the kind. Allingham* has frequently performed this operation, and speaks very favourably of it. He mentions one case of great interest, in which he found by this means, and completely stretched, a band of false membrane or peritonæum which was binding down the bowel as it crossed the brim of the pelvis. The obstruction was removed, and the patient recovered.

This examination can be much more readily practised on women than on men; but even in the latter, if the hand is tolerably small, and due caution be used, much information can be gained. The smaller the operator's hand is, the better; the limit assigned by Simon being a tolerably liberal one, 25 cm. (9¾ in.) greatest circumference. After the operation slight incontinence sometimes is experienced, but it seldom lasts over a few days. If an incision has been rendered necessary it is better to pass a couple of deep sutures to keep the sides of the wound in apposition, and so hasten healing.

* "Diseases of Rectum," p. 9. Fourth edition.

CHAPTER II.

THE CONGENITAL MALFORMATIONS OF THE RECTUM AND ANUS.

VIEWED either from an embryological or surgical standpoint, congenital malformations form one of the most interesting of all the subjects connected with the contents of the present volume. And although it is true that the number of such cases coming under the observation of any one surgeon will probably be small, as they are all somewhat rare, still it is essential that all medical men should be thoroughly familiar with the varieties and the treatment required, as the saving of life in some of the forms is mainly dependent upon the promptitude with which surgical aid is invoked.

In order to arrive at an understanding of the method of production of the various forms, it will be necessary for us briefly to refer to the development of the intestinal tract; it will, however, be possible here only to refer to the more important facts. For fuller details the reader must be referred to the systematic works on Embryology.

The mesenteron or central portion of the alimentary tract is formed from the hypoblast, and consists, in the first instance, of a simple tube which ends at the anterior extremity of the embryo in a blind sac, while at the posterior extremity a similar cul-de-sac is formed. This tube of hypoblast represents only what is to constitute the mucous membrane of the alimentary tract, the other coats of the intestine being subsequently formed by this hypoblastic portion becoming enveloped in a layer of mesoblast, which differentiates into two portions, the outer forming the peritonæal covering, while the inner develops into the muscular and connective tissue elements of the intestinal wall.

An invagination of epiblast at the anterior extremity of the embryo meets and communicates with the anterior portion of the mesenteron; this, which is called the stomodæum, constitutes the mouth; while a similar depression of the epiblast at the posterior extremity, called proctodæum, forms the anal orifice, and communicates with the mesenteron. It will be necessary to study this proctodæum a little more in detail. If a human embryo about the sixth week be examined, it will be found that immediately in front of the coccygeal eminence, which at that period of development is relatively very prominent, a slight elevation surrounded by a furrow is to be seen; from this eventually will be developed the anus and generative organs, but as yet they are not differentiated. At about the eighth or ninth week the anus will be separated from this cloacal opening, and the rudiments of the perinæal septum will have been formed; and in an embryo of about ten weeks, when the genital organs will have been so far developed that it is possible to determine sex, the anus also will have been fully formed and separated by a distinct septum from the structures of the anterior perinæum. The anal depression continues to deepen until it reaches the mesenteron, with which it becomes continuous, and so the patency of the extremity of the intestinal tract becomes established. In the higher vertebrates this intercommunication becomes so complete that no permanent trace of the junction is left, except a change in the character of the epithelium. According to Balfour,* a permanent fold marks the hypoblastic section of the cloaca from the proctodæum proper in birds, and in this class

* "Comparative Embryology," vol. ii. p. 641.

of vertebrates the proctodæum is further complicated by the development from it of the *bursa Fabricii*. The perinæal septum, which is such a characteristic feature of the higher mammalia, in the human subject forms a very complete barrier between the anterior and posterior perinæum. This is practically illustrated by the way in which extravasation of urine is sharply limited to the front portion of this space, while for the same reason abscess connected with the rectum seldom makes its way forward. Most of the malformations of the rectum are to be referred to arrested development or irregular growth of the proctodæum or mesenteron; and it is a remarkable fact how frequently, when any of these malformations are present, there is also a failure of development of the perinæal septum, and a tendency for abnormal communication to occur between the intestinal tube and the genito-urinary system. This possibly is due to the fact that in early fœtal life the proctodæum is portion of the genito-urinary tract, from which it subsequently becomes differentiated.

The proportion of children born with congenital defects of the termination of the intestine is small. The statistics of Zöhrer at the Vienna Lying-in Hospital, and of Collins at the Dublin Lying-in Hospital, jointly reach to a record of 66,654 deliveries, and of these only three were born imperforate. Other observers show a rather larger proportion, but sufficient has been said to demonstrate that the abnormality is by no means common. The proportion of males to females is 241 to 184.*

The most practical classification of rectal malformations is that given by Papendorf, and with slight modifications adopted by Bodenhamer, Mollière, Esmarch, Kelsey, and other writers. He gives the following nine varieties.

* "Leichtenstern's Statistics."

First variety.—Preternatural narrowing of the anus at the margin, occasionally extending some distance above this point.

Second variety.—Complete occlusion of the anal aperture by simple membrane or by the common integument.

Third variety.—No anus whatever, the rectum being partially deficient and terminating in a cul-de-sac at a greater or less distance above its natural outlet.

Fourth variety.—The anus in this variety is normal, the rectum at a variable distance above it being either obstructed by a membrane or completely occluded for a greater or less distance.

Fifth variety.—The anus opens at some part of the perinæal or sacral region, instead of at its normal position.

Sixth variety.—The rectum opens into bladder, urethra, or vagina, sometimes forming a complete cloaca, the normal anus usually being absent.

Seventh variety.—Rectum normal, with the exception that either the ureters, the vagina, or uterus, open preternaturally into it.

Eighth variety.—The rectum is entirely wanting.

Ninth variety.—Where rectum and colon are both absent. In these cases there is sometimes an opening leading to the intestine at some other part of the surface of the body.

It will be seen that in the first three varieties the development of the proctodæum is principally at fault, while in the others the malformation is of a more complicated character.

First variety: congenital stenosis of the anus.—Of this malformation but few cases are recorded. According to Bodenhamer, however, it is more common than generally supposed, as many of the minor cases are not considered worth recording, and it is probable that others undergo spontaneous cure, the efforts of the child to pass fæces being sufficient to produce the necessary dilatation.

It is possible that this condition might be mistaken for the effects of congenital syphilis; but the series of symptoms indicating the latter disease are usually sufficiently characteristic.

Varieties.—The anus may be simply too narrow and incapable of sufficient dilatation. When this is the case the symptoms usually do not develop until the child is some months old, the anus, unless the constriction be extreme, being sufficient to permit the soft and semifluid fæces of early infant life to pass. When, however, the child begins to take solid food, and the fæces become harder, an accumulation is likely to take place in the rectum. I have seen a case of this kind in a child aged three years, in whom nothing abnormal was noticed until the patient was seized with great straining and slight catarrhal dysentery. Upon examination it was found that the anus was incapable of dilating to more than the size of about a No. 8 catheter, and that the rectum was plugged with a mass of hardened fæces; a slight incision and subsequent mechanical dilatation rapidly cured this case.

In other instances the anus may be partly occluded by a membrane resembling the hymen, the opening in which may be so small that it will only admit a fine probe, or the membrane may simply exist as a fold which does not interfere with the opening of the anus to any appreciable extent. Such folds, manifestly of congenital origin, are not unfrequently met with during the examination of the rectums of adults for other affections. When so slight as this they are of no surgical significance, and the patients are probably unconscious of their existence.

The **treatment** of congenital anal stenosis must be conducted on the same lines as those laid down for acquired stricture. There is, however, a much better prospect of doing permanent good by mechanical dilatation than in those cases in which the stricture is due to cicatricial contraction. The best of all bougies is the mother's index finger introduced daily into the bowel.

Mr. Morgan* records two remarkable cases in which a curious congenital malformation produced a partial occlusion of the anus. In the first case, "a male child, aged six months, was taken to the hospital on account of the pain which he suffered whenever motions were passed. The pain was such as to cause the child to cry continuously before and after the bowels were relieved. The body was well formed and otherwise healthy, but on examining the anus, which was of usual size and in proper position, there was found to be a band of tissue passing from a point corresponding to the apex of the coccyx to the median raphe of the scrotum, with the posterior extremity of which it was continuous. The band was about three-quarters of an inch long, and was attached at both ends, the remainder forming a thick free cord which lay below the aperture of the anus ; while from the centre of the band there ran a small branch of similar tissue which was attached to the skin of the left buttock, and was about half an inch in length. The skin which covered the central band exactly resembled that of the scrotum, shrinking and contracting upon stimulation, and it was so placed that any matter passed per anum must cause it to be stretched, thus accounting for the pain which attended every relief of the bowels. The whole band was removed by cutting the attached ends with the scissors. The wound healed at once, and the child

* *Lancet*, Oct. 22, 1881.

was relieved of all pain." Mr. Morgan's second case was somewhat similar. "On examining the parts there was seen a small thick band, passing from the median raphe of the perinæum in front to the depression between the buttocks posteriorly, and broadest behind. At a spot corresponding to the anus on either side of the band was a depression; that on the right side was patent, and admitted a probe to pass into the anus; that on the left side, though similar in appearance, proved to be only a cul-de-sac. Motions passed freely, but caused much distress. The band was snipped off, and the child was relieved of pain." Mr. Cripps states* that he has seen a similar case under the care of Mr. Willett.

Second variety: complete occlusion of the anal aperture by simple membrane or by more fully formed skin (Fig. 5).—In this variety the anus may simply be occluded by a structure which is in many respects similar to the hymen. And its presence is not inconsistent with the complete development of the sphincter and other portions of the termination of the rectum. It would appear to be due to an adhesion or skinning over of the surface of the anus, the rest of the proctodæum being normally formed.

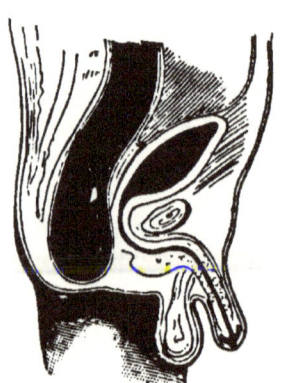

Fig. 5.—Second Variety of Congenital Malformation. Complete occlusion of the anal aperture.

Diagnosis. — It will be noticed that the child passes no meconium; and that it is constantly straining, and apparently in pain. An inspection of the anal region shows that there is a protrusion at the anus, especially if the child

* Diseases of Rectum and Anus," p. 25. London, 1884.

strains; and if the covering be very thin the colour of the meconium can be distinctly seen through the membrane: fluctuation is also distinct. The cure of this case is as simple as the diagnosis. A crucial incision with snipping off the angles is sufficient to give exit to the meconium and to cure the child, at all events when supplemented by the occasional passage of the tip of the mother's finger as a bougie.

Third variety : no anus whatever, the rectum being partially deficient, and terminating in a cul-de-sac at a greater or less distance above its natural outlet (Fig. 6).—In this variety there are many different degrees, the simpler forms resembling the more marked examples of atresia ani, the rectal cul-de-sac descending well down into the pelvis, and reaching to within a short distance of the skin; while in the more severe forms the rectum may terminate high up in the pelvis. The appearance of the perinæum varies somewhat in these cases. In some a ring of darker-coloured skin points out the position where the anus ought to have been; while in others a little button of skin is to be found in the same place, or even a pendulous tumour. In others, again, the ridge of skin forming the raphe of the anterior perinæum is continued in an unbroken line back to the coccyx, no mark whatever being visible to indicate the site of the anal orifice. Associated with this latter condition other abnormal relations of the pelvic structures are sometimes met with. The external genital organs may be nearer the tip of

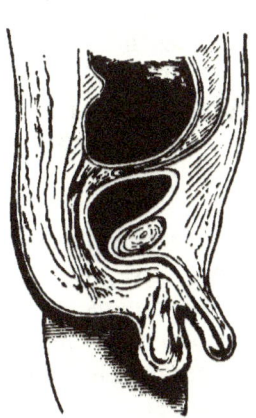

Fig. 6.—Third Variety of Congenital Malformation. No anus whatever, the rectum terminating in a cul-de-sac.

the coccyx, and otherwise abnormally developed; the tubera ischii may be abnormally close together; and the pelvis altogether manifestly smaller than natural. When any of these latter conditions are observed, it renders the prognosis very formidable, as the probability is strong that the rectal cul-de-sac is situated a long distance above the perinæum. This form of malformation is usually early diagnosed, as the attention of the parents or nurse is attracted by the fact that the child's bowels have not moved, and a mere inspection serves to establish the cause of the constipation. If unrelieved, the child soon becomes restless; it is manifestly in great pain, straining a great deal, and, at a later period, fæcal vomiting comes on, and death soon follows. In some rare instances, however, life may be prolonged for several weeks, the fæcal vomiting alternating with the swallowing of milk.

Treatment.—The duty of the surgeon in these cases is clear: in the first place he should attempt to reach the rectal cul-de-sac by perinæal incision, and the sooner this is attempted after the diagnosis is made the better, as in all cases of intestinal obstruction delay means a fatal result. If we wait until the abdomen is distended with gas, and the little patient is much run down, death will be the inevitable sequence. In selecting the form of operation it is of the utmost importance to determine the position of the rectal cul-de-sac as far as possible. Unfortunately this is seldom possible with any degree of accuracy. If the pelvis is of normal shape, and the genital organs in natural position, if on crying or straining there is a distinct protrusion in the anal region, then the probability that the pouch is within easy reach is strong. The protrusion may usually be incited by tickling or pinching the skin of the perinæum, but the administration of purgatives for such purposes is to be absolutely interdicted. It has been recommended by some surgeons to delay the operation for a day or two if the symptoms are not very urgent, in order to give an opportunity for the rectum to become more distended and prominent. Such advice is quite erroneous, as, in the first place, the meconium in the bowel becomes less by the absorption of fluid, and, what is more important, while we are waiting, the time may slip away when alone a chance of success exists. Before undertaking any operation it is well briefly to review some of the more essential anatomical arrangement of the pelvic viscera, as success will in great measure depend upon the care exercised by the surgeon. The rectum descends in the hollow of the sacrum, closely applied to that bone, and, except at its upper part, is uncovered by peritonæum posteriorly. In front the peritonæal pouch descends to a much lower level, while its close relations to the genito-urinary organs anteriorly would prove an additional reason for selecting the posterior aspect for exploration. The following are the measurements of the infant's pelvis, as given by Bodenhamer. From one tuberosity of the ischium to the other, 1 inch to 1 inch 1 line; from tip of coccyx to symphysis pubis, 1 inch 1½ lines to 1 inch 3 lines; and from the tip of the coccyx to the promontory of the sacrum, 1 inch 1 line to 1 inch 2 lines. The object of the surgeon in operating on these cases is to reach the bowel by direct incision from the perinæum. Formerly it was the custom to use a trocar and cannula, which was driven into the perinæum in the direction where the rectal cul-de-sac was supposed to be. No doubt, occasionally, the bowel was opened, and exit given to a small quantity of meconium, but even when this result was attained the opening was too small to serve as a permanent vent, while, frequently, fatal injuries were inflicted, the peritonæum, bladder, and even the common iliac vein, having been opened by the instrument.

Method of operating.—The little patient being held in the lithotomy position, an incision should be carried from a little behind the root of the scrotum right back to the tip of the coccyx, taking care to keep accurately in the middle line. The pelvis should be from time to time during the progress of the operation explored with the finger, in order to try and feel the bulging of the rectal pouch, and the incision carried deeper at its posterior extremity than in front. It has been recommended to keep a sound in the bladder of the male, or in the vagina of the female, and, occasionally, this may be of use in recognising the parts where the dissection is carried deeply, but, under ordinary circumstances, if care be taken to keep well back in the hollow of the sacrum, the danger of wounding the genito-urinary organs is trivial. When the light is good the proximity of the rectum has been sometimes determined by the colour of the meconium showing through, but, generally, the finger will be the best means of detecting the gut. Having found and opened the intestine, the meconium should be evacuated, and, if possible, the recommendation of Amussat followed, to bring down the bowel, and carefully stitch the edges of the intestine to the skin wound, and, where necessary, the bowel should be cleared well from its pelvic attachments in order to allow this to be done. In stitching the intestine to the wound, great care is necessary to pass the sutures so deeply through the surrounding structures that no cavity can possibly be left outside the rectum in which fluids could collect; the danger of septic periproctitis, in great measure, depending upon the amount of success with which the suturing is done. If it be found impossible to close the deeper parts of the wound completely by sutures, one or two small drainage tubes should be passed to the bottom of the incision. The method of bringing down the mucous membrane, and suturing it to the skin, introduced by Amussat, is a vast improvement upon the older method, by which the incision was made simply into the rectum, and the fæces allowed to escape over the open wound. Not alone were the immediate dangers of fæcal extravasation into the areolar tissue of the pelvis great, but, even should the patient survive, difficulty more or less great is almost certain to result from the contraction of the cicatrix producing stricture. It will still, however, occasionally happen that the termination of the rectum is situated so high up that it may be found impossible to bring down the mucous membrane to the skin without an amount of traction which would tear the gut. Under these circumstances it is necessary to keep the wound open by means of a metal tube, or other similar contrivance; but such cases are not likely to terminate successfully. In order to render the dissection more easy, Amussat recommended the removal of the coccyx;* and the procedure is still more strongly advocated by Verneuil.† As the latter has shown, not only does the removal of the coccyx assist much in the search for the occluded rectum, but it also enables the intestine to be more readily attached to the wound, as the latter is by this means much nearer to the cul-de-sac. In this variety of malformation there sometimes exists a cord, consisting of the outer tunics of the bowel, leading from the cul-de-sac to the normal site of the anus, which will be more particularly described when discussing the next variety. Should this be present, it may, as has been pointed out by my colleague, Professor Bennett,‡ form a valuable guide to the pervious portion of intestine. If it be found impossible to

* Quoted by Bodenhamer, "Malformations of the Anus and Rectum," p. 147. New York, 1860.
† *Med. Times and Gazette*, July 5th, 1873.
‡ Transactions of Academy of Medicine in Ireland, vol. L p. 150.

reach the rectum by carefully dissecting up in the direction of the promontory of the sacrum, and should a careful digital examination of the rest of the pelvis, as far as it can be explored, fail to indicate any fluctuating tumour resembling the distended intestine, unquestionably the proper surgical course is to perform colotomy, and most surgeons are now agreed that laparo-colotomy is preferable in these cases to the retro-peritonæal operation in the lumbar region. I would refer the reader to the chapter upon colotomy for a fuller discussion of this subject. Dr. McLeod has suggested * that after the performance of laparotomy, instead of bringing out the intestine at the abdominal wound, the perinæal incision should be continued so as to open the peritonæal cavity, and the intestine should then be forcibly drawn down, opened, and fastened to the skin of the perinæum, the abdominal wound being closed; the object being to make the artificial anus in the perinæum instead of in the anterior abdominal wall. It does not appear that this operation has been practised; nor do I think that in the majority of cases it would be practicable. In a few recorded cases it has been attempted to establish a perinæal anus subsequent to colotomy, after the wound has healed, by passing a sound down the rectum, and attempting to cut upon this from the perinæum; the results, however, have been far from encouraging. What I would propose doing in these cases where perinæal exploration has failed, is to open the abdomen, preferably in the left linea semilunaris, introduce a finger into the abdominal cavity, and trace down the rectum; if it is found to descend well into the pelvis, the search from the perinæum should be again instituted, and with the additional guide of a finger in the peritonæal cavity it may probably be successful, in which case the

* *British Medical Journal*, vol. ii. p. 657; 1880.

abdominal exploration will not have much increased the risk; if, on the other hand, the pervious bowel terminates above the pelvis, all hope of making a perinæal anus should be for ever abandoned, and a colotomy completed.

Fourth variety : in this form the anal portion, although apparently normally formed, ends in a cul-de-sac, the rectum terminating at a variable distance above this point (Fig. 7).—Sometimes the tubes are separated by a more or less thick membrane, while at other times there is a considerable length of intestine impervious. Of this malformation there are several interesting varieties; in one form the obstructions are multiple, and the important bearing that these cases may have upon operative treatment is obvious, the obstruction continuing after the apparently complete division of a membranous septum. Of these Bodenhamer has collected a number of examples.* Thus, Friedberg mentions a case of a new-born female child whose

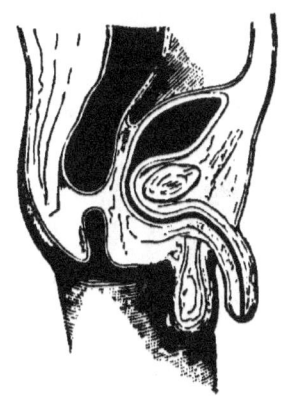

Fig. 7.—Fourth Variety of Congenital Malformation: anal portion ending in a cul-de-sac, rectum terminating in a blind extremity.

anus was well formed, but the anal canal was closed a little above the sphincter. The attempt made to open it by puncture produced no evacuation of meconium, and the child died six days after birth. At the autopsy the walls of the intestine were found adhering to each other and closely united in two different places. Schenck records a case in which the rectum above a natural anus was closely united

* *Loc. cit.*, p. 162.

at two places, as if glued together, and at two other points the rectum was occluded by two annular membranous septa; while Voillemier reports a case in which the rectum was divided by membranous septa into four distinct compartments, the anus remaining normal; and Dr. Bushe, of New York, and Goeschler, of Prague, both give cases in which a double obstruction existed. In other cases, instead of the two constituent portions of the rectum abutting one against the other, they may pass parallel to one another, the lower portion passing usually in front. Cases of this kind are recorded by Amussat,* by Godard,† and by Curling.‡ In Godard's case the rectal portion was attached to the coccyx, while the anal portion passed up in front of the prostate gland. These cases are especially interesting, as they unquestionably indicate that the malformation is due to the fact that the proctodæum has failed to meet the mesenteron, and that they cannot possibly be due to an obliteration of the rectum after complete development. In connection with these cases must be noted the classical case recorded by Amussat (Fig. 8),§ in which the anal portion communicated with the vagina, while the rectum ended in a cul-de-sac. I have not met with a description of any similar instance to this, which is given with great detail by Amussat. Not only is this case interesting from the nature of the

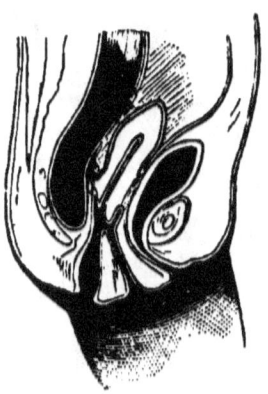

Fig. 8.—Congenital Malformation: rectum ending in a cul-de-sac, anal portion opening into vagina (Amussat).

* "Sur la possibilité d'établir une ouverture artificielle." etc. Troisième Mémoire. Paris, 1843.
† *Gaz. Méd. de Paris*, 1856.
‡ "Diseases of Rectum," p. 200.
§ *Loc. cit.*

malformation, but also for the success of the treatment adopted. An incision was made behind the anus ; and the rectum having been reached, it was separated from its connections, brought down, and stitched to the skin, this being the first occasion upon which this procedure, now so well known as "Amussat's operation," was performed. The anus was left untouched, and the communication with the vagina not interfered with. The patient made a good recovery, and was married at the age of twenty-one years.

Two theories are put forward to account for the usual condition met with in the fourth variety of malformation : one that it is due to failure of meeting between the mesenteron and the proctodæum; and the other that it is due to an intra-uterine inflammatory occlusion of the fully developed intestine. The supporters of the latter view point to the fact that in many of these cases there is to be found a cord manifestly continuous with the outer tunics of the bowel connecting the two pieces of gut. Unquestionably this cord is very frequently present, but it by no means follows that its presence presupposes a pervious intestine. On the contrary, its presence can be shown with much greater probability to have a developmental origin ; the mesenteron which originates from the hypoblast, as before mentioned, forms the upper portion of the rectum, but from it the mucous membrane alone is developed, a layer of mesoblast subsequently surrounding the tube to form the muscular and other external portions of the intestinal wall ; consequently, when the development of the cul-de-sac of mesenteron becomes, from any cause, arrested, it does not follow that the growth of the other tunics originating from the mesoblast should be arrested also ; and when there is no mucous coat to be surrounded, it can be readily understood how this portion of mesoblast can form itself into the rounded cord.

Again, we must remember how exceedingly rare it is for a mucous canal to be obliterated by inflammation, unless attended with a very considerable superficial loss of substance. The only instance that I know of in which a mucous canal is obliterated during the process of development in the human subject is that of the urachus, but even in this case evidence of the mucous membrane, and even small mucous cavities, are still found in the cord which forms the remains of this fœtal structure. I have recently had an opportunity of carefully examining a case of this kind from a patient under Professor Bennett's care in Sir Patrick Dun's Hospital, in which, after failure to meet the rectum by perinæal incision, a colotomy was performed, but the result was fatal. In this instance there was a very firm and strong cord extending from the cul-de-sac to the anal portion; a microscopical examination of this cord showed it to be composed entirely of muscular and connective tissue, without a trace of mucous membrane. I was also able to determine another important point in this case. If the anal depression is composed alone of proctodæum, it is obvious that, as it originates entirely from the epiblastic layer of the embryo, it should have its surface covered with scaly and not columnar epithelium. I consequently obtained a small piece from the fundus of the anal depression, and made sections of it. There was not a trace of glandular epithelium to be seen in it, so that, in this case at any rate, the conclusion was unavoidable that the malformation was due to the fact that the mesenteron did not descend low enough for the proctodæum to meet it; and that, I believe, is the explanation of the majority, if not all, of these cases.

The **diagnosis** of the fourth variety of malformation is not so likely to be made at an early date as in the previous form, a mere inspection of the anus revealing nothing abnormal. And it is only when continued constipation exists, attended possibly with vomiting and meteorism, that the surgeon's attention is usually directed to the case. If an attempt is made to pass a probe or the tip of the little finger up the rectum, the condition will at once be recognised, the anal portion of the intestine seldom extending more than a very few lines from the outlet. An attempt should be made to feel, if possible, the upper cul-de-sac, so as to form an opinion as to the distance intervening between the two portions of the bowel.

As these cases are seldom diagnosed until grave symptoms have become developed, it is obvious that the prognosis is usually more unfavourable than in the last variety.

The **treatment** of this variety does not differ in any essential respect from what has been already described for the third variety. In all cases an incision should be made from the anal depression back towards the coccyx, and the rectum sought for and opened, and stitched to the site of the normal anus. Even in those cases in which a septum alone intervenes between the two portions of intestine, the incision should be preferred, as where an attempt has been made to puncture with a trocar and cannula a valvular and inefficient opening has been formed in some cases, while in others serious injury has been inflicted on the peritonæum or other pelvic viscera. As in the last instance, if the surgeon fail in finding the rectum, colotomy is the sole resource for saving the infant's life.

Fifth variety : the anus is completely absent, or rudimentary, and the rectum terminates in a cutaneous opening at some other situation than the normal one.—As might be anticipated, there is a very considerable amount of variety met with in the situation at which the abnormal anus opens, while in some cases more than a single orifice exists. It has been suggested that these cases are due to the formation of fistulæ, similar to the fistulæ found in connection with stricture of the rectum, but this view may at once be disposed of by the fact that no evidence of suppuration exists; and the channel is lined with mucous membrane instead of granulation tissue, clearly demonstrating that these cases are congenital malformations in the true acceptation of the word, and not of pathological origin. Bodenhamer, who has made these cases a distinct class from the following, in which the abnormal anus is situated in one of the other mucous tracts, has collected a number of interesting varieties, but as space will not allow me to go into details, the reader, for further particulars, must be referred to Bodenhamer's work.* The positions at which these openings have been described are indeed various, as may be gathered from the following list of the more important situations: (1) within the prepuce, the rectum being continued as a narrow channel under the raphe of the perinæum and under surface of the penis (Fig. 9); (2) at the symphysis pubis; (3) at the root of the penis; (4) at the posterior portion of the scrotum; (5) at the fourchette in the female; (6) at various portions of the perinæum; (7) in the right gluteal region; (8) in the loin; and (9) through a perforation in the sacrum. These last cases, when viewed in the light of modern

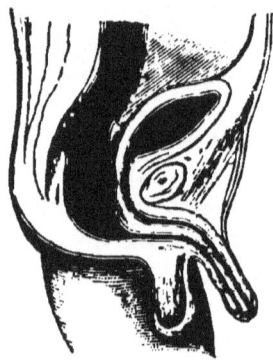

Fig. 9.—Congenital Malformation: the rectum ends in a narrow channel which opens beneath the prepuce (Cruveilhier).

* "Congenital Malformations of the Rectum and Anus." New York, 1860.

embryological knowledge, possess a special interest. Bodenhamer mentions three of them: two recorded by M. De La Faye,* and one recorded by M. la Coste.† Unfortunately there does not appear to be any allusion in these cases to the relation of the abnormal anus to the termination of the neural canal. In the lower vertebrates there is a distinct communication between the mesenteron and neural canal known as the neurenteric canal, which is a very distinctive feature during the early periods of development, and which, when the communication with the neural canal is cut off, which occurs at an early period, constitutes a prolongation backwards of the intestine behind the anus. This, which is termed the post-anal gut, has also been observed in birds and mammals; and, according to Kölliker,‡ in the young embryo of the rabbit, the post-anal gut, or *pars-caudalis intestini*, is a very conspicuous structure, terminating in a small vesicle posteriorly. As in the case of the lower vertebrates, this caudal intestine undergoes a very rapid obliteration during the process of development. Balfour§ attaches great weight to the presence of the neurenteric canal and post-anal gut, as throwing light upon the ancestral form from which the *chordata* have been evolved, and he considers that the anus as at present found in the vertebrates is of comparatively recent evolution; for, as he states, the post-anal section of the alimentary tract cannot always have been devoid of function, as is shown by the following facts: (1) by the constancy and persistence of this obviously now functionless rudiment; (2) by the greater development in the lower than in the higher forms; (3) and by

* " Principes de Chirurgie," p. 358. Paris, 1811.
† " Bulletin de la Société Médicale d'Émulation de Paris," p. 417; October, 1822.
‡ " Embryologie," p. 878.
§ " Comparative Embryology," p. 267. London, 1881.

its relation to the formation of the notochord and subnotochordal rod. He therefore concludes that in the ancestral type the anus was situated at the termination of the post-anal gut. I have alluded at some length to this subject, in the hope that should any examples of these (cases where the rectum perforates the sacrum) be met with, they may be made the subject, if possible, of a detailed anatomical examination.

The **prognosis** of these cases in which the anus is only abnormally situated but remains pervious, is manifestly very much better than those in which occlusion exists; and in some cases it may be unnecessary to contemplate any operative interference, the abnormal anus being quite sufficient for all purposes. But in other instances, and apparently in the majority of the recorded cases, the termination of the rectum and the aperture were quite too narrow to permit of efficient evacuation of the bowel at a later period of life, though possibly allowing the semifluid meconium to pass off in early infancy. Operation can, however, here be undertaken with a very much greater chance of success, as it is no longer necessary to search for the rectum, a probe through the abnormal opening indicating the position of it. If the probe indicates clearly that the pouch of the rectum descends down close to the normal position of the anus, the best course will be to make an incision down to the rectum, clear it round, and bring it down, after the method of Dieffenbach, and stitch it to the skin. The abnormal passage should now be closed, if possible, as if it is left open there will be a constant tendency for nature to close up the artificial anus, and continue the use of the abnormal one. For this purpose the use of the actual and potential cauteries has been recommended, but they do not appear to have answered the purpose properly ; probably the best course is to dissect out the narrow abnormal tube, and bring the parts together by deep sutures. Vicq D'Azyr recommended the introduction of a director into the rectum from the abnormal anus, and the incision of the parts, up to the normal situation of the anus, a metal tube to be retained, and the rest of the wound allowed to heal. Possibly where the abnormal anus was situated close to the normal position, this method might be with advantage adopted. When the rectal pouch cannot be proved to be near the perinæum, as is likely to be the case in those instances where the abnormal anus is situated at a considerable distance from the usual site, no attempt should be made to reach the bowel from the perinæum, but, if absolutely necessary, the abnormal opening might be dilated. And in the rare form in which more than one anus is present the recommendation of Ashton* may be followed with advantage; not to interfere if either is sufficiently large to provide for efficient evacuation; but if both orifices are very minute an incision through the septum, as in the treatment of rectal fistula, will probably prove an efficient form of treatment.

Sixth variety: the anus being absent or only rudimentary, the rectum terminates in some of the mucous passages of the genitourinary system.—The subdivisions are: (1) where the communication is with the bladder (*atresia ani vesicalis*); (2) where it opens into the urethra (*atresia ani urethralis*); and (3) where the abnormal anus is situated in the vagina (*atresia ani vaginalis*). It is remarkable the frequency with which the sixth variety of malformation is met with. According to Leichtenstern's statistics,† out of a total number of 375 cases of malformation of the rectum 40 per cent. were of this variety; and in

* "Diseases of the Rectum," p. 327.
† Ziemssen's Cyclopædia, vol. vii. p. 485.

Bodenhamer's statistics, out of a total of 287 cases 85 were of the sixth form. This large proportion is exceedingly remarkable when we remember that in the adult the separation between the rectum and anterior perinæum is so exceedingly definite ; and I am not aware that any satisfactory explanation has been offered to account for this marked tendency of the rectum to communicate with the genito-urinary apparatus in cases where the anus is undeveloped, unless, indeed, it be due to the method of development of the proctodæum (*see* page 21), or a tendency to reversion to the cloacal types of the birds and lower mammalia.

Atresia ani vesicalis (Fig. 10).—The communication between the rectum and bladder in these cases consists usually of an exceedingly narrow canal having its termination at some part of the trigone of the bladder by a minute orifice, while in very rare cases the communication has been near the fundus, and when this is the case the opening has been found to be more free.

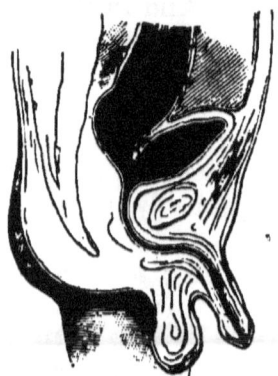

Fig. 10.—Congenital Malformation: rectum ending in bladder (atresia ani vesicalis).

The **diagnosis** of these cases is usually sufficiently simple. With the symptoms dependent upon the occluded rectum will be found a small quantity of fæces passed with the urine, sometimes only in sufficient quantity to stain the water of a green colour. The fact that the meconium is intimately mixed with the urine, and that it only appears during urination, will at once distinguish this variety from the atresia ani urethralis. The prognosis, as might be expected in these cases, is eminently unfavourable. From the fact that the bladder is in the young infant an abdominal viscus, it will be readily understood that the termination of the rectal pouch is probably situated in such a position that any attempt to reach it from the perinæum will prove a failure, although two successful cases of this operation are mentioned by Bodenhamer.* If, however, nothing is done, death will be the inevitable result, as in the majority of cases the opening is so small that but a small quantity of meconium can pass; and even where the opening is somewhat larger, although possibly life might be prolonged while the fæces remained semi-fluid, a fatal cystitis will eventually be produced. M. Martin, of Lyons, has suggested† an operation for the relief of this form, viz. to incise the neck of the bladder as in the operation for lithotomy, and then to continue the incision on into the rectum. Granting that this operation might in a few cases be feasible, the result would be of very questionable utility, as if successful it would leave the patient with a urinary and fæcal fistula. I believe that the true scientific procedure in these miserable cases would be to perform a laparo-colotomy, and then to cut the colon completely across, as recommended by Madelung; closely suture up the lower end after inversion of the serous membrane, and bring the upper end out at the wound to make an artificial anus. Although this procedure would leave the patient with an abdominal anus, it would have the advantage of restoring the functions of the bladder.

Atresia ani urethralis (Fig. 11).—In this form the canal leading from the rectum, instead of opening into the bladder, communicates with some portion of the urethra. As in the former case, the opening is generally by an exceedingly minute orifice, so that but a small quantity of meconium can escape. In this

* *Loc. cit.*, p. 232.
† "Dictionnaire des Sciences Médicales," tome xxiv. p. 127.

variety it comes more or less in the intervals between urination. When the urine is passed, although the first that is evacuated may be stained, it afterwards comes quite clear.

In the **treatment** of this form, the probability of reaching the bowel from the perinæum is much greater, and this should always be attempted; moreover, the presence of a subsequent recto-urethral fistula is a much less serious condition than a communication between the rectum and bladder, the former being frequently curable; or, at any rate, not directly threatening life, while the latter is generally beyond the range of surgical aid to cure, except by establishing an artificial anus, and is likely to prove fatal. This is the most suitable place to consider a form of rectal malformation which I have not hitherto seen described. It is one of atresia ani in which a diverticulum from the rectum passes forward, and becomes related to the urethra without opening into it. Of this remarkable malformation I have only the following case to record. In the spring of the year 1886 I saw, in conjunction with my colleague Professor Finny, a medical man in Dublin, who, after having suffered for some days from an intermittent fever, developed an inflammatory swelling at the root of the penis and deep in the scrotum. He told us he had suffered from some pain while the bowels were being moved and asked me to delay examining the rectum till he was under æther. Upon making an incision into the swelling, a gangrenous and abominably fœtid abscess was opened, apparently in connection with the left

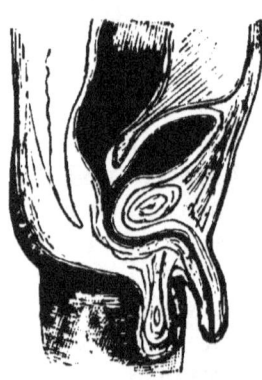

Fig. 11.—Congenital Malformation: rectum ending in urethra (atresia ani urethralis).

crus penis. Passing my finger now into the rectum, I was surprised to find a membranous stricture immediately within the anal verge, through which, however, the finger readily passed, and above it I could detect a diverticulum of the rectum passing off in the direction of the situation of the abscess. I forcibly dilated the anus and membranous stricture; subsequently fæcal matter in small quantity passed from the wound at the root of the penis. Although he developed septic pneumonia and other symptoms of wound infection, he eventually, after a protracted and severe illness, made a good recovery. After the operation we were informed by his mother that he had been born imperforate, and was operated upon by the late Dr. Fleetwood Churchill. There was never any mixture of fæces with the urine. He is now quite free from any trouble, and I believe the diverticulum has been obliterated by the inflammatory process.

Atresia ani vaginalis (Fig. 12).—In this form the rectum terminates at some part of the posterior wall of the vagina, either having a tolerably large and free aperture, quite sufficient for the evacuation of the intestinal contents, or more rarely the rectum ends in a cul-de-sac, the communication with the vagina being by means of a narrow pipe-like tract. The points at which the rectal orifice may open are various, either immediately within the fourchette (Fig. 13)

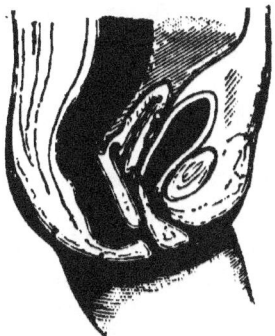

Fig. 12.—Congenital Malformation: rectum ending in vagina (atresia ani vaginalis).

at the entrance of the vagina, or high up in that canal; and cases have been recorded by Papendorf and Ainsworth,* in which the communication between

* Quoted by Bodenhamer, *loc. cit.*, p. 227.

the rectum and vagina was by a double orifice. If the rectum ends by a true vaginal anus, as is most frequently the case, the patient experiences but little discomfort; and many cases are on record in which women have married and borne large families while suffering from this malformation. One such case has come under my notice. The woman was the mother of six children, and did not suffer the slightest inconvenience. The anus opened into the lower portion of the vagina, and was so far provided

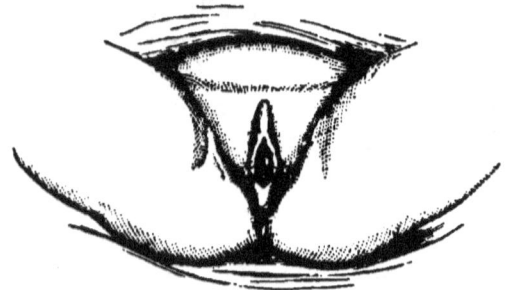

Fig. 13.—Vulvar Anus in a Child aged 19 months. The opening is immediately within the fourchette, and is well provided with a sphincter.

with a sphincter that the top of the finger, when introduced into the rectum, was tightly grasped. There was not the least incontinence, and the bowels acted regularly every day. It will thus be seen that this variety of malformation is the one in which the prognosis is most favourable, many of the cases, as in the one above recorded, requiring no operative interference, while in others the fact that a more or less free outlet exists for the meconium relieves the case of its urgency, and enables the surgeon to act with deliberation and choose his own time for operative interference, when the little patient has gathered some strength.

Amongst the operations which have been suggested for the treatment of vaginal anus, the following may be mentioned : M. Vicq d'Azyr has recommended a similar procedure to that stated in discussing the last variety, namely, to make an incision from the vaginal opening to the site of the normal anus, and then to introduce a cannula into the rectum, and allow the anterior part of the incision to heal up. Velpeau simply made an opening into the rectal pouch, without dealing in any way with the vaginal portion. Dieffenbach and Barton[*] recommend that the rectal cul-desac, when present, should be dissected free by perinæal incision, and stitched to the skin margin, thus shutting off the vaginal portion from the intestinal tube altogether; and then closing the vaginal orifice, or not, according to circumstances. By far the best plan, however, that has been hitherto suggested, is that of Rizzoli.[†] An incision is carried from the lower margin of the vaginal anus through the perinæum backwards towards the coccyx, care being taken not to open the intestine. The termination of the rectum, with its vaginal orifice, is now to be carefully dissected out, and the abnormal anus is to be transplanted to its natural situation; the perinæal and vaginal wounds are to be brought together by deep sutures, thus restoring the recto-vaginal septum. The great merit claimed for this highly ingenious procedure is that it retains intact the outlet which Nature has formed, and which, therefore, will probably have sufficient sphincter action to avoid incontinence of fæces, and is less likely to be followed by stricture than where any other operation has been performed. The formation of a proper recto-vaginal septum is also an important matter in preventing uterine trouble in after-life. Dr. Aveling has recorded a very interesting case of this malformation, in which, after a series of operations, a very brilliant result was obtained.[‡] The malformation

---

[*] *Medical Recorder*, vol. vii. p. 357; Philadelphia, 1824.
[†] "System of Surgery," Gross, vol. ii. p. 605. Sixth edition.
[‡] *Lancet*, Dec. 20, 1884; p. 1085.

remained undetected until the child was five weeks old, when Sir Prescott Hewett operated by making an opening at the natural situation, and attempting to close the vulvar orifice by caustic, without success. At the age of seventeen she came under Dr. Aveling's care, and he endeavoured to close the vulvar anus by plastic operation. In the first instance this object was defeated by the passage of a hard mass of fæces, which tore away the sutures. A second operation of similar nature proved more successful, the opening being reduced to the size of a goose-quill. A third operation was performed, a small cylindrical speculum being introduced into the artificial anus to ensure the exit of flatus by that route, and so keep the vulvar wound at absolute rest. The result of this operation was completely successful in shutting off the intestinal tube from the vulva. As hardened fæces still accumulated in the diverticulum which existed in front of the artificial anus, and as this portion of gut was unable to evacuate its contents, a still further operation became necessary for its obliteration. An incision was carried through the skin of the perinæum and vulvar mucous membrane up to but not including the site of the previously occluded anus. The mucous lining of the diverticulum was carefully dissected away, and the perinæum closed by deep sutures. In this way a firm recto-vaginal septum was formed, and the patient completely cured. This case demonstrates very forcibly the absolute necessity of adopting efficient means to obliterate the abnormal channel, as well as establishing an opening at the natural situation; and I believe that in similar cases it would be well to do both operations at the same time.

Seventh variety: in which the rectum and anus are normal in situation and calibre, but in which the ureters, vagina, or

uterus open into the intestinal tube.—Bodenhamer has collected seven cases in which the ureters are described as terminating in the rectum, the bladder being absent. In these cases the position at which the opening took place was at the reflection of the peritonæum. Such cases are, at present at any rate, beyond the reach of useful operative interference. The same authority has also collected nine cases in which the vagina terminated in the rectum. In several of these pregnancy and successful parturition took place; but in one only is it stated that operation with a view of establishing a separate vagina was attempted and successfully carried out. It does not appear, however, that there would be greater difficulty in operating upon these cases than in those of the converse condition already described, where the rectum opens into the vagina.

The eighth and ninth varieties of Bodenhamer, in which the entire rectum is absent either alone or conjoined with more extensive congenital malformation of the intestinal tube, are beyond the scope of this volume to discuss.

CHAPTER III.

PROCTITIS.

INFLAMMATION of the rectum may be the result of injury, or it may be due to various specific influences. In Great Britain, however, its occurrence from these latter causes is not by any means frequent. The mucous membrane of the last part of the intestinal tube is more liable to injury from matter contained in the fæces than any other part of the gut, because, in the first place, owing to absorption of the more fluid parts, the fæcal mass has now become more solid and resisting, so that when the rectal wall contracts firmly upon it in the expulsive effort, any hard particles which project, such as pieces of bone, glass, and such like, are liable to lacerate the delicate mucous membrane. Again, as the calibre of the tube so suddenly diminishes at the junction of the rectal pouch with the anus, hard substances projecting from the fæcal mass are more liable to penetrate this portion than any other of the entire intestinal tract. I have seen cases of stercoral abscess following perforation at this situation by a pin which was swallowed accidentally, and also by a fish-bone; and most authors on rectal surgery allude to similar cases. The introduction of foreign bodies intentionally or accidentally through the anus (which in a later chapter will be more fully alluded to) is an obvious cause of local traumatism.

Inflammation of the rectum may be due to any of these injuries, and when arising in this way has a tendency to spread past the limits of the bowel constituting the condition elsewhere described as periproctitis (page 57).

The varieties of inflammation not directly due to injury are the catarrhal, the dysenteric, and the gonorrhœal.

Of simple catarrhal inflammation we may recognise two varieties, acute and chronic.

Acute catarrhal proctitis presents many symptoms in common with dysentery, and the difference is rather one of degree than kind. I shall, therefore, confine the term to those cases in which the inflammatory process is limited to the rectum, and in which the abdominal pain and constitutional symptoms of typical dysentery are absent. As such we not infrequently meet with cases, especially in children. The symptoms are great tenesmus; with the frequent passage of small quantities of bloody mucus, at first mixed with fæces, and then alone; at the same time there is vesical irritation, and general sense of heat and weight about the pelvis, resulting from the inflammation of the mucous and submucous tissue, œdema is present, and frequently, in consequence, a partial prolapse takes place, resembling the chemosis met with in acute inflammation of the conjunctiva (*ectropion recti ;* Roser).

Catarrhal proctitis is in all respects analogous to the localised inflammation of other parts of the intestinal tube, such as gastro-duodenal catarrh, typhlitis, colitis, etc., and it may terminate in several ways. In the vast majority the disease subsides completely in a few days, but it may, if severe, be accompanied by inflammation of the structures outside the rectum (*periproctitis*), which may eventuate in abscess and fistula.

Or, again, the disease may merge into the chronic form, or be followed by ulceration more or less deep.

The frequency with which inflammation of the intestinal mucous membrane is accompanied with considerable bloody discharge has been noticed by Cohnheim;* he suggests that, from the fact that during the process of digestion the chyle vessels of the mesentery always contain red blood corpuscles, it may be inferred that the intestinal mucous membrane is one of the regions of the body where the passage of blood corpuscles through the walls of the vessels takes place with special facility.

Amongst the causes which give rise to this disease, epidemic influences are undoubtedly occasionally noticeable, but in many cases direct local irritation can be made out, such as the presence of large numbers of oxyurides, neoplasms in the rectal wall, intussusception, or other forms of rectal disease and fæcal

* Leube ; Ziemssen's Cyclopædia, vol. vii. p. 363.

accumulation. The latter acts either by its direct mechanical effect or by chemical action, the result of putrefactive changes. In some persons, certain articles of food, as high game or old cheese, always give rise to a slight amount of irritating rectal catarrh; or the use of large quantities of purgative medicines may produce a like result. Esmarch has pointed out the fact that the long-continued action of damp and cold may produce this disease, and states that it is common amongst coachmen, who sit for a long time on wet seats. As a complication of child-bed and various forms of uterine disease, it may also be observed.

Chronic catarrhal proctitis.—Where acute catarrhal proctitis has merged into the chronic form, the symptoms become somewhat modified; the acute pain and tenderness give place rather to a sense of weight and fulness than actual pain. The discharge also becomes altered; instead of consisting of a tolerably intimate mixture of blood and mucus, as is found in the acute form, it becomes more purulent, and if blood is present it exists as streaks in the pus, which have evidently arisen from ulcerations of the mucous membrane rather than from a general oozing from the inflamed surface. On inspection the mucous membrane appears more thickened and indurated, but the œdema is less. Ulceration of the surface is also more frequent in the chronic form.

There is a condition of the rectum occasionally seen, which, although unattended by any overt evidences of inflammation, appears to be most fitly considered here. It is that in which there is a constant and abundant flow of clear glairy mucus resembling white of egg in appearance; it is highly alkaline and causes considerable irritation of the skin surrounding the anus, and is a source of great annoyance to the patient. I have seen three well-marked cases—two males and one female—of this condition;

in all, the most careful examination of the rectum, both by finger and speculum, failed to detect any abnormal appearance, except the continued flow of this glairy fluid, which the sphincters were unable to retain. All were attended with an extreme amount of nervous depression, and stated that they were rendered absolutely miserable by the discharge. Treatment in the shape of astringent injections and suppositories did not appear to influence the disease much, but all of them got well after some months. I have no idea what the pathology of this distressing affection can be.

Dysenteric proctitis.—Of the specific inflammations, dysentery is much the most important, and although not always confined to the rectum, this portion of the intestinal tube is more or less affected in all cases, and sometimes the typical lesions are only to be found here. But few opportunities offer themselves to civil surgeons in Great Britain of investigating this disease during its acute stages; but amongst Anglo-Indians and others who have resided for a long time in tropical climates, the chronic form and some of the important sequelæ are not unfrequently met with.

For a full account of *dysentery* I must refer the reader to the systematic treatises of medicine, as a detailed investigation of this subject would be quite beyond the intention of the present work.

Gonorrhœal proctitis.—As a result of gonorrhœa the rectum occasionally becomes the seat of acute purulent inflammation. In women this may occur in consequence of the discharge escaping from the vagina and trickling over the anus, the liability to secondary infection being greatly increased by the presence of prolapse of the mucous membrane, piles, fissure, or relaxation of the sphincter from any cause. In men, and in women in whom the anus is normal, the disease probably only originates as a result of the direct introduction of virus within the sphincter. Rollet* reports a case in which a man who was suffering from gonorrhœa inoculated the rectum by the introduction of a finger for the purpose of causing a motion. Boniere,† on the other hand, states that, although the anus is easily infected, the express introduction of gonorrhœal virus through a tube into the rectum showed that the mucous membrane was only slightly susceptible to this form of inflammation. According to Lebert‡ the symptoms are as follows : gonorrhœa of the rectum causes congestion and swelling of the mucous membrane, and is attended by a *purulent* discharge. The constant sense of pressure, burning, and itching in the anus is much increased on each evacuation of the bowel, and sometimes there are very severe spasmodic attacks of pain in the anus, extending even to the bladder. Excoriations and fissures are very liable to form in the folds of the anus, and render the evacuation of the bowels still more painful. Gonorrhœa of the rectum has, however, happily a decided tendency to get well in a few weeks. Exceptionally the inflammation spreads to the submucous connective tissue, and in this way gives rise to the formation of abscess or even fistula in the neighbourhood. Much more frequently we observe slight erythematous irritation of the skin around the anus. Sometimes papillary outgrowths may afterwards develop. The diagnosis of gonorrhœal proctitis, as distinguished from other forms of inflammatory affection of the rectum, presents obvious difficulties, but can generally be arrived at from the circumstances of the case—the co-existence of urethral discharge ; the

* " Dict. Enc. des Sciences Méd."
† *Arch. Gén. de Méd.*, April, 1874.
‡ Ziemssen's "Cyclopœdia," vol. viii. p. 808.

profuseness and purulent nature of the discharge from the rectum; and the symptoms of extreme rectal irritation alluded to above.

Before, however, making an absolute diagnosis, the existence of the specific microbe of gonorrhœa (*Micrococcus gonorrhœæ*, or *gonococcus*) in the pus discharged should be verified. The detection of this organism by the microscope is now a tolerably easy matter;* and in any case of suspected gonorrhœa of the rectum its presence should be sought for.

Diphtheria of the anus is noticed by Trousseau† as occurring during the course of diphtheria in the pharynx and trachea; but is only secondary to the more usual form, and is an indication of profound implication of the system by this terrible disease.

Having now discussed the various forms of inflammation of the rectum, I would speak of their **treatment.**

In dealing with acute proctitis the first essential is to evacuate the canal, and this should be accomplished by the administration of purgatives, not by enemas, owing to the pain that the latter occasion and the danger of spreading infection up the intestine. Some saline will usually prove most suitable, preferably sulphate of soda, or sulphate of magnesia in some form, such as the more potent mineral waters; or the effervescent sulphate of soda will be found an easily taken and satisfactory aperient. Such powerful drugs as gamboge, jalap, etc., must be carefully avoided. Where it can be taken without much nausea, castor oil will fulfil every requirement. Absolute rest in bed is essential, and the occasional use of a hot hip-bath will give relief. The diet should be

---

* Klein, "Micro-Organisms and Disease," p. 77. Third edition, 1886.
† "Clinical Med.," New Sydenham Soc., vol. ii. p. 515.

carefully regulated. All food leaving a copious fæcal residue is to be avoided ; and during the acute stage the patient should be restricted to milk, strong meat-soups, and eggs. If there is much tenesmus, injection of two ounces of mucilage of starch with a few drops of tincture of opium may be used, but as a rule it is better not to use opium or morphia to any great extent, as, owing to its tendency to produce constipation, the disease may be aggravated. A suppository, consisting of 5 grains of iodoform, with ½ grain extract of belladonna, made up with oil of theobroma, will be found useful. In dysentery, in addition to the foregoing, the use of ipecacuanha in large doses is indicated ; and, where there is much exhaustion, the free use of stimulants.

Where inflammation of the rectum has become chronic, astringents must be employed. Nitrate of silver, 5 grains to two ounces of water, may be introduced into the bowel, followed in a few minutes by a large enema of warm water to wash out the rectum ; or liquor bismuthi with glycerine of starch, in the proportion of two drachms of the former to two ounces of the latter, will sometimes check the discharge of pus ; and where the discharge is fœtid, the following formula may be tried :

  ℞ Liquor. carbonis detergens . . . ℨ ii
    Tr. krameriæ . . . . . ℨ iv
    Mucilag. amyli . . . . ad ℥ iv
    An ounce to be injected night and morning.

For the sequelæ of proctitis the reader is referred to the chapters on Periproctitis, Ulcer, and Stricture.

CHAPTER IV.

PERIPROCTITIS.

INFLAMMATORY changes originating in the mucous membrane may spread to the loose areolar tissue surrounding the rectum; or the inflammation may commence primarily in the tissues on the outside of the bowel; giving rise in one of these two ways to the affection that is termed periproctitis. Of the former we can recognise two very distinct varieties: first, a diffuse form of septic infection consequent on the direct absorption from a wounded surface of poisonous matter similar to that which produces, in the recently-delivered female, parametritis and puerperal peritonitis; and, secondly, circumscribed phlegmon, resulting from spreading of a limited inflammation through the rectal wall, or more commonly subsequent to a perforation of the gut, the result of an ulcer or of direct traumatism.

Septic periproctitis is a very dangerous disease, and is not infrequently the cause of death after rectal operations. The clinical features of this condition are in many respects similar to those of puerperal septico-pyæmia, and, like that affection, may present themselves in various degrees of severity.

In the more serious cases, after a stage of incubation lasting a few days the patient is seized with rigors, quickly followed by general and considerable febrile disturbance, with high temperature and rapid pulse. Profuse perspirations occur at irregular intervals, causing the temperature to fall temporarily; pain is complained of in the region of the pelvis, and in many cases this rapidly increases and involves the lower portion of the abdomen; the patient lies on the back, with legs drawn up;

meteorism is present, and the face presents an aspect of great anxiety, a train of symptoms clearly indicating involvement of the peritonæal serous membrane; when this has occurred, death, in the great majority of cases, soon terminates the patient's sufferings. Other symptoms indicating the septico-pyæmic state are sometimes to be met with, as an erysipelatous blush on the buttocks; the deposit of diphtheritic patches on the mucous membrane of the rectum; the involvement of some of the synovial cavities or of the pleural and pericardial sacs; and the occurrence of metastatic abscesses in many parts of the body. Owing to the fact that the veins of the rectum are tributaries to the vena portæ, these metastatic abscesses are more frequently met with in the liver than when the source of infection is more directly connected with the systemic circulation.

In the milder cases of septic periproctitis the disease may be somewhat more limited in extent, and consist in a diffuse pelvic cellulitis, without involvement of the peritonæal cavity or evidence of septic manifestations in other parts of the body. This condition is analogous to the parametritis or pelvic cellulitis of the puerperal state; and, like it, may end in resolution, but much more frequently terminates in somewhat diffused suppuration. Matter formed in this way accumulates in considerable quantity in the loose areolar tissue situated between the recto-vesical fascia and the peritonæum, and may point near the anus, finding its way down posteriorly between the two levatores ani muscles; or it may escape through the sacrosciatic notch; or, passing upwards, appear as an iliac abscess. In other cases suppuration commences in the pelvic lymphatic glands.

Of the pathology of this affection but little can positively be stated. A large volume might be filled with an account of the various theories and discussions that have occupied obstetricians as to the aetiology and pathology of the septic febrile conditions subsequent to childbirth, but the probability is that in most instances at any rate the first important lesion is a lymphangitis of the pelvic lymphatic vessels. In those cases in which the disease is limited to a diffuse pelvic cellulitis, the limitation is due either to the blocking up of these vessels by thromboses, as Virchow has pointed out; or to the filtration power of the glands situated in the hollow of the sacrum being efficient. When neither of these conditions serves to stop the spread of the disease, the implication of the great lymph-space of the peritonæal cavity rapidly ensues.

The more important features of a case of mine that died many years ago after the operation of linear proctotomy, as a case typical of the lesions we have been considering, I will briefly describe. The patient, a man aged sixty-three years, was admitted into Sir Patrick Dun's hospital with piles, fissure, and a history of obstruction in the rectum. Digital examination revealed the presence of a tumour occupying the hollow of the sacrum, and compressing the rectum towards the bladder. As the finger could be passed above the upper limit of the growth, an attempt was made by linear proctotomy to remove the tumour, or at any rate to relieve the obstruction, the symptoms being so urgent that either this operation or colotomy was indicated. The incision was made in the usual manner, and, by means of the finger and scoop, the tumour, which proved to be in part cystic, was broken down and as much as possible removed, the obstruction being completely overcome. As hæmorrhage was free the rectum was plugged. On the following day the plug was removed and rectum and wound were irrigated with carbolic lotion. This treatment was directed to be repeated every three hours. The case went on well till the morning of the fourth day after operation, when the patient complained of pain ; he had a rigor, followed by a rise of temperature to 102·8° Fahr. ; the lower part of the abdomen became tumid and tender ; later on the same day he was restless, with high temperature alternating with profuse perspirations. He died on the sixth day after operation. At the autopsy the usual signs of purulent peritonitis were present, the peritonæal cavity of the pelvis being full of thin serous pus, and the coils of intestine being glued together with recent lymph ; the loose areolar tissue of the pelvis, more particularly that situated between the folds of the meso-rectum, being infiltrated with pus. Several of the lymphatic glands of the pelvis were also much inflamed, and in two instances suppurating. The extreme limit of the operation wound was separated by a considerable distance from the peritonæum. The tumour in this case will be again referred to in the chapter on Neoplasms, as there are some points of interest about it.

That the cause of death in this case was due to septic infection there can be no reasonable doubt, and the post-mortem appearances were in all respects similar to those found in the majority of instances of puerperal septico-pyæmia.

As far as **treatment** is concerned we can do but little in the more severe forms of infection. Where matter forms in the neighbourhood of the rectum, free incision, followed by irrigation with some antiseptic solution, and free drainage, should at once be adopted. Where the peritonæum has become implicated, opium in full doses is indicated, with the double object of relieving pain and lessening peristaltic action. The application of an ice-bag or cold water compress to the abdomen will sometimes relieve suffering ; and again in other cases, warm moisture in the shape of stupes and poultices will be found more comfortable.

The prophylactic treatment, however, holds out to us better hopes of success. We frequently find surgeons say that there is no use in adopting antiseptic measures in rectal operations, owing to the necessary escape of fæcal matter into the wound. It is true, certainly, that where fæces are extravasated into areolar tissue, and have no *free exit*, considerable irritation and suppuration are set up; but a stercoraceous abscess formed in this manner is essentially different from the more diffuse inflammation and involvement of the venous and lymphatic systems characterising the extensive septic poisoning. If the extravasation of fæces has free exit, so that it merely passes over a wounded surface, such as is left after the division of a fistula, no undue inflammation is produced, and we find in these circumstances wounds implicating the rectum healing as well and as rapidly as in other parts of the body. In the minor operations, such as the removal of piles by ligature or cautery, septic periproctitis is of extreme rarity; but antiseptic precautions should not on that account be dispensed with. In the more severe operations, such as extirpation of portion of the bowel, or extensive incisions implicating the rectal wall, septic infection is not infrequent, so that it is necessary for the surgeon to adopt the fullest antiseptic precautions. In the first place, thorough drainage is essential, and this is to be ensured by having the external wound sufficiently free, and, in cases where sutures are necessary, the introduction of rubber drainage tubes between them; this will be found especially necessary in operations where only a small portion of the lower rectum is removed, and in other operations where the mucous membrane is sutured to the skin. In some cases where deep sutures can be passed, so as completely to obliterate the cavity of the wound, drainage tubes may be dispensed with. In the next place, thorough and frequent cleansing of the wound must be attended to; this may most effectually be accomplished by irrigation several times in the twenty-four hours,* with a solution of corrosive sublimate (1 in 2,000). A light pad should now be applied over the anus and wound, and retained in position by a T-bandage, care being taken that the pressure is not so great as to retain discharge. The pad may be made of any of the usual antiseptic and absorbent materials, salicylic wool fulfilling all the indications admirably. All plugging of the wound should be avoided, except when applied in the lightest manner, to keep the edges of a wound from uniting too rapidly, as in fistula, or when hæmorrhage renders plugging a necessary evil; and in these circumstances the most extreme precaution must be taken that the materials used are not only aseptic, but powerfully antiseptic.

Improved methods of wound treatment have largely diminished the number of cases in which septic periproctitis follows even such severe operations as excision of the rectum; but unfortunately cases are still to be met with, the difficulties of obtaining a complete asepsis being obviously so much greater here than in other parts of the body.

Circumscribed periproctitis, which is secondary to changes in the rectal wall, is undoubtedly due in the great majority of cases to perforation, either the result of ulceration or direct traumatism; but sometimes an abscess may occur external to the rectum without any perforation, as a direct result of acute inflammation of the mucous membrane. We have examples of abscess produced in this way in other organs; for instance, in the neighbourhood of the urethra in gonorrhœal inflammation; and iliac abscess, the result of appendicitis following inflammation of

---

* Volkmann recommends continuous irrigation, but this is difficult to carry out satisfactorily.

the mucous membrane of the vermiform appendix. Abscess so formed may burst externally, or both through skin and mucous membrane.

Where perforation of the gut has taken place extravasation of a minute quantity of fæces is sufficient to produce a considerable amount of inflammation, and as a result suppuration. Abscess produced in this way discharges from time to time a little matter into the bowel, constituting an *internal rectal sinus*, and in this way may continue for a considerable time in a passive condition; but much more frequently it makes its way to the skin as well, and, opening there, forms a complete fistula.

Inflammation originating externally to, but in the neighbourhood of the rectum, in which the bowel escapes, or is at any rate only secondarily affected, presents itself under several conditions. Pott,* in his classical treatise on fistula in ano, enumerates three distinct varieties of this inflammation, or phyma, as he calls it: the *phlegmonoid*, the *erysipelatous*, and the *gangrenous;* and clinically these three forms may easily be distinguished. As I believe this classification to be founded on a correct pathological generalisation, I adopt it.

The *phlegmonous abscess* is the commonest of these varieties, the loose, fatty, areolar tissue that fills the ischio-rectal fossa being peculiarly prone to suppuration, the great movability of the parts no doubt also predisposing to this result. Although met with at all ages, ischio-rectal abscess is more particularly a disease of middle life, and more common in men than in women. It occurs most frequently in those who from some reason are run down and out of health, as the result of long-continued and emaciating illness, overfatigue, intemperance, etc.; and we may often be able to trace an exciting cause, such as an injury, exposure

* Earle's edition, vol. iii. p. 49.

to cold, or the straining necessary to pass a hard fæcal mass. A very frequent cause is slight external fissures, similar to the chapped nipples that so commonly give rise to mammary abscess.

The formation of acute ischio-rectal abscess is attended in the majority of cases with some fever, and the constitutional disturbance is generally out of all proportion to the local disease. The pain is severe and is increased by defæcation ; there is frequently irritability of the bladder, strangury, or retention of urine being not uncommon symptoms. Its situation is most frequently on the lateral aspect of the anus, and it presents itself as a prominent, tense, red, and shining swelling, in which fluctuation is distinctly to be felt.

As to the **treatment** of ischio-rectal abscess, it must differ according to the stage and intensity of inflammation. During the earlier stages, a bladder filled with ice, and moulded to the anal region, will be found to give considerable relief ; but when matter has formed, incision should at once be performed. This is of great importance in these cases, as if done early enough in abscesses that have originated outside the rectum, we may be enabled to prevent the formation of fistula ; or in those cases in which we cannot stop the formation of an internal opening, or in which such is already present, we may be enabled to limit to a considerable extent the amount of burrowing of the matter, and its extension upwards underneath the mucous membrane. Speaking of the prevention of fistula by early opening, Allingham,* whose experience in the treatment of these diseases is so great, speaks as follows : " If the patient will allow me to act in my own way I can almost guarantee that no fistula shall result. The following is the method to be adopted : The patient must take an anæsthetic, as the operation is

---

* "Diseases of Rectum," p. 15. Fourth edition.

very painful. I first lay the abscess outside the anus open from end to end, and from behind forwards, *i.e.* in the direction from the coccyx to the perinæum. I then introduce my forefinger into the abscess, and break down any secondary cavities or loculi, carrying my finger up the side of the rectum as far as the abscess goes, probably under the sphincter muscles, so that only one large sac remains. Should there be burrowing outwards, I make an incision into the buttock deeply at right angles to the first. I then syringe out the cavity, and carefully fill it with wool soaked in carbolic oil, 1 in 10 or 12. This I leave in for a day or two, then take it out and examine the cavity, and dress again in the same manner; but, in addition, I now use, if I think it necessary, one or more drainage tubes. In a remarkably short time these patients recover; the sphincters have not been divided, and the patient therefore escapes the risk of incontinence of fæces or flatus, which sometimes occurs when both the sphincters are incised."

The important element in the above treatment is undoubtedly the free opening and the breaking down the loculi into one cavity. For this purpose I prefer to use a flushing scoop (Barker's), with which the entire granulation surface can be satisfactorily removed. When this is complete the filling of the interior with boric acid and the immediate introduction of drainage tubes, together with the application of an antiseptic absorbent pad, instead of plugging with wool and carbolic oil, will be found, I think, to give the best results.

Instead of ischio-rectal abscess running the acute course above described, we sometimes meet with a chronic or cold abscess in this locality. This is mostly found in strumous individuals, and is attended with but little pain or uneasiness of any kind. The treatment to be adopted may be similar to the above; or, as in one case which was recently under my care, hyperdistension of the cavity with carbolic lotion, as recommended by Mr. Callender, left nothing to be desired.

A much less severe form of abscess than the ischio-rectal just described, is that known as the *marginal abscess* of French authors. This may originate in several ways, but the most frequent is by the suppuration of an external pile; in other instances it is more like a form of furuncle, originating in some of the mucous follicles of the anal margin, resembling a hordeolum in the eyelid; and again in other instances the suppuration starts from the base of a small external fissure or tear. These minute suppurations, owing to the extreme sensitiveness of the anal margin, are attended with considerable pain, but are not otherwise generally of consequence. Incision is the proper treatment, and the best way to do it is to pass the index finger into the rectum, and, with the tip of the thumb outside, the little tumour can be perfectly steadied, and the point of a lancet or small knife passed through it. As it is painful, local anæsthesia by ether spray should be used.

The marginal abscess, originating in a pile, sometimes leaves a small superficial fistula, which will be alluded to farther on, but the furuncle does not usually leave any fistulous trace behind it.

Erysipelatous periproctitis is a diffuse inflammation starting in the skin surrounding the anus, and presenting the characters of this disease in other parts of the body. As it presents no essential peculiarity in this region, we need not further allude to it here, except to mention that a subcutaneous abscess may result from it requiring the above treatment.

Gangrenous inflammation of the tissues surrounding the rectum is a very formidable disease, and may be attended with fatal result, or, where recovery takes place, there is a great destruction of tissue, with the consequent troubles attending the contraction of cicatrices. This disease may commence in the skin and rapidly spread to the underlying connective tissue, or the skin may be only secondarily affected, the deeper structures being first implicated. It occurs amongst the well-to-do, who consume large quantities of animal food, and drink too much alcohol, and is almost exclusively confined to the male sex, exposure to wet and cold being sometimes noted as an exciting cause in the recorded cases. There is a livid tumefaction in the posterior portion of the perinæum attended with considerable pain; if the disease spreads deeply into the pelvis, pressure on the sacral plexus may give rise to reflected pain. Fever of a low type is present, with brown furred tongue, quick weak pulse, and in severe cases delirium supervenes. The skin covering the swelling sloughs generally at several points as in ordinary carbuncle, to which this disease bears great resemblance; as the cutaneous slough separates, masses of dark gangrenous tissue are brought into view, discharging an offensive ichorous fluid rather than normal pus. After separation of the gangrenous tissues, a great chasm is left, which heals but slowly; and relapses, with extension of the disease, are common. According to Furneaux Jordan, who has published an interesting clinical lecture on this disease,* perforation of the rectum is rare. No doubt this is due to the very slight anastomosis which exists between the vascular supplies of the rectum and the surrounding structures.

A well-marked case of this kind bears a strong

* *Brit. Med. Journ.*, Jan. 18, 1879; p. 73.

resemblance to the gangrenous inflammation produced by extravasation of urine, and in one case I know so great was the resemblance that a surgeon advised incision of the perinæal urethra. In a case of doubt the urethra should be explored with a sound. If no obstruction is experienced, extravasation of urine is improbable.

The **treatment** should be prompt; early and deep incisions must be made in lines radiating from the anus; frequent syringing with solution of chlorinated lime, or corrosive sublimate (1 to 2,000), or the application of charcoal poultices to keep down fœtor, will be of use. And where any tendency to spread is recognised, chloride of zinc, or some other powerful escharotic, may be employed. At the same time the patient's strength must be supported by liberal allowance of milk, strong beef-tea, eggs, etc.; and the administration of alcohol is usually indicated.

In making the diagnosis of abscess in connection with the rectum the surgeon must bear in mind that suppuration, symptomatic of other diseased structures, may present in the posterior perinæal region. This is especially the case with urinary abscess, and with collections of matter originating in the female generative organs. Suppuration in connection with disease of the bony pelvis or vertebræ may also point beside the rectum.

CHAPTER V.

RECTAL FISTULÆ.

OF the diseased conditions consequent on the affections detailed in the last chapter by far the most important are : **Rectal fistulæ**, which present themselves to our consideration in considerable variety, and from their great frequency, and the discomfort they produce, constitute a subject of great interest to the surgeon; it is, therefore, to be little wondered at that, in looking through the voluminous literature on this subject, the names of most of the great masters of surgical art are to be found. According to the statistics of St. Mark's Hospital, as given by Allingham,* out of 4,000 cases taken consecutively, and without selection, from the out-patient department, there were 1,057 persons suffering from fistula, and 196 from abscess of which 151 eventuated in fistula, so that the disease constitutes considerably over one-fourth of all the cases presenting themselves as out-patients. Mr. Allingham also tells us that an examination of the records of the in-patients at the same hospital, during several years, shows that two-thirds of those operated on were cases of fistula. As the author very justly points out, statistics of this kind must be taken with some reservation, as we frequently find fistula co-existing with other forms of rectal disease; in some instances the accompanying disease being of a trivial nature, as external piles, etc.; while in others the fistula may be only secondary to more grave pathological changes, such as malignant disease or simple stricture. Another element of error which must not be lost sight of is that St. Mark's has

* "Diseases of Rectum," p. 12. Fourth edition.

a special reputation for the cure of fistula, so that many patients suffering from this disease go there, and so the list is swelled. At the Dublin general hospitals, although fistula is common, it is by no means the commonest of rectal diseases; and, in my own practice, it has not furnished more than one-sixth of rectal operation cases.

The use of the term fistula (a pipe), so common now in general surgery, originated probably in the special variety now under consideration, and amongst the older surgeons the term was confined to those cases in which the tissues immediately surrounding the ulcerating tract had become much indurated and thickened, the effects of long-continued inflammatory changes. Now, however, it is used in a broader sense, and is applied to any abnormal communication between a mucous canal and the external skin, or between two mucous surfaces; the term "incomplete fistula" being used where the fistulous tract has only one aperture, either mucous or cutaneous.

The exciting cause of rectal fistula may occasionally be a penetrating wound, but very much more frequently it is the result of suppuration, any of the varieties of abscess described in the last chapter commonly eventuating in this condition. Why an abscess situated in this region is more likely to become fistulous than abscess situated in other parts of the body, is a question of considerable interest, but the true explanation is not far to seek. In the first place, the looseness and free motion of the tissues in the ischio-rectal fossa tend to prevent rapid closure of an abscess cavity, in the same way that we see sinuses remaining for a long time unclosed in the axilla or female breast; and where a cutaneous opening alone exists this must be taken as the principal cause, but where there is a mucous aperture, a very much more potent influence is brought to bear;

namely, the constant trickle of mucus and thin fæculent matter along the track of the fistula. We must also bear in mind the peculiar vascular supply of the rectum, the veins being without valves. Hence it is that as man occupies the erect position, a considerable amount of congestion of the lower end of the bowel is frequently present, which we well know seriously impedes the process of cicatrisation.

Of the complete muco-cutaneous fistulæ connected with the rectum, the best practical division is according to the position of the internal opening; and we find there are three situations in which this opening may be found: in the first and most simple case, the fistula is subcutaneous, the inner opening being situated superficial to the external sphincter, in the anal canal. To this alone should the term *fistula in ano* be confined. Next we may find the opening situated between the two sphincters; this is probably the most common variety. And, lastly, we have the fistula in which the opening into the rectum is situated above the internal sphincter, traversing, therefore, the so-called "superior pelvi-rectal space of Richet." The first of these three varieties corresponds with the *fistules sous-tégumentaires* of Mollière, and the latter two with the *fistules sous-musculaires*. The superficial fistula has its origin in the marginal abscess before alluded to, and is generally very minute, and productive of little inconvenience; it may be altogether superficial, or extending through a few fibres of the external sphincter: as, however, it does not open directly into the cavity of the rectum, the escape of flatus and fluid fæces does not take place, consequently secondary suppuration is uncommon. So little is the irritation, that occasionally we see the internal surface of the fistulous tract cicatrise, and then it remains as an epidermis-lined tube, like the hole for an ear-ring, and ceases to give annoyance.

The ordinary form of **complete rectal fistula** is the result of abscess in the ischio-rectal fossa, either primary, or secondary to changes in the rectal wall, and the position occupied by the internal opening is tolerably constant, being situated between the two sphincters. The external opening is subject to more variety, but is generally within one inch from the anal verge, though occasionally at a considerably greater distance, as in the groin or thigh. The position of the internal opening in this variety is a subject of very considerable importance, as, unless acquainted with its usual situation, the surgeon may fail to find it; the reason for this being, that in most cases the mucous membrane of the bowel is separated from the muscular wall for a variable height above the orifice, so that the opening is not really at the highest point to which suppuration has extended, as one might be led to expect. In eighty cases which M. Ribes* had an opportunity of examining post mortem, in none was the internal opening situated at a greater height than 5 or 6 lines from the verge of the anus, and in many it was not so high. Velpeau tabulates the result of thirty-five post-mortem examinations as follows : In four the internal opening was one inch and a half from the anus, in one it was three inches, and in thirty a few lines from the anal outlet. Sir B. Brodie goes so far as to say : " The inner orifice is, I believe, always immediately above the sphincter muscle, where fæces are liable to be stopped, and where an ulcer is most likely to extend through both the tunics." Sir B. Brodie was of opinion that a primary perforating ulcer was always essential to the production of that form of abscess liable to terminate in fistula, and he did not admit the possibility of abscess from without causing fistula by secondary perforation of the rectum;

* *Revue Médicale Historique et Philosophique*, liv. i. p. 174. Paris, 1820.

however, Syme held the opposite view, and expressed the opinion that the abscess formed before the internal opening, as has already been pointed out in the preceding chapter. There is no doubt that both these causes may operate. Where the abscess is the primary affection, the point at which it will penetrate the muscular wall of the rectum depends chiefly, in my opinion, upon its relation to the levator ani muscle. If superficial to this muscle it is prevented from reaching the gut at a point above the insertion of the levator ani muscle, as it cannot well perforate the fascial structures that cover this structure upon both sides; it therefore makes it way through the insertion of this muscle.

A clear understanding of the levator ani muscle and the constitution of the pelvic diaphragm will much facilitate the comprehension of this subject. On each side the levator ani arises from the posterior aspect of the symphysis pubis, and from a tendinous thickening of the obturator fascia, stretching from this point to the spine of the ischium; from this extensive origin it descends and is inserted, posteriorly, to the coccyx, thence to the ano-coccygeal ligament; next the fibres become blended with the longitudinal muscular coat of the rectum, and with them pass in between the sphincter muscles, while the anterior fibres pass over the prostate gland, and blend with the muscle of the opposite side, thus constituting a complete diaphragm, the continuity of which is perfect around the entire circumference of the pelvis. As the upper surface is covered everywhere with the strong recto-vesical fascia, while the under surface is similarly covered with the more delicate anal fascia, it will be seen that a very efficient barrier is thus interposed between an abscess in the ischio-rectal fossa and the upper regions of the pelvis. Once having passed the muscular wall of the rectum an abscess distends the loose submucous tissue; but although considerable separation of the mucous membrane may take place, this structure is generally perforated in the neighbourhood of the situation indicated. Although measurements from the anus have been given above, it is more correct to say that the opening occurs immediately above the anal *canal*, what is described as the anal verge being the lower termination of that canal, and varies immensely with the motions of that canal.

The internal opening usually corresponds closely with the point at which the fibres of the levator ani muscle blended with the longitudinal external muscular coat pass between the two sphincter muscles. Where, however, the abscess originates above the levator ani muscle in the superior pelvi-rectal space, the matter may separate the rectum from its loose connections for a considerable distance upwards, and open into the lumen of the bowel at almost any point. In these cases of pelvi-rectal fistula the cutaneous orifice is usually situated posteriorly, the abscess having passed between the posterior portions of the levator ani muscle; whereas in ischio-rectal fistula the cutaneous opening is most frequently lateral. Again, in examining a case of fistula by means of probe with the finger in the rectum, the diagnosis can be made between the superior pelvi-rectal and the ordinary (ischio-rectal) fistula in which the orifice is high up, by the fact that in the latter the probe can be felt to pass, in the greater part of its course, immediately under the mucous membrane, while in the former the entire thickness of the gut is felt to intervene between the finger and probe.

Of the incomplete fistulæ two varieties are described; to these the terms "blind internal" and "blind external" have been applied. These names are most confusing, and, were it not that they have received the sanction of long usage, would not be referred to here. It is necessary to state, however, that the blind internal fistula is that in which there is a mucous, but no cutaneous opening; and the blind external fistula is merely a sinus with a cutaneous aperture, but no opening in the mucous membrane; although this is the generally accepted way in which these terms are used, Chelius and a few other authors reverse this arrangement. I have very often found students transpose the names, a mistake which, considering the nomenclature, is scarcely to be wondered at. In the following pages the terms "internal and external rectal sinus" will be used, as more clearly indicating the conditions referred to.

Internal rectal sinus may therefore be defined as a suppurating tract communicating with the cavity of the rectum by means of an opening in the mucous membrane; its origin may be due to similar causes to those which produce the complete fistula, and, indeed, it is frequently only the preliminary stage of the latter condition, an internal rectal sinus, if left to nature, sooner or later forming an opening in the skin, and thus forming the complete fistula.

The **external rectal sinus** is the result of an abscess which has commenced in the structures external to the rectum, not, of course, stercoral in its origin, which has broken or been opened externally, and which, from causes previously enumerated, is prevented from healing in the ordinary manner. The direction of the sinus may be towards the rectum, and sometimes ulceration will transform this also into a complete fistula, but in other cases the tract leads away from the rectum into the upper portions of the ischio-rectal fossa. Owing to the difficulty that is experienced sometimes in detecting the internal opening of a complete fistula, the diagnosis of external rectal sinus has been more often assumed than its existence proved. The study of dissected specimens has, however, completely determined the existence of this condition; but it is very much rarer than the complete fistula.

The tract of a fistula is seldom straight, and the calibre generally varies at different parts of its course. This irregularity is one of the most important reasons why difficulty is experienced in making a probe traverse the full extent of the fistulous channel. We sometimes find a sharp angle in the course, the cutaneous extremity at first leading away from the rectum, and then taking a sudden turn towards the bowel. This is probably due to the fact that the changed course is rendered necessary by the fistula following for some distance the direction of the levator ani muscle. And, again, it is common to find pouches, or diverticula, communicating with the main tract. These diverticula sometimes become blocked up, forming the foci of fresh suppuration which may perforate the skin at a distance from the original orifice, so that we sometimes meet with cases of fistula in which numerous external openings exist. I have seen in one case as many as twenty-two external openings, the whole of the buttock on both sides being riddled with fistulous tracts. Increase in the number of internal openings is extremely rare, and it never exists to the same extent as is the case with the cutaneous orifices, but sometimes two or more openings may be recognised in the mucous membrane. In the case above mentioned only one internal opening was discoverable.

Of these *complex fistulæ*, as they have been called by Hamilton, one important variety remains for consideration, namely, "the horse-shoe fistula." In this the external and internal openings are on opposite sides, the tract running round the margin of the anus subcutaneously; this may be due to abscesses starting in both ischio-rectal fossæ simultaneously; or it may simply be the result of the burrowing of matter from an abscess which was in the first instance unilateral. It is to be recognised by the discovery of the external and internal apertures upon opposite sides, and by the induration which marks the path of the fistula.

Histologically considered, a fistula resembles in structure an indolent ulcer. Its surface is covered by unhealthy granulations in part, and in others granulations are wanting; frequently we see at the external orifice a flabby sentinel granulation protruding, which, when present, renders the opening much more obvious. Immediately surrounding the granulation is a layer of fibrous tissue, which is sometimes of extreme density, feeling almost of cartilaginous hardness. This is found only in fistulæ of long standing, and is due to the proliferation of the connective tissue elements at the base of the granulations in abortive attempts at healing. It is remarkable the rapidity with which this hardness disappears when the fistula is healed. ...The most marked example of this condition that ever came under my notice was in a man who had had fistula for fifteen years. He had a hard warty growth about one inch from the anus, and ¾ inch in diameter, which was attended with but little discharge; from this point a hard ridge could be felt passing deeply. This wart was so firm and elevated that it suggested to more than one surgeon who saw it the idea that it was epitheliomatous; a probe, however, could readily be passed into the rectum, and simple incision sufficed to effect a permanent cure. I saw this patient a year after the operation, and the cicatrix was as soft and elastic as the neighbouring skin, no trace of the warty protrusion remaining. The older surgeons looked upon the thickening as a new growth, which it was necessary to remove. Hence arose the term of cutting *out* a fistula, and so freely was this done that in many cases incontinence of fæces was produced, and the patient was rendered very much more miserable by the *cure* (!) of his fistula.

The symptoms of which fistulous patients complain vary according as to whether there is active suppuration going on or not. The abscess from which the fistula originates is, if acute, extremely painful : though sometimes, as before noticed, a cold or chronic abscess may occur in this locality, in which case the initial pain is slight, and the first symptoms noticed by the patient are discharge or involuntary escape of flatus, and, more rarely, a small quantity of fæces. With a fistula fully established little pain is experienced, but considerable annoyance is occasioned by the soiling of the linen with discharge, which is generally accompanied with an unpleasant odour. There is generally some tenesmus, and the fæces may be stained with blood, etc., except when the case is one of external rectal sinus.

A person suffering from fistula is always liable to attacks of secondary suppuration, due to blocking of the tube by small particles of fæces or exuberant growth of the granulations. This, of course, is attended with pain, until a new opening forms or is made by the surgeon, or until, as sometimes happens, the passage of the original fistulous tract becomes re-established.

Fistula in some persons, more particularly those of a nervous temperament, produces an amount of depression and constitutional disturbance altogether out of proportion to the local disorder; their minds being impressed with a feeling of physical weakness, rendering them miserable and unhappy.

The diagnosis of fistula is generally attended with but little difficulty, although sometimes the small size of the external orifice renders it difficult of detection; and in others the great induration and prominence may give rise (as in the case above detailed) to the suspicion of epithelioma. Cases of the latter kind show how important, as an adjuvant to diagnosis, the positive detection of an internal opening may be.

A difficulty may possibly arise in determining whether a fistula is connected with the bowel or with the genito-urinary organs; but in the majority of cases no such difficulty should arise. Rectal fistulæ usually open in the posterior perinæum, while those in connection with the urethra open anteriorly; at any rate, a probe introduced into a fistula will show to which side of the perinæal septum the tract lies. This perinæal septum is a very definite structure, and is seldom perforated by suppuration, so that it may be generally assumed that abscess posterior to it is associated with the rectum.

In urinary fistulæ the escape of urine along the tract will serve to establish the diagnosis. The fistula of the superior pelvi-rectal space may be difficult to recognise. If the opening into the rectum can be found with the probe and finger in the bowel, or if fæcal matter escapes at the orifice, the diagnosis is obvious; but in other circumstances it may be impossible to distinguish this form from sinus following parametritis or that connected with diseased bone.

The **method of examination** for the purpose of diagnosis of rectal fistula is one of the most important points in connection with the discussion of this subject.

In order to examine a patient with suspected fistula he should be placed lying on the side, preferably that on which the external opening is situated, with the legs well drawn up towards the abdomen. If the external orifice is prominent, or if there is a sentinel granulation present, it will be obvious; but where small and situated between folds of skin, pressure with the top of the finger will usually cause a little drop of matter to exude, and so demonstrate its position.

Careful feeling with the finger will also frequently tell us the direction which the tract takes. A probe should be now passed along the fistula, and in doing this considerable care is requisite, and the utmost gentleness should be observed, always remembering that the probe is to be directed by the channel it is passing through, and not by the hand of the surgeon. It must also be remembered that fresh suppuration not unfrequently follows the use of a probe, no matter how gently handled, and against this, complete sterilisation of the probe and the use of antiseptics is not always efficacious. The fistula contains highly septic pus, from which the surrounding structures are protected by a layer of granulations. If this delicate granulation tissue is bruised and lacerated by the probe, inoculation and suppuration may readily be started afresh. It is well to be provided with one of the flexible spiral metal probes, and also with a very fine silver probe, for the investigation of tortuous and very narrow fistulæ ; sometimes, also, by bending the point of the ordinary probe, it will be found to pass with greater facility. The probe having been passed, the finger should be introduced gently into the rectum ; and it is a matter of considerable importance that this should be subsequent to the passage of the probe, as, if the finger is introduced into the rectum in the first instance a spasm of the sphincter muscles is set up, which will greatly increase the difficulty of passing the probe. It will be frequently found that, as the top of the finger passes the sphincter muscle, the end of the probe is felt free in the cavity of the rectum, thus demonstrating the fact that the fistula is complete. In other cases the mucous membrane is felt to intervene between the tip of the finger and the probe, and the latter can be passed freely over a considerable surface, showing that the mucous membrane has been separated to a large extent from the muscular walls of the rectum. In these cases the internal opening generally exists, but it may be hard to find; the usual fault being that it is looked for too high up. Careful palpation with the pulp of the finger will sometimes, if the sense of touch is delicate and has been educated, determine the orifice, while at others it may be brought into view by simply everting the mucous membrane of the anus. Should a doubt still exist as to the completeness of the fistula, injection of milk may be had recourse to, a speculum having previously been introduced into the rectum, when of course the appearance of milk in the bowel will set the question at rest; or, with the finger in the rectum, a few drops of a strong solution of iodine may be injected along the fistula; and if there is an internal orifice present, a brown stain will be perceived on the finger. If none of these methods demonstrate the inner opening, the case must be looked upon as one of external rectal sinus, but when the fistulous tract passes through the muscular wall and separates the mucous membrane from the outer coat, a mucous orifice may nearly always be found. Should this, however, not be the case, the failure will be of little moment practically, as the same plan of treatment applies to both. Where the entire substance of the rectal wall intervenes between the finger and probe, or where the probe is guided away from the rectum along the anal fascia to the upper portion of the ischio-rectal fossa, the case is one either of external rectal sinus, or of fistula originating in the superior pelvi-rectal space. We must, therefore, in these cases go farther, and try to find the cause, such as diseased bone, etc.; and in the female a vaginal examination may show us a uterine or ovarian origin. Where there are numerous external openings it is necessary to carefully probe all these so as to determine whether they are all connected and the direction which they take. The upper limit of the separation of the mucous membrane should also be made out, and search should be made for more than one internal orifice, if such is likely to be present.

The diagnosis of internal rectal sinus is more difficult generally than the forms we have been considering. In this form the fæces will be smeared with pus, or blood, and a boggy swelling is to be felt at some portion of the anal circumference. If the internal orifice can be felt, or seen, through a speculum, a bent probe may be introduced into it, and made to protrude in the perinæum.

Having made the diagnosis of complete fistula, we must, before proceeding to treatment, satisfy ourselves as to the presence or absence of other important diseases, and, of these, changes in the rectum itself hold a first place. The co-existence of fistula with other pathological conditions in the lower bowel has been already alluded to; but I cannot too fully impress on the junior practitioner the necessity of examining for stricture, malignant disease, hæmorrhoidal and other tumours, in every case that comes under his notice. A thorough physical examination of the chest should also be made, the co-relation between phthisis and fistula being of extreme importance. Any serious disease of the kidney should be eliminated before recommending operation; albuminuria in this, as in almost all other surgical procedures, being a very unfavourable indication.

Where the inconvenience produced by fistula is but slight, there is sometimes considerable difficulty in getting a patient to consent to the treatment necessary for cure; but it may be taken as a general rule that the longer a fistula is left without being efficiently dealt with, the more tortuous and complicated it becomes, and the more difficult will any subsequent procedures be.

TREATMENT WITHOUT INCISION.

The **treatment** to be adopted may be conducted upon two principles : (1) by trying to excite healthy action in the fistulous tract, and so produce its obliteration ; (2) by the division of the structures intervening between the rectum and the fistula by some means.

Of the former, but little can be said in its favour. The process is slow and the result uncertain; whereas, by division of the septum the result is generally eminently satisfactory. In exceptional cases, however, where the patient has a very great dread of any cutting operation, and when he is willing to give up a sufficient time to the treatment, a cure by the milder means may, in the first instance, be attempted; but for this to be attended with any reasonable prospect of success the fistula must be of somewhat recent origin, and unattended with any considerable amount of induration. Indeed, the occasional occurrences of spontaneous recovery, or cure resulting from the slight amount of irritation produced by the passage of a probe, show that operation is not always essential; but so rare are these cases, that the surgeon should recommend the more certain and speedy method by division except under the circumstances above stated, and where the fistula is uncomplicated, or is obviously an instance of external rectal sinus. The essential elements of treatment, where the attempt is made to obliterate the fistulous tract without division, are : rest, free drainage, and the use of antiseptic and stimulating applications to the suppurating tract. The patient should be confined to the horizontal position, either in bed or on a sofa, this being a most important condition, as it relieves congestion. A firm pad, well supported by a T-bandage, should also be worn for the purpose of limiting the motions of the anus due to alternate contraction and relaxation of the levator ani. For the purpose of clearing out the fistula dilatation should be employed by the introduction of a piece of sea tangle, or some other species of dilator, which should pass to the bottom of the tract. This not only renders the application of antiseptics and the introduction of a drainage tube easy, but the stretching of the walls affords a useful stimulus, and frequently initiates the healing process. Allingham recommends the thorough swabbing out of the fistula with strong carbolic acid, and the passage of a full-sized soft rubber drainage tube right to the bottom. As the case progresses, the drainage tube may be shortened down and finally removed, or, instead of the tube, a bone shirt-stud, perforated through the centre, may be buttoned in to ensure the patency of the external orifice. The application of carbolic acid may have to be repeated, or some other antiseptic may be used with advantage, such as ethereal solution of iodoform, corrosive sublimate solution, nitrate of silver, or tincture of iodine; but although success may at first appear to follow this line of treatment, division will have very frequently to be adopted subsequently. Allingham, who has had such extensive experience in fistula, and who has tried this plan of treatment more fully probably than any other surgeon, states his experience thus : * "Altogether I find I have had twenty-one successful cases, and a considerable number in which I have failed to effect a cure after a prolonged attempt, therefore I cannot say the prospect is very encouraging."

To Pott is undoubtedly due the credit of having demonstrated the fact that a simple *incision*, converting the tubes of the fistula and rectum into one, is generally sufficient to effect a permanent and complete cure. Previous to his writings, if any operation was attempted, it was one of partial or complete *excision* of

* *Loc. cit.*, p. 26.

the entire fistula; for instance, Cheselden recommended that one blade of a polypus forceps should be passed into the fistula, and the other into the rectum, and then the entire portion included was cut out with scissors. Such a proceeding is wholly unnecessary, and justly merits the term "barbarous," which has since been freely applied to it. Since the classical treatise on fistula in ano of Pott appeared, considerable ingenuity has been displayed in attempts to improve on the means to be adopted for dividing the structures intervening between the rectum and fistula; but before considering these, it may be useful to discuss the rationale of the treatment. In the first place, the effect of incision is to ensure free drainage, and the prevention of any accumulation of pus in the deeper parts, in the same way that we see incision cure sinuses in the breast, or those following bubo, etc., but a like result can be obtained by the careful introduction of a drainage tube, so that we must look for other and more potent reasons for the necessity of incision in this disease, and these are to be found in the action of the sphincter muscles. If we undertake the cure of urethral fistula the result of stricture of that tube, every surgeon knows that his efforts will be futile until the stricture is removed, the urine trickling through the fistula, and so keeping it open; but once the stricture is fully dilated, the fistula can generally be closed without much difficulty. Now, the sphincters of the anus produce a similar effect in the rectum : as they are tightly closed except during an attempt at defæcation, there is a frequent discharge of flatus, rectal mucus, and occasionally slight amount of fæcal matter through the fistula, effectually keeping it open. Incision through the portion of muscle below the tract, by stopping for a time all retention of gas and fluid, tends to promote healing. And, again, the mechanical action of the sphincter muscles in constantly pulling on and moving the fistula, prevents its repair in the same way that we see a sinus in close relation to a tendon in the hand or foot kept open by the constant action of the muscle. If a splint is so applied that the tendon is kept at rest, the sinus will soon close : in the same way incision quiets for a time the sphincter, and allows repair to go on. Forcible dilatation of the sphincter, no doubt, produces a temporary paralysis sufficient to prevent retention of gas, and to keep the muscle at rest, and may, therefore, in exceptional cases, be found a useful adjunct to the treatment by drainage tube, but as it requires the administration of an anæsthetic, it is better to proceed at once to the more certain plan of division.

Division of the intervening septum between the fistula and the bowel may be effected in a variety of ways : the knife, the elastic ligature, the thermocautery, the galvanic écraseur, and the wire écraseur of Chassaignac, all having their advocates. The operation which is recommended by Pott, and which is practised still upon a large scale, is performed with a curved blunt-pointed bistoury ; the knife having been passed through the fistula, and the index finger of the operator's left hand introduced into the rectum, the probe point is felt for ; if it has passed through the internal opening into the cavity of the rectum, it is rested against the pulp of the finger, and both withdrawn together with a sawing motion, the intervening structures being thus divided ; if, however, the internal opening cannot be made out, the blunt point of the knife being felt outside the mucous membrane, a very little pressure will suffice to make it penetrate this soft structure, and then the operation can be completed as before. This operation has been modified in many ways, as accidents may happen during the performance of it : the surgeon is apt to wound his finger; and sometimes, when the fistula is very hard and callous, the knife breaks. All risk of the former may, however, be obviated by substituting for the finger a speculum, a gorget, or even a tallow candle. A preferable mode of operating where the internal opening is situated low down, is by passing along the fistula a probe-pointed flexible director (Fig. 14), the end of which can be hooked out at the anus by means of the left index finger, and the parts divided by running a knife along the groove of the instrument. The pulling down the director gives the patient a little pain if general anæsthesia has not been induced, but this is more than counterbalanced by the

Fig. 14.—Brodie's probe-pointed Fistula Director.

fact that the parts situated on the director can be frozen with ether spray previous to division. Allingham figures and describes an instrument which he has devised for performing this operation.* It consists of a director and spring-scissors. There is a button on one blade of the scissors, which runs in the groove of the director, and from which, when once introduced, it can be only removed at the end next the handle. He uses this for fistula situated high up, where difficulty would be experienced in bringing the end of the director out at the anus. I think that where simple division is required a knife is much to be preferred, except under special circumstances, and in no case of fistula that I have ever met with did I consider the use of scissors called for.

When either of the foregoing cutting operations is decided upon, the patient should be prepared by having his bowels well cleared the day before the

* *Loc. cit.*, p. 42.

operation, care being taken not to provoke a diarrhœa, and the rectum should be washed out with an antiseptic enema immediately before he is placed on the table. He should lie on the left side, with the knees well drawn up to the abdomen, and if an anæsthetic is used, it is well to tie up the right knee by a band passing round the neck and thigh: a strong calico bandage folded lengthwise three or four times answers the purpose well, and will be found to aid very materially in retaining the patient in proper position. A still more convenient plan is to retain the patient in the lithotomy position by means of Clover's crutch. (*See* page 11.) In making the section the surgeon should take care that the incision passes at right angles to the anal margin, as if the cut goes obliquely through the muco-cutaneous tissue, leaving the edges bevelled, healing will be much retarded, and incontinence more likely to result.

In dealing with **complex fistulæ** the surgeon must be guided by the individual peculiarities of each case. Where there are numerous external openings it is the safest plan to slit up all the sinuses at the same time, but where more than a single internal opening exists the case is different. If numerous incisions be made in the external sphincter, more or less incontinence will be a probable result; so that the safer plan of treatment to adopt in these cases is to incise the sphincter on one side only in the first instance, and to attempt to cure the other fistulæ by the application of caustics and use of drainage tube, the physiological rest afforded to the part by the incision already made lending powerful assistance to the treatment, so that success may be looked forward to with a considerable degree of confidence. Should, however, any fistulous tract remain unhealed after the one incised has cicatrised, a secondary operation will weaken the sphincter less than if the sections are made simultaneously.

Where the mucous membrane is separated from the muscular tunics of the rectum for a considerable distance above the internal opening, professional opinion is somewhat divided as to the proper course to pursue. Allingham states most positively that the incision must extend to the uppermost limit of the suppurating tract.* Curling advises a similar procedure; while Syme, Brodie, and many others, state that it is sufficient to divide the fistula between the two openings, and that the upper portion will heal. I believe the best rule of treatment lies between these extremes. If in the first instance the ordinary incision is made, the surgeon can watch the progress of the case, and it will frequently be found that, as granulation proceeds, the pouch between the mucous membrane and muscular wall will fill up; but if it is not healing satisfactorily, slight pressure will be sufficient to make a few drops of matter exude, indicating its position, and, by passing a fine probe-pointed knife up, division of the mucous membrane can be effected with almost no pain to the patient, as the mucous membrane above the external sphincter is possessed of but very slight sensibility. I have on several occasions adopted this plan with excellent results. The careful watching for small suppurating tracts communicating with the original wound, constitutes, in my mind, one of the most important parts of the after-treatment of these cases.

Where the edges of the fistula are very indurated and callous, it will promote more rapid healing if the tract is scraped with a sharp spoon, or the plan of operation which has been termed the "back cut" by Mr. Salmon may be adopted sometimes with advantage. He recommends that, after division in the usual way, a linear incision should be made through that portion of the sphincter muscle which lies *outside* the tract of the fistula, hereby adopting the

* "Diseases of the Rectum," p. 93.

operation customary for the treatment of ordinary anal fissure.

Where the edges are much undermined, or where, after division, there is a tendency to curve inwards, it is advisable to cut off the overhanging margin of skin.

In the treatment of **horse-shoe fistula** just referred to, it will be found advisable to incise the sphincter only on the side of the internal opening; and open up the sinus extending round the anus, so as to allow of its being dressed from the bottom throughout its entire extent.

The treatment of **external rectal sinus** must vary according to the direction and extent of the suppurating tract. If a probe is found to pass through the muscular wall, so that the mucous membrane alone intervenes between the finger introduced into the rectum and the probe passed along the sinus; then, by perforation of the mucous membrane at the usual situation of the internal opening, the sinus should be transformed into a complete fistula, and the ordinary operation for division of the sphincter performed. If, however, the external sinus, when it impinges against the levator ani muscle, be directed by that structure away from the rectum, instead of the usual course down between the sphincters, division of the sphincter is wrong in theory; and futile in practice. The proper course will undoubtedly be to enlarge freely, by incision, the cutaneous orifice, and by dressing carefully from the bottom a satisfactory result will in all probability be obtained.

Internal rectal sinus always demands operative interference, and the sooner this is adopted after the diagnosis has been made, the better for the patient; for, if left alone, the matter burrows, and may produce a considerable amount of separation of the rectum from the surrounding structures before the skin gives way, converting it into a complete fistula. In making an external opening into the suppurating cavity, difficulty may be experienced from the fact that even slight pressure may empty the matter into the rectum, and so the indication of fluctuation may be lost : it is, therefore, better to introduce a suitably bent probe from the rectum into the abscess, and then the blunt extremity of the probe can with ease be cut down upon from without, and the case treated as one of complete fistula.

The **deep fistula** which runs through the muscular pelvic diaphragm, formed of the two levatores ani, and then, traversing the superior pelvi-rectal space, perforates the muscular coat of the rectum at a considerable distance from the anus, presents much more difficulty in treatment than those which we have hitherto been discussing. If the intervening tissues between the rectum and fistula are divided with the knife, the risk of serious hæmorrhage is encountered, as it involves the incision of a considerable portion of the highly vascular intestinal tunics : it will be well, therefore, in these cases, to substitute for the knife the cautery, the galvanic or wire écraseur, or the elastic ligature. Gerdy has recommended a form of clamp founded on the principle of the *enterotome* of Dupuytren, by which the division of the deeper parts can be gradually accomplished. It is claimed for this instrument that should a fold of peritonæum exist in the structures to be divided, a limiting peritonitis would be produced, and so the danger of opening the peritonæal cavity obviated. In this respect, no doubt, it has advantages over the knife, cautery, or écraseur, but not over the elastic ligature, which is more certain and more rapid in its action, and can be applied with less pain and annoyance to the patient. (*See* page 97.)

When a sinus opening beside the anus is found to be connected with diseased bone, or is due to pathological changes in the genito-urinary apparatus, the treatment must be conducted on general surgical principles, which it would be out of place for us to discuss further here.

As to the **after-treatment of division of the sphincter**, there are some important points for consideration. Hæmorrhage is seldom troublesome where the sphincter has been divided for ordinary fistula, but occasionally it is somewhat free. This is generally the case, as Gross has pointed out, where the edges of the fistula are very much indurated, and is no doubt due to the fact that the divided vessels are unable to retract and contract as they do in normal tissues. A parallel example is to be seen in the very free bleeding frequently following the division of the thickened cicatrices left after a severe burn. If any vessel is to be seen spouting, it should, of course, be secured with a ligature, but in general the application of cold and pressure, or the application of the thermocautery will be found sufficient.

Van Buren recommends that the surface of the wound left after incision should be dusted over with ferrous sulphate, which not only stops bleeding, but, according to him, forms an adherent crust over the surface, so imitating the natural healing under a scab. It is well, too, I think, to combine the ferrous sulphate with iodoform: a mixture of equal parts of these drugs dusted over the surface forms the best possible dressing that can be applied. A piece of iodoform gauze or salicylic cotton should be *lightly* placed in the wound, a morphia suppository having previously been introduced into the rectum. A pad of some antiseptic absorbent should be now applied; the sanitary pads made of wood wool, impregnated with corrosive sublimate, and sold for the purpose of absorbing menstrual discharge, are admirably suited for use after this as after many other rectal operations. All is fixed with a T-bandage, applied with sufficient firmness to limit the motion of the anus, and so keep the parts at rest.

If the bowels have been well cleared out before the operation, and no discomfort is complained of, it will be well to let matters rest until the third day, when a mild unirritating aperient should be prescribed, such as compound liquorice powder, some of the saline mineral waters, or, where the patient has not a very great dislike to it, castor-oil. Immediately before the bowels move, it will conduce much to the patient's comfort if he have a hot hip-bath, so that the dressings may be thoroughly well softened and loosened. As soon as the rectum is evacuated, the wound should be gently and completely cleansed, and a fresh dressing applied, consisting of a little antiseptic cotton. After the first dressing, however, it will render the subsequent treatment easier if the cotton wool is covered with iodoform ointment. The patient should be kept in bed, and a nourishing but unstimulating diet ordered, in which milk forms a large item. Until the process of cicatrisation is well established, the wound should be dressed daily, preferably immediately after the bowels move; and a careful watch for any suppurating tract discharging into it should be maintained. The process of healing is, as a rule, somewhat slow, and is liable to sudden arrest after it has hitherto progressed favourably. The causes of this delayed healing are many. In the first place, the frequent distension and soiling of the wound by the passage of fæces is, of course, impossible to obviate, but much may be done by ensuring a soft motion at regular intervals, at the same time taking care not to establish a diarrhœa. Another important factor is the congestion of the lower portion of the rectum, so common in these cases, and due, as pointed out by Verneuil, to the anatomical disposition of the veins. These vessels, which are destitute of valves, pass obliquely through the muscular walls of the rectum by openings which are unprotected by tendinous rings, so that they are compressed by the muscles when contracting to empty the bowel. Where this congestion is considerable, healing is much retarded, the wound resembling a varicose ulcer of the leg, which is sometimes so difficult to heal. Precisely the same treatment will be found to apply to both, the most essential element of which is rest in the recumbent position : this, by relieving the vessels of the weight of the contained column of blood, materially assists the healing. After the first three or four days it is not necessary to keep the patient in bed, but the recumbent posture should be maintained on a couch for the greater part of the day, until cicatrisation is completed. Stimulating applications will be found of great service in these sluggish cases which become indolent : such as balsam of Peru, compound tincture of benzoin, elemi ointment; or sometimes lotions of sulphate of zinc or ferrous sulphate will answer better. If not progressing satisfactorily, it is a good plan to vary these applications; or more potent stimulants may be tried, such as red oxide of mercury, nitrate of silver, or even nitric acid. When other means have failed, change of air, particularly to the seaside, will frequently cause these indolent sores to assume a healthy surface and rapidly heal. This is especially true of hospital patients sent to a convalescent home in the country.

When the fistula has been the result of a marginal abscess, and when the tract passes through the fibres of the external sphincter, a small ulcer may remain unhealed after treatment by division, presenting the character of the **painful fissure** of the rectum. It is to these cases that the " back cut " of Mr. Salmon is so eminently suitable in the primary operation ; but if this has not been done, and a fissure results, the proper line of treatment is to incise the floor through the fibres of the sphincter, as in the case of a spontaneous fissure.

Should **tuberculosis** be present, as evidenced by livid edges, undermined and infiltrated skin, a thorough scraping with a sharp spoon, and liberal use of solid nitrate of silver is indicated.

Under the term "trichiasis recti," Gross* has described a condition which retards healing of these wounds. It is, as the name implies, analogous to the trichiasis of the eyelid, the margins of skin becoming turned inwards. The hairs which grow in the neighbourhood of the anus are directed against the wound, and so a source of irritation is kept up. The treatment should consist in the removal of the cutaneous margins, or, where this is not considered advisable, careful epilation.

Secondary suppuration occurring in the proximity of the wound will of course delay the healing process, and the surgeon should always be on his guard for this complication. Any complaint from the patient of pain about the incision or any increase of the discharge calls for the most minute examination of the part, and if any secondary focus of suppuration be found during the after treatment, it should be opened from the original wound without delay.

An unpleasant sequela to the operation for fistula is a certain amount of **incontinence of fæces** when fluid, or of flatus. This is happily of rare occurrence, and only follows extensive operations, such as those required for the superior pelvi-rectal fistula, or where more than one division of the sphincter has been rendered necessary. Where it exists to any extent it is productive of great annoyance to the patient, possibly more than the original fistula, the cure proving worse than the disease. Much can be done to remedy this by the judicious use of the cautery; if the sharp

* "System of Surgery," vol. ii. p. 602. Sixth edition.

point of Paquelin's thermo-cautery be applied to the cicatrix of the operation wound, contraction will follow, giving tone and increased power to the sphincter and decreasing the size of the anal outlet. In this way this troublesome complication can generally be relieved; but care is obviously necessary not to cauterise the anal margin to too great an extent, as an intractable form of stricture might be the result.

Ligature for the cure of rectal fistula where the material used is inelastic has but little to recommend it; it is impossible to tie a silk ligature so tight that it will completely cut through and strangulate all the tissue which requires division in an ordinary case of fistula; and many plans have been devised to try and improve this method of treatment, such as hanging weights to the ligature, and the use of the tourniquet of Mr. Luke, by means of which the ligature could be screwed up tighter every day. These plans entailed so much suffering on the patient that they are now completely discarded; and the elastic ligature has come to be the only form that is now used in the treatment of this disease. Although occasionally practised by other surgeons, the use of an indiarubber band for effecting division of tissue in surgery has been prominently brought under the notice of the profession by Professor Dittel of Vienna; the great difference between division effected by this means and that produced by the inelastic ligature being that the latter, if including a large amount of tissue, sinks into it, and soon ceases to constrict the vessels, the central portion soon regaining its suspended vitality, separation thus being stopped. With the elastic band, on the contrary, the pressure is maintained, as the cord cuts its way through the constricted structures, so that a comparatively thick mass of tissue can be divided by this means with one application of the indiarubber band. What, then, are the relative merits of the knife and the elastic cord in the treatment of rectal fistula? The knife possesses many advantages, which may be briefly enumerated as follows: Division is quickly accomplished; pain can be obviated by anæsthesia; where the fistula is complex the secondary channels can be at the same time laid open; and the after treatment is almost painless. On the other hand, if the fistula is a deep one, the use of the knife may be followed by considerable hæmorrhage.

The advocates for the **elastic ligature** claim for it that there is no hæmorrhage. This is a matter of considerable importance where the fistula penetrates deeply, and also in those rare cases of so-called hæmorrhagic diathesis where severe bleeding is apt to follow a trivial incision. It is stated by those who have had large experience in this plan of treatment, that contrary to what one might expect, the pain attending the ulceration of the band through the soft parts is slight, the patient being often able to go about his occupation during treatment. Too rapid closing in of the wound by primary union of the parts divided is impossible. In my experience, this danger when the knife is used is more to be read of in books than seen in actual practice. The difficulty that the surgeon has to experience is more frequently due to a want of reparative action than to too great activity in the process of cicatrisation. The time occupied in treatment by elastic ligature is not longer than when simple incision is practised. The average time the elastic ligature took in cutting out, as estimated by Allingham from an experience of ninety cases, was about six days; but nevertheless the total time of treatment is not increased, as healing goes on in the deeper portions of the wound as the rubber band sinks into the constricted tissues. Another advantage claimed for it is that patients who have an insuperable dread of any cutting operation will sometimes submit to the application of an elastic cord.

If, then, we try to define the cases that experience has proved are most suitable for treatment by elastic ligature, we find that its use is somewhat limited. In the first place, the fistula must be simple, *i.e.* uncomplicated with sinuses. This, it may be remarked, is a matter not always easy of diagnosis before commencing operation. Probably the most suitable cases are those in which the patient is debilitated by phthisis or other chronic disease, or those in which the patient is unable to lie up during treatment, or in which there is known to be a hæmorrhagic tendency. Where the fistula is deep elastic ligature may with advantage be combined with the knife, the latter being used for the superficial structures and the elastic cord for the deeper parts where vessels of considerable size exist, and where difficulty might be experienced in securing them if divided by the knife.

The application of the elastic ligature is best effected in the manner recommended by Allingham.

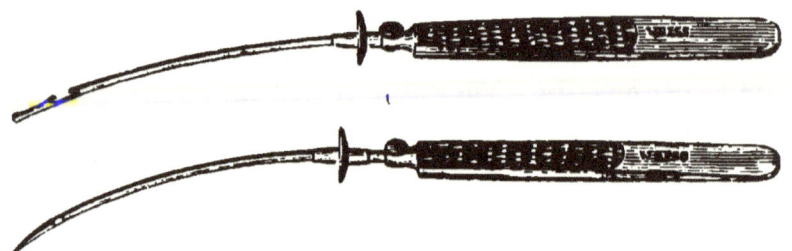

Fig. 15.—Allingham's Instrument for the Introduction of Elastic Ligature.

He uses solid rubber cylindrical cord $\frac{1}{10}$ of an inch in diameter, and so strong that it cannot readily be broken by pulling. For its introduction he has devised a very ingenious instrument (Fig. 15), which consists of a curved blunt-pointed probe with a deep notch near the extremity. This is fitted in a handle, and a cannula slides on the probe, so that once a loop of the ligature is caught in the notch the cannula is slipped up and the ligature prevented escaping until released by the cannula being drawn down again. The instrument can be fitted with a sharp-pointed probe instead of the blunt point where it is necessary to transfix tissues. In order to apply a rubber cord to a case of simple complete fistula, this instrument, fitted with the blunt point, is passed through the fistula, and the index finger of the left hand introduced into the rectum, carrying a loop of elastic ligature stretched over the tip. The point of the instrument can now readily be passed between the tip of the finger and the ligature, and the latter slips into the notch. The cannula is now pushed up and the instrument withdrawn; by this means the ligature is drawn through the fistula, one end hanging out of the rectum and the other out of the external orifice. Both ends are now passed through a small pewter ring made for the purpose, and by pulling, the cord is stretched to the required extent. The pewter clip is compressed with a strong forceps, the excess of cord cut off, and the operation is finished.

This method is incomparably superior to the other plans previously devised, such as that recommended by Dittel, which is to introduce a grooved director up the sinus, and along this a needle threaded with the ligature is passed, by means of the left index finger in the rectum a loop of the elastic cord is to be hooked down, and the needle withdrawn along the fistula. The pewter clips answer admirably for fastening rubber cord, as if it is knotted it is apt to slip. Should, however, these clips be not at hand, the elastic band may be firmly held by a piece of strong silk applied in the following manner : It should be laid on the skin between the anus and external orifice of the fistula, and the ends of the elastic made to cross it at right angles.

The ligature is now pulled quite tight, and the silk tied firmly round the ends of the elastic band where they cross. This answers well, but as it requires the help of an assistant to tie the silk while the rubber is extended, it is not so handy as the clips.

After the application of an elastic ligature, but little after-treatment is required; it will be frequently found that by the time the cord separates the wound is superficial.

Mr. Reeves[*] has suggested a plan of division of the septum in rectal fistula, which he designates by the not very happy term "immediate ligation." He recommends that a strong ligature should be passed with one end through the fistula and the other out at the anus: the ends then being firmly held, by a to-and-fro motion the ligature is made to saw its way out. This manœuvre resembles the obstetric operation by which, in impacted transverse presentation, the neck of the fœtus is sawn through with a strong piece of whip-cord. Mr. Reeves claims for this that it is free from hæmorrhage; and that healing from the bottom progresses favourably.

The same author has also made a suggestion[†] which may, in some cases, be of service, more especially where the external opening is situated at a long distance from the anus, and that is, after thoroughly dividing and scraping the fistula, to bring the sides together by sutures passed deeply in the hope that by obtaining primary union the tedious process of healing by granulation may be obviated.

Dr Stephen Smith[‡] speaks very favourably of this method. He advocates the complete extirpation of all the granulation tissue, and then bringing the opposed surfaces accurately together by means of (1)

---

[*] *Med. Times and Gazette*, June 15, 1878; p. 649.
[†] *Brit. Med. Journal*, p. 917; 1881.
[‡] *New York Med. Journal*, June 12, 1886.

a saddler's stitch of carbolised silk passed about three-quarters of an inch from the margin deeply through the sides, and extending the entire length of the wound; (2) a few points of interrupted suture passed *underneath* the wound; (3) a fine catgut continuous suture to the edges of the incision.

In a number of cases I have tried this plan, and in uncomplicated cases of fistula consider it a very great improvement. If we fail in obtaining primary union the fact will be indicated by an escape of pus from the deeper parts of the wound about the third or fourth day. In these circumstances all the sutures should be removed at once, any adhesions broken down with a probe, and the wound dressed from the bottom. The patient will then be in the same position as he would be if no attempt had been made to obtain primary union; whereas, if we succeed in obtaining healing, the gain is a very considerable one. I am so impressed with the value of this method of treatment that I invariably adopt it in uncomplicated cases, and am confident that if the curette is thoroughly used and the fresh surface rendered efficiently aseptic, union can be obtained and the patient cured in ten days. In my own practice 80 per cent. of simple cases heal by primary union.

The **thermo-cautery** of Paquelin (Fig. 30, page 215) is an instrument which has rendered very great service to rectal surgery. Indeed, no operation in this special department, where the occurrence of hæmorrhage is possible, should be undertaken without having this most useful appliance at hand. In principle it depends upon the well-known power which platinum (in common with some other metals) has of causing, when heated, the rapid oxidation of volatile hydrocarbons, and it consists essentially of hollow platinum points, of various shapes, fitted upon metal tubes. In the interior of each of these is a smaller tube, ending inside the platinum point, where it is covered with a small piece of fine platinum-wire gauze. By means of a rubber hand-bellows air is made to pass over benzole, contained in a bottle, and the air now charged with benzole vapour is conveyed by means of rubber tubes to the inner tube of the instrument. If, now, the platinum point be heated in the flame of a spirit lamp, combustion of the benzole vapour will take place in the interior of the cautery, the products of combustion passing down the outer tube. As long as the supply of air and benzole vapour is maintained the point will remain hot, and the degree of heat can be regulated to a nicety, from a bright white to a dull red or black heat, according to the rate at which the air is propelled through the benzole bottle. In using this instrument care must be taken that it is not *too* hot, a dull red heat being the most suitable, as if it is more intense than this the tissues will be protected by a thin film of steam, in the same way that a drop of water assumes the spheroidal state and will not boil, when placed on a white-hot platinum dish. The result of this will be that either the tissues will not be sufficiently divided, or if the point is firmly pressed it cuts through without charring the divided vessels enough to stop bleeding.

The treatment of fistula by this cautery appears to meet with more favour in France than in Britain, both Verneuil and Pozzi speaking highly of it; and in deep fistula it can be used with great safety and certainty to open up thoroughly the ramifications of a suppurating tract in complicated cases. In simple cases, however, the methods before alluded to will generally be found more satisfactory in practice. When the knife is used the cautery may prove a useful addition, either to check bleeding from vessels divided by the incision, or to continue the section through deeper parts where serious hæmorrhage is to be apprehended.

The **galvanic écraseur** can be used for the treatment of simple fistula, but it is not suitable for making the secondary sections necessary in complex cases; and the requisite cumbrous battery limits greatly its utility. When, however, it can be readily obtained, it does its work cleanly and expeditiously.

The wire or chain écraseur is difficult to apply, and its only advantage, freedom from hæmorrhage, can be quite as well or better ensured by the elastic ligature or cautery; so that I do not recommend its use in the treatment of rectal fistula.

Fistulous communications between the rectum and other mucous tracts (**fistulæ bimucosæ**) may be conveniently classified into three divisions : (1) In which some portion of the bladder or urethra is penetrated; (2) where the female genital organs are implicated; and (3) where a fistulous channel is established between some other portion of the intestinal tract and the rectum.

Recto-vesical and recto-urethral fistulæ may result from a variety of causes, of which the following may be enumerated : Direct traumatism. This will most commonly result from, or be occasioned by, an accident attending some surgical operative proceeding. Thus, perforation of the bladder through the rectum for the relief of retention of urine entails of necessity this condition; but in the great majority of cases this perforation closes rapidly when once the normal channel for the evacuation of the bladder is restored. But where it has been found necessary to tie in a cannula, or where the obstruction in the urethra continues for long, a fistulous communication may be established. Penetration of the vesical wall may also result from the forcible introduction of an enema pipe or other foreign body through the anus. The most frequent traumatic origin, however, is undoubtedly the accidental wounding of the rectum in the operation for lateral lithotomy.

Suppuration originating in the structures surrounding the rectum or urethra may, by opening both these mucous tracts, result in the formation of a fistulous communication; such is occasionally the case with prostatic abscess.

Malignant disease, whether originating in the tissue of the rectum or in its immediate proximity, is undoubtedly the most frequent cause of an abnormal communication being established between these two organs. This subject, however, will be more fully considered in the chapter upon Neoplasms.

Primary ulceration of the bladder of a non-malignant type must be admitted as a cause, though it is an extremely rare one; a few cases having been recorded in which urinary calculi escaped by this means into the rectum, and were discharged through the anus.

In the female, recto-vesical fistulæ are, as might be anticipated from the anatomical relation, excessively infrequent. Simon has noted one case* which was subsequent to a difficult labour, and was attended with occlusion of the upper two-thirds of the vagina. In other instances it would appear that the cause of the abnormal communication has been the simultaneous opening of an abscess into both rectum and bladder. Two cases of this nature have been put upon record by Simpson.†

The **symptoms** of this condition are extremely distressing. The escape of urine into the rectum causes excoriation of the anus and great irritability, but if the opening into the bladder is sufficiently large to permit the entrance of fæces, the suffering of the patient is usually almost intolerably severe, cystitis of an aggravated form being the inevitable result. I have, however, seen one case of the kind in which, although a small quantity of fæces passed into the

* *Arch. klin. Chir.*, Bd. xv. p. 111.
† "Obstet. and Gynecolog. Works," pp. 814–816. Edinburgh, 1871.

cavity of the bladder, but little irritation was produced, the diagnosis of the condition being first made by the appearance of striped muscular fibre as a urinary sediment; this was accounted for by the escape from the bowel of small particles of incompletely digested meat.

The **treatment** of this condition must vary with the extent and nature of the fistulous communication. Where the size is moderate and the surrounding tissues normal, an attempt should be made to close the opening by a plastic procedure; but where there has been very extensive destruction of the recto-vesical septum, or where the surrounding tissues are infiltrated, especially if that infiltration has the clinical features of malignancy, the proper course is undoubtedly to divert the fæcal current from the rectum by the establishment of an abdominal artificial anus.

The operation of closing recto-vesical fistula is practised much in the same way that the cure of vesico-vaginal fistula is effected. The bowel having been well cleared out, the patient is placed in the semi-prone position of Marion Sims; æther administered; and the sphincter thoroughly well dilated. This last is a most essential step in the performance; indeed, it is only by its means that the operation is rendered possible. Should sufficient room not be available, more may be gained by an incision carried backwards through the sphincter towards the coccyx, and a large-size duck-bill speculum introduced, so as to bring the rectal wall well into view. The edges of the fistula are now thoroughly pared, care being taken that the whole circumference of the orifice is vivified, and the edges brought together by a number of sutures passed through the entire thickness of the recto-vesical septum. Afterwards the bladder should be frequently evacuated by the catheter if necessary, so as to prevent over-distension; and the sutures removed on the fourth or fifth day. The bowels may be kept confined for a week or ten days by the use of opium, and with this object in view the diet should be so regulated that excrementitious matter be reduced to a minimum: it is better, indeed, that it should consist almost entirely of milk. The recumbent position ought to be maintained until union is firm.

Dittel[*] has advocated an exceedingly ingenious operation for the cure of recto-vesical fistulæ in the male, the steps of which are, to a certain point, similar to his operation for the removal of calculus from the bladder where lithotrity is contra-indicated, and when the condition of the prostate gland renders its section inadvisable. The steps are as follows: A sound having been introduced through the urethra, he makes a transverse incision in the perinæum in front of the anus, and carefully dissects between the rectum and prostate and then between the rectum and bladder. When the fistula is met, it is divided, and the rectal and vesical orifices are separately sutured. This operation would appear to promise excellent results, but as far as I can learn it has not been performed sufficiently often to justify conclusions as to its merits.

I have personal experience of but one example of this operation. The patient, a man, aged 40 years, was admitted into Sir Patrick Dun's Hospital under the care of my colleague, Professor Bennett. He had slipped down a haystack and impaled himself upon a pitchfork, one prong of which entered the anus, penetrating the rectum and membranous urethra; the other prong slipped up beside the scrotum, without doing any injury. The result was a tolerably large recto-urethral fistula. Dr. Bennett operated by the method above detailed; the fistula having been divided, the urethral and rectal orifices were separately vivified and sutured, and a drainage

[*] *London Medical Record*, p. 139; July 15th, 1878.

tube was introduced into the wound; in this instance, however, probably owing to the large amount of cicatricial tissue about the fistula, the posterior wall of the rectum and anus unfortunately sloughed, and the operation failed in closing the abnormal communication.

For the consideration of recto-vaginal fistulæ the reader is referred to the special works on obstetric surgery.

Fistulous communications between the rectum and some other **portion of the intestinal tract** are extremely rare. Esmarch* describes the formation of an *anus preternaturalis in ano* as a sequela to extensive prolapsus recti. Where the prolapse is considerable, a pouch of peritonæum descends through the anus at the anterior aspect of the protrusion, and in this pouch a bundle of small intestine is not infrequently situated. If now this becomes sufficiently constricted to induce gangrene, a fistulous communication between the rectum and small intestine may be a possible result. An artificial anus so formed differs in no material respect from a similar condition elsewhere. If the discharge is light in colour and contains a large proportion of but-little-digested food, and if there is progressive marasmus, the probability is that the portion of small intestine implicated is high up, and consequently of a kind where, if unrelieved, a fatal termination must be looked for soon. An attempt might be made to cure it either by means of the enterotome of Dupuytren, or by resection of the intestine.

According to Schroeder† a direct communication between the small intestine and rectum has been produced intentionally by the surgeon in order to cure an

* "Pitha u. Billroth, Handbuch der allgemeinen und speciellen Chirurgie," Bd. iii. p. 149.
† Ziemssen's "Cyclopædia of the Practice of Medicine," vol. x. p. 531.

artificial anus in the posterior vaginal cul-de-sac, the idea being to divert the discharge from the vagina into the rectum, where the power of retaining fæces will exist; and the method adopted is with an intestinal shears, one blade of which is passed into the rectum and the other into the small intestine. The communication having been established, the vaginal opening may be closed. A far preferable plan, however, of treatment, which, according to Schroeder, has been successfully performed by Heine, is to re-establish communication between the two ends of small intestine.

A few cases have been noted of communication having been formed between other parts of the intestine and the rectum. McCarthy* records a communication between the vermiform appendix and rectum; and recently I have had a case of this nature at Sir Patrick Dun's Hospital. A man, aged forty, was admitted with symptoms of violent peritonitis, of which he died, and at the post-mortem it was found that the vermiform appendix was firmly adherent to the upper portion of the rectum. An abscess had formed in the adhesions, which ruptured, and gave rise to the fatal peritonitis.

CHAPTER VI.

THE RELATIONS OF PULMONARY PHTHISIS TO RECTAL FISTULA.

THE connection between phthisis and rectal fistula is a subject of extreme importance, and well deserving of separate consideration. A considerable amount of difference of opinion, however, prevails amongst authors as to the relations that exist between these two diseases. Physicians, in treating of phthisis, rarely mention fistula

* Path. Trans., Lond., 1876; p. 161.

as a common complication, while, on the other hand, surgeons who have written on the subject of fistula have generally noticed its frequent co-existence with pulmonary disease. This, no doubt, is due to the fact that patients often undergo treatment for phthisis without mentioning to the physician the presence of fistula, whereas, if the surgeon is consulted for the rectal disease by a person suffering from phthisis, the latter condition will, in the majority of cases, be sufficiently apparent. Allingham has noticed patients who have come to St. Mark's Hospital to be treated for fistula while they are attending at other hospitals for their cough, having said nothing about their rectal trouble at the latter institutions.

We meet with fistula in phthisical persons under two distinct conditions, which we may distinguish as the tuberculous fistula; and the simple fistula occurring in the tuberculous patient : the former being the result of a true tuberculous ulceration of the lower end of the rectum, and the latter occurring as a result of emaciation and absorption of fat in the ischio-rectal fossa, similar to that which we find in other debilitating diseases.

The researches of Koch upon the ætiology of tuberculosis have thrown important light upon tubercular ulceration of the intestinal tract. According to him,* intestinal tuberculosis may exist either as a primary disease, or secondary to pulmonary phthisis, the first arising by the introduction into the intestinal tract of the flesh or milk of tuberculous animals, and the latter in consequence of the person swallowing sputum containing tubercle bacilli. It is obvious that persons suffering from pulmonary tuberculosis who have any expectoration must frequently swallow some of the microbes, but yet only a few of these

* "Mittheilungen aus dem Kaiserlichen Gesundheitsamte," Band ii.

individuals are affected with intestinal ulcers. Koch suggests as an explanation of this difficulty, that this may be due partly to the fact that the intestinal mucous membrane is not very susceptible to infection by the bacilli, and also to the slow development of the tubercle bacillus, he having cultivated the microbes of intestinal tuberculosis only nine times in a period of six months. Experiments with *bacillus anthracis* have shown that the fully-developed bacilli were destroyed by the gastric fluids, but that the spores passed through and subsequently developed; and Koch concludes that possibly the same may be true of the tubercle bacillus; and that it is only in those cases where the spores are delayed, from some cause, for a sufficiently long time in the intestinal canal, that the bacilli develop and form intestinal ulcers. When once developed, the tuberculous ulcers contain vast numbers of the bacilli. In one case[*] which he records they were found in great quantities in the floor of intestinal ulcerations, although they existed only in small numbers in the pulmonary cavities. According to Lichtheim and Gaffky,[†] the fæces of persons suffering from intestinal tuberculosis contain numbers of bacilli. The latter of these authorities has drawn the following conclusions from a number of observations on this subject. In health, and in non-tubercular illness, no tubercle bacilli could be found in the fæces; in phthisis, where the sputum contains bacilli, none were found, except where symptoms of intestinal ulceration were present, and then they were abundant. He also found in fæces two other forms of bacilli, which were stained blue by Ehrlich's method, but they differed in shape from the tubercle bacillus.

Professor G. Sormani has published some interesting

[*] Koch, *loc. cit.* Case II., p. 35.
[†] Koch, *loc. cit.*, p. 34.

experiments on the artificial digestion and other conditions influencing the life of these organisms.* The stomach of a pig recently killed, and kept without food for forty hours before death, was the source of the gastric juice employed. Complete physiological digestion not only destroys the vitality of the bacillus of tubercle, but its form also. The destruction of the bacillus is not among the first phenomena of digestion, rather among the last to happen; that is to say, these organisms are, among organised substances, the least easily attacked by the digestive juices. A digestion of too short duration, or of little activity from scarcity of gastric juice, or from insufficient acidity, does not attack the bacillus of tuberculosis, and in such case it maintains its virulence nearly unaltered.

The knowledge of this, according to Sormani, helps to explain the frequent association of intestinal tuberculosis with pulmonary phthisis. The stomach of the tubercular patients, little active, as a rule, from catarrh due to the fever, and possibly to the remedies, is of so weak digestive power that it readily allows these bacilli to pass unaltered, so that they may subsequently become the foci of intestinal tuberculosis.

The **symptoms** of tubercular fistulæ are tolerably diagnostic: the internal opening is easily felt; is large, sometimes being the size of a threepenny piece, or even larger; it is irregular, and the mucous membrane surrounding it is infiltrated, and feels to the finger elevated and knobby. The external opening is also large, and the surrounding skin is much undermined; the skin itself is congested and livid, and the edges have a tendency to curl inwards. The discharge is thin and curdy.

The occurrence of tubercular fistula as a primary

* *Annali Univ. di Medicina*, Aug., 1884; and *Lond. Med. Record*, March 16, 1885.

affection, without pulmonary or other manifestation of the disease, is extremely rare. I have, however, seen one case, which I believe to have been of this nature. A boy, aged ten years, came under my care in January, 1878, with a complete fistula : the inner opening, which was situated in the usual position, was large and rugged ; the edges were indurated and elevated ; the external opening just admitted the tip of my little finger, and the skin was undermined for a distance of three-quarters of an inch from the margin ; it was thinned and livid. The discharge was somewhat profuse, but thin and curdy. His general health was otherwise very good, and there was not the slightest indication of pulmonary mischief. I divided the sphincter in the usual way, and then thoroughly scraped the granulations with a sharp spoon, cut away the thinned margin of skin, and cauterised the surface with nitrate of silver. The wound healed rapidly and perfectly, and five years after the operation I learned that the boy had remained well.

Since the investigations of Koch, we should, I think, establish the fact of the presence of the bacillus before accepting the diagnosis of tubercular fistula in any given case, its existence in discharge being possible to demonstrate by the same methods as are employed in investigating sputum.[*]

Should a phthisical patient be operated on for fistula?—Very different opinions are held by surgeons upon this point. Some say, " Do not operate, because, if you cure the fistula, the pulmonary mischief will be aggravated by stopping the discharge;" while others say, " If you operate, the wound will not heal."

The former objection is one which is now properly disregarded as belonging to the period of surgery

[*] Smith, *The Lancet*, June, 1883 ; p. 1108.

when issues, setons, and moxæ, were in high repute; but the latter is sometimes an important consideration. If the fistula is tubercular, it is obvious that simple incision will fail. Something more must be done, as in the instance above described, and in cases of simple fistula, where the pulmonary mischief has much advanced; and where cough is frequent, failure would be the probable result, the constant cough causing such frequent movement of the anus, that the rest so necessary to repair is impracticable. On the other hand, when the lung disease is slight and not extending, and where the patient is not much run down, it is well to operate, especially where the fistula is simple; and the removal of a source of irritation and suppuration by this means will frequently result in very great benefit. No definite rule can, however, be laid down in these cases. The surgeon must be guided in recommending operation by the individual peculiarities of each case that comes under his notice.

CHAPTER VII.

ULCERATION OF THE RECTUM.

ULCERATION of the rectum is a tolerably common affection, and we find that in addition to the causes which tend to produce this destructive disintegration in this situation in common with other parts of the body, several special ætiological influences act more particularly upon the lower bowel.

Of predisposing causes the most potent is congestion, and just as we see ulceration following varicose veins in the leg, so enlargement of the hæmorrhoidal veins is the occasional precursor of rectal ulceration. In order to fully appreciate this, we must take into consideration the anatomical arrangement of the bloodvessels of the lower bowel. Of the arteries, which are distributed to this portion of the intestinal tract, the most important is the superior hæmorrhoidal. This vessel, which is a branch of the inferior mesenteric, passes down between the folds of the mesorectum, where it divides into two branches, which form loops, curving downwards, and from which the parallel rectal arteries described and delineated by Quain, are given off. These vessels pass down in the columns of Glisson, which are vertical folds of mucous membrane, from six to eight in number, but liable to some variety in size and distinctness, terminating immediately inside the anal margin. The blood supply of the mucous membrane is almost entirely derived from these vessels.

The veins which return the blood from the rectum pass with the arteries upwards to the inferior mesenteric vein, and so their blood finally passes into the portal circulation. In passing through the rectal wall, the hæmorrhoidal veins penetrate small openings in the muscular coat, about four inches above the anus, unprotected by any tendinous ring; and Verneuil, who has drawn special attention to this arrangement, considers that the pressure to which these veins must of necessity be subjected during defæcation is a fertile cause of hæmorrhoids and congestion. In common with other radicles of the portal system, these veins are destitute of valves, so that in the erect position which man occupies the weight of the column of blood contained in the vessels tends powerfully to produce stasis in the smaller terminal branches; and again during defæcation the passage of the fæcal mass along the rectum, tightly contracted around it, and in a direction opposed to that by which the current of blood is returning, must be admitted as a frequent although temporary cause of congestion. Lastly, obstruction to the portal circulation through the liver must not be overlooked.

It will thus be seen that the anatomical arrangements tending to produce congestion of the lower bowel are numerous and important, so that a very trivial exciting cause may start an ulceration which may prove very intractable. Undoubtedly, one of the most frequent exciting causes is a direct traumatism, and this may be produced in a great variety of ways. Injury sufficient to induce the ulcerative process may originate from without, as from the introduction of foreign bodies through the anus, but very much more frequently the initial laceration is caused by the contents of the intestine, and where the bowels have become much confined, the effort to extrude the hardened fæcal mass may result in a tearing of the anal margin, which, instead of healing healthily, remains as a more or less permanent ulcerated surface; or, as Bushe has pointed out, a fold of mucous membrane may be prolapsed, and so pressed upon by the indurated fæces, that it loses its vitality, and sloughs off. Pieces of bone, nutshell, and similar hard substances contained in the fæces, may also injure the delicate mucous membrane sufficiently to induce ulceration. Another traumatic origin which has been described is the pressure of the fœtal head during child-birth. This cause has been assigned for the greater frequency of non-malignant stricture, resulting from ulceration, in the female. Various operations, as the removal of a hæmorrhoid, or the division of fistula, prove to be the direct exciting causes of ulceration not unfrequently, especially where the patient is allowed up too soon, or when the bowels are permitted to become constipated during after-treatment.

The rupture or sloughing of an internal hæmorrhoid may also be the starting-point of this process.

The classification of ulcers (at all times a matter of difficulty and vagueness) cannot be more definitely arranged in the rectum than in other parts of the body ; there are, however, some forms which appear sufficiently characteristic to warrant special notice. Of these the more important are : The hæmorrhoidal, the follicular, the tubercular, the lupoid, the dysenteric, the irritable, and the syphilitic.

The hæmorrhoidal ulcer.—Under the above heading Rokitansky* describes ulceration due to congestion, in which, no doubt, some traumatism is the immediate exciting cause. This form resembles, in all respects, the varicose ulcer of the leg, and is characterised by marked chronicity, elevated irregular edges, and a tendency to bleed. Its situation (Plate II., Fig. 2) is usually well within the external sphincter, and confined to the mucous membrane; it does not implicate the anal verge. The amount of pain that the patient suffers is slight, in this respect contrasting markedly with the irritable ulcer or fissure; the reason is, no doubt, that in the first place it is situated out of the grasp of the sphincter; and secondly, the sensory nervous supply of the mucous membrane is very much less than at the margin of the anus.

The complications with which this form of ulceration may be associated are in the first place the perforation of the gut, and the establishment of an internal rectal sinus or complete fistula as a result; secondly, if the destruction of the deeper coats of the bowel be considerable, a stricture may be formed during the process of cicatrisation; or, again, the ulcer may, by passing across the anal margin, and exposing some of the delicate nerve fibres, be associated with a true irritable ulcer.

* "Path. Anat.," vol. ii. p. 107.

Symptoms.—The amount of discomfort occasioned by hæmorrhoidal ulcers is slight. There is tenesmus with frequent passage of a small quantity of fæces, generally more noticeable in the morning, with pus, and occasionally bleeding, more or less severe. The educated finger introduced into the rectum will, without difficulty, determine the number and extent of these ulcers; or, after dilatation of the sphincter under an anæsthetic, a good view can be obtained of them with a rectal speculum.

The treatment of this form of ulceration is not very satisfactory. Rest in the recumbent position is necessary for the same reason that we prescribe it in varicose ulcer of the leg. The bowels should be kept regular, preferably by some of the saline mineral waters given in sufficient quantity to ensure one soft motion daily. At the same time the diet should be so regulated that the fæces may be as unirritating as possible; indeed, an exclusive milk diet may sometimes be advantageous. For local treatment an injection of liq. bismuthi, ʒss.; liq. morphiæ, min. xv.; mucilag. amyli ad ʒii., night and morning, will be found of service; or, where a more powerful astringent appears indicated, a solution of nitrate of silver in water, 2 or 3 grains to the ounce, may be tried. In other cases the introduction of iodoform ointment, by means of Allingham's instrument (Fig. 16), will answer

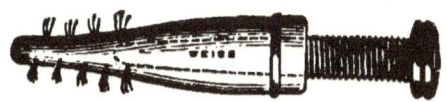

Fig. 16.—Allingham's Screw Ointment Introducer. (Scale ¼.)

better. Should iodoform irritate, or the smell be much objected to, an ointment containing finely-powdered boric acid may be substituted with advantage.

In cases where the amount of ulceration is extreme, and where the above treatment fails to relieve, colotomy should be contemplated, the object being in this case to afford physiological rest to the part, and so enable the healing process to progress, by diverting the fæcal flow from the rectum. This is undoubtedly sound surgery. Where a joint is diseased we stop its functional activity by a splint. And in cases of intractable ulceration of the larynx, Bryant has advocated the performance of tracheotomy for a like purpose, with excellent results. We should remember that when performed in this way as a remedial measure the prognosis is much more favourable than when the operation is called for in a patient worn down by long-continued intestinal obstruction or open rectal cancer; and should the ulceration in the lower bowel heal, steps may be taken to close the artificial anus.

Allingham in his table of ulcer and stricture of the rectum gives two cases (Nos. 7 and 25) in which, after colotomy, the rectal ulceration healed; and in the latter he states that the lumbar opening was subsequently closed, the fæces passing normally; and at the meeting of the International Medical Congress at Copenhagen, Bryant strongly advocated colotomy as a curative treatment of rectal ulceration. Dr. Bridge and other surgeons have recorded similar cases.

An interesting variety of ulceration, probably more nearly related to the hæmorrhoidal than any of the other varieties enumerated, is that which originates within the lacunæ (sinuses of Morgagni) just above the external sphincter, probably from the lodgment and decomposition of fæcal *débris*. These little pockets are situated between the columns of Glisson, and are formed by the anal valves, which have a faint resemblance, as pointed out by Ribes, to the semilunar valves and sinuses at the aortic and pulmonary openings. The lacunæ are very variable both in number and extent, in some cases constituting diverticula (*the encysted rectum*); while in others they are scarcely recognisable. Physick first drew attention to the occurrence of this form of ulceration.*

A very instructive instance of this disease is quoted by Kelsey, which was recorded by Dr. Vance.†

"A lady, aged eighteen, had suffered for more than a year from all the symptoms of fissure; had been frequently examined to no purpose, and was reduced to a very miserable state. On examination, the integumentary folds were congested, thickened, and œdematous, doubtless as a result of scratching; but there was no trace of anything like a fissure. The lining membrane was searched with the utmost care, but no lesion of any sort was revealed, except slight hypertrophy of the sphincter. A second painstaking review of every part of the rectum gave the same result, and the author was about to abandon the hope of finding any local lesion when, as a matter of form (for there was no evidence of disease about them) he determined to pass a probe into each of the pouches. The probe could not be forced into the first one, and with the second he fared no better; but with the third, after an ineffectual attempt, the probe passed into the sacculus. No sooner had it entered, however, than the patient screamed with pain, there was a spasmodic retraction of the levator ani and sphincter muscles, and the part was forcibly withdrawn from view. The site of the sacculus felt as if a buck-shot had been embedded in the tissues, so hard and swollen was the part. A small probe-pointed tenotome was carefully passed along the canal, and as soon as the sensitive point was touched the handle was brought down, and the edge of the knife made to sever the inner wall of

* "American Ency. of Pract. Med. and Surg.," Art. Anus. 1836.
† *Medical and Surgical Reporter*, Aug. 14th, 1880.

the sacculus and expose the diseased point. This done, the cause of the suffering was revealed. On the left side of the anus, and at a point where there had been no unusual sensibility, an indurated ulcer had formed

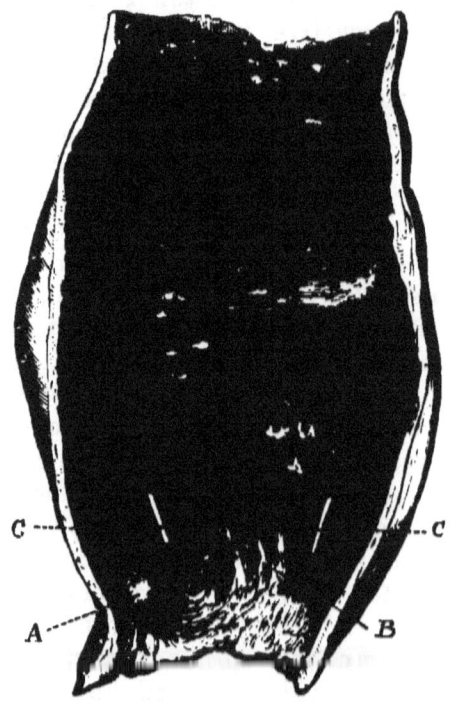

Fig. 17.—Ulcers originating in the Sinuses of Morgagni.
A, Small internal pile; B, three small ulcers; C, glass rods indicating position of unaffected sinuses.

within one of the little pouches. When the sacculus was opened and the ulcer exposed, it seemed very much like an ordinary fissure of the anus; but before cutting it open there was no evidence whatever, save the symptoms the patient complained of, to indicate the existence of such a lesion."

Ulceration of this kind, although extensive, may produce but trivial symptoms, as is shown in a case upon which I made a post-mortem examination. The patient, a man aged sixty, was for several weeks in the medical wards of Sir Patrick Dun's Hospital, suffering from Bright's disease, from which he eventually died. During his stay in hospital he never complained of any uneasiness about the rectum, and as there was no indication to excite suspicion, no rectal examination was made. I injected the vessels of the rectum for anatomical purposes, and on slitting up the bowel was surprised to find three well-marked ulcers immediately inside the external sphincter. They had manifestly commenced in the little sinuses, two of which remain unaffected, and are indicated by pieces of glass rod. It will be observed, on reference to the woodcut (Fig. 17), that none of these ulcers invaded the anal margin, with its numerous sensory nerves; hence the complete absence of symptoms in this case, as contrasted with the extreme agony attending the ordinary irritable ulcer.

In connection with this case it may be well to inquire whether there is anything more than an accidental relation between ulceration in the lower bowel and disease of the kidneys. Bartels* says degeneration of the blood-vessels of the intestinal mucous membrane is certainly a very common condition with amyloid disease of the kidney, and extensive ulceration destructive of the mucous membrane is by no means a rare consequence thereof. He considers that intestinal ulceration has a direct sequential relation to diseased kidney, and he states that follicular ulceration of the large intestine is not infrequently associated with amyloid changes. Dickenson also, in the Croonian Lecture for 1876, mentions intestinal ulceration as one of the rare complications of renal disease.

Follicular ulceration may occur in any part

* Ziemssen's Cyclopædia, vol. xv. p. 522.

of the large intestine, but the seat of election of this disease is undoubtedly the rectum and sigmoid flexure, sometimes affecting only a few follicles, while at others the ulcers are very numerous and extensive (Fig. 18). They arise as a result of inflammation of the mucous membrane either catarrhal or dysenteric, and are immediately due to the breaking down and necrosis of infiltrated and swollen solitary follicles. As to the pathological anatomy of this disease, when seen in the early stage, before ulceration sets in, the follicles appear tumid, and raised above the surface of the mucous membrane, due to the proliferation of the cell contents. This proliferation proceeds with such rapidity that sphacelus of the follicle takes place, and as the slough separates an ulcer is left. At first the ulcers are of a small size, and are of a more or less circular outline; the margin is considerably raised, and the base is formed of the submucosa. Occasionally, by increase and coalescence of two or more follicular ulcers, a considerable loss of substance results, but more frequently the dimensions remain limited. Cicatrisation progresses somewhat slowly, and where the

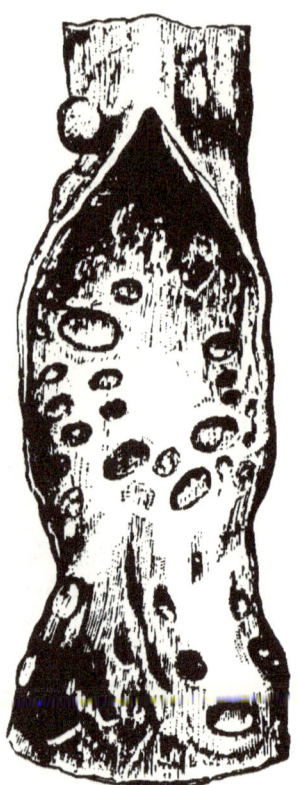

Fig. 18.—Follicular Ulceration of Rectum.*

* Museum Royal College of Surgeons in Ireland.

Plate II

Fig. I.

Fig. II.

E. Burgess ch lith

PLATE II.

Fig. 1.—Lupoid Ulceration of Rectum.
Fig. 2.—Hæmorrhoidal Ulceration of Rectum.

destruction of tissues is considerable, may lead to stenosis of the bowel.

The diagnosis of these ulcers during life must, from their very nature, be difficult and obscure, but their presence may be suspected when the symptoms of intestinal inflammation persist for a long time, especially when there is frequent hæmorrhage. When low down in the rectum they may be felt with the finger, or seen through a speculum. In some instances we find in the discharges masses of inspissated mucus, which represent the form of the ulcer. These are compared by Bamberger to frog-spawn or boiled sago-grains, and have been shown by Virchow to consist, in part at any rate, of particles of imperfectly digested starch.

In the case of a gentleman who had spent twenty-five years in India, and who came to me complaining of dysenteric-like symptoms, the surface of the rectum as far as it could be felt with the finger was studded with these small follicular ulcers. Various plans of treatment were adopted, but without causing any appreciable change.

Tubercular ulceration.—Intestinal tuberculosis and ulceration may be either a primary disease, or it may be secondary to similar changes in the lungs. In the first instance the cause is, in all probability, a direct inoculation by the ingestion of bacilli in the food; and in the second the sputum which the patient swallows, as pointed out by Klebs, is the probable source of infection. When implicating the rectum, tubercular ulceration may, in the generality of cases, be recognised by the following characters: the size of the ulcer is usually considerable, and in position it is found in the rectal pouch, or invading the anus; the shape is generally irregularly oval, the long axis being parallel to the vessels, which in this part of the bowel are vertical,

124 THE RECTUM AND ANUS. [Chap. VII.

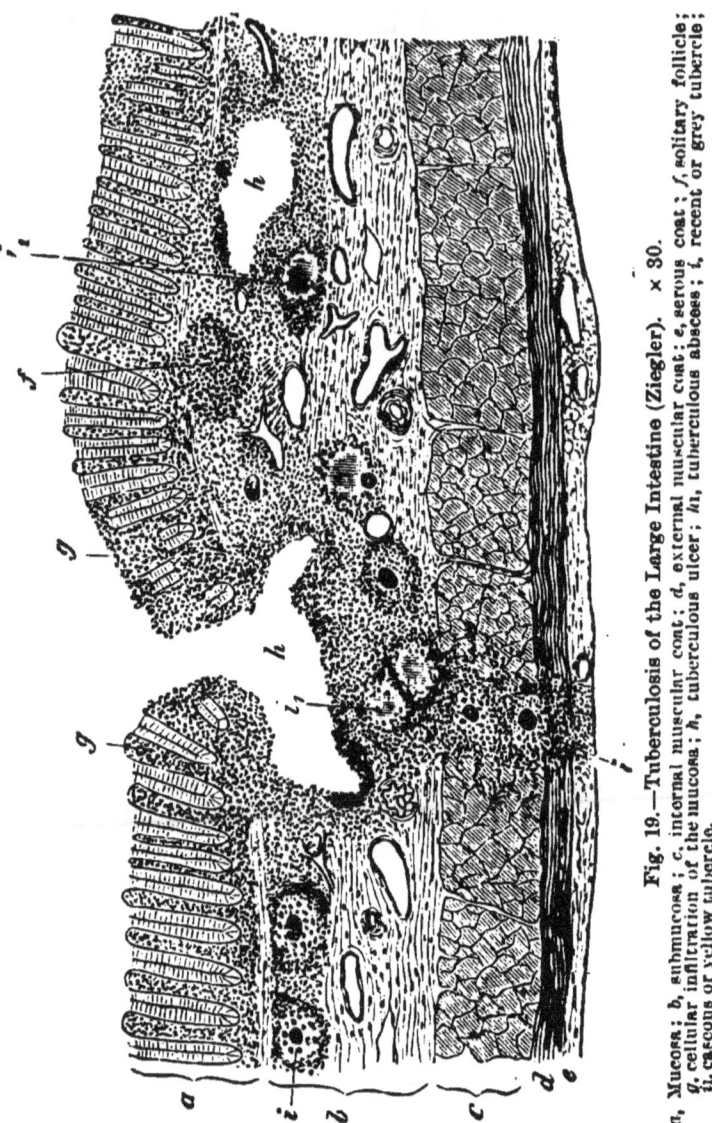

Fig. 19.—Tuberculosis of the Large Intestine (Ziegler). × 30.

a, Mucosa; *b*, submucosa; *c*, internal muscular coat; *d*, external muscular coat; *e*, serous coat; *f*, solitary follicle; *g*, cellular infiltration of the mucosa; *h*, tuberculous ulcer; *h*₁, tuberculous abscess; *i*, recent or grey tubercle; *i*₁, caseous or yellow tubercle.

whereas in the other parts of the intestine they are transverse. The edges of the sore are much undermined and ragged, and there is considerable infiltration of the mucous membrane in the immediate neighbourhood. It is impossible to draw any definite line between the limited tubercular ulcers met with in this region and the more extensive destruction of tissue which has been described as lupoid ulcer. Tubercular ulcers of the large intestine manifest a strong tendency to perforate the bowel (Fig. 19). Hence it is that fistula is such a common result as a sequela to this disease when situated in the lower portions of the rectum; and similarly, when higher up, perforation may not unfrequently originate a fatal peritonitis. For a further consideration of rectal tuberculosis the reader is referred to the chapter on the connection between fistula and pulmonary phthisis.

Lupoid ulceration.—Under the names "rodent ulcer," "lupus of the vulvar-anal regions," and "l'esthiomène," a number of cases have been described, in which the essential element is a chronic intractable form of ulceration in the neighbourhood of the anus and genital organs.

The first description of lupus in this region, more particularly implicating the vulva, and the differential diagnosis of this condition from syphilis on the one hand and carcinoma on the other, was by Huguier, in 1848, who, under the term *l'esthiomène*, described a number of cases that he had met with.

Like lupus in other parts of the body, we find considerable variety in the tissues affected and the mode of progress of the disease. For practical purposes it may be well to classify these under two heads—the superficial, or serpiginous, and the hypertrophic. Amongst the published cases, the greater number appear to have been of the latter variety.

As an example of the great ravages of this disease, the following case is given in a valuable paper by Dr. Angus McDonald* :—

* *Edinburgh Medical Journal*, April, 1884; p. 910.

"The case came under my notice after it reached an extreme degree of advancement; it had then lasted some six or eight years; the destruction of tissue was terrible in extent. I have reason to know that it is the same case as that referred to by Duncan, in 'Duncan and West,' p. 656. Of it (at the time when he saw the case, which at least was a year before I was introduced to the patient) Dr. Duncan says: 'A case to which I was called some years ago is, so far as I know, so unprecedented in the amount of destruction as to be worth describing. I only saw it once in consultation. The disease was at one time regarded as cancerous. The patient, aged about forty, had had the disease for at least five years, and she lived many years after my visit. While the disease was already extensive, she bore a child. On the hips, just beyond the ischial tuberosities, were long scars, thin and bluish, of healed ulcers. The entire ano-perinæal region was gone, there being a hollow space as big as a fœtal head. The urethra was entire, as well as the mucous membrane between it and the cervix uteri, which was healthy. Except the anterior portion of the vagina, no trace of it, or of the anus or rectum, was discoverable; behind the cervix uteri the bowel opened by a tight aperture, just sufficient to admit a finger; when the fæces were hard she could keep herself clean, but only then. Although the extent of ulceration was severe, the patient was attending to her household duties.' To this graphic description of the case I can fully subscribe, with this addition, that latterly the ulceration went still higher up into the pelvis, leaving the bowel hanging loose for some distance from the upper level of ulceration, giving it the appearance of the torn sleeve of a coat. This patient lived two and a half years after the time referred to by Dr. Duncan, and died of exhaustion and diarrhœa. Notwithstanding this shocking amount and prolonged continuance of ulcerative action, there was no involvement of inguinal or other glands."

Allingham uses the term "rodent," or "lupoid," for an intractable form of ulceration, of which his records show several cases similar to, although scarcely so extensive as, the case detailed by Dr. McDonald.

Through the kindness of Dr. R. McDonnell I had an opportunity of seeing a case of this kind, which was under his charge in Steevens' Hospital, Dublin, in the person of a policeman, aged thirty-five years. In this case an extensive and deep ulceration extended round the margin of the anus, with the exception of a small part on the left side. Spreading more in an antero-posterior direction than laterally, it passed up into the interior of the rectum for a distance of about one inch and a half, completely destroying the entire thickness of the bowel (Plate II. Fig. 1). A thorough scraping with the sharp spoon and cauterisation was only followed by temporary benefit and the ulcer was afterwards excised. A careful microscopic examination of the parts removed showed several giant cells, but no tubercle bacilli were found.

In a second case which was under my care in Sir Patrick Dun's Hospital, 1890, the appearances were remarkable and unlike any of the ordinary forms either of lupus or rodent ulcer. The patient was a lad aged sixteen years, apparently in robust health, who stated that the ulceration had commenced two years previously. It extended up the entire length of the anal canal, but did not encroach much on the mucous membrane; it extended out into the right buttock for about one inch and a half, and reached from the coccyx nearly to the scrotum. The whole ulcerated surface was raised about one-eighth of an inch above the surrounding structures, the margin remarkably

128   THE RECTUM AND ANUS.   [Chap. VII.

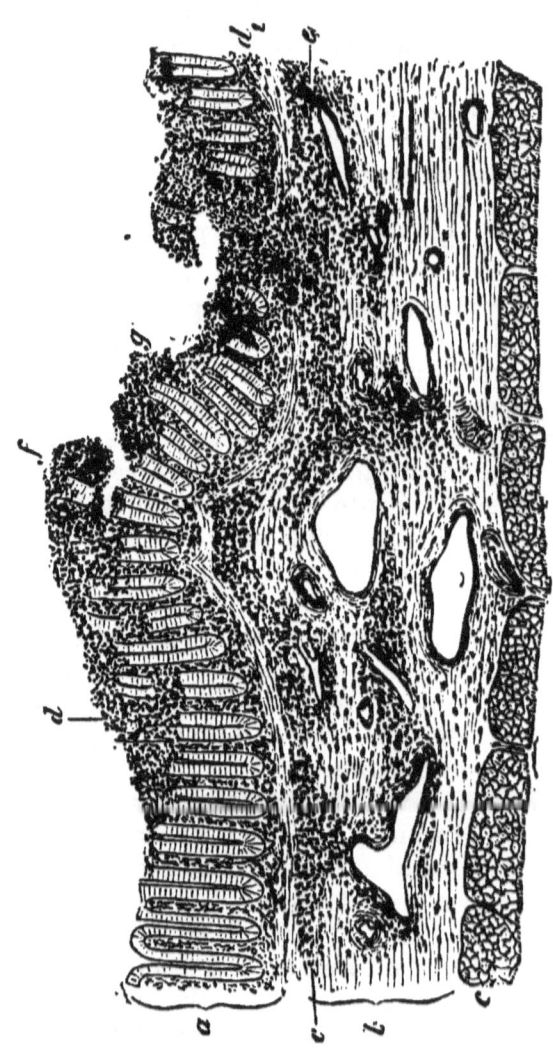

Fig. 20.—Section of the Colon from a Case of Dysentery (Ziegler). × 25.

a, Mucosa: *b*, submucosa; *c*, muscular coat; *d*, infiltration of the mucosa between the crypts; *e*, infiltration of the submucosa; *f*, infiltration of the superficial glandular layers which are in process of desquamation; *g*, ulcer with infiltrated floor.

convoluted and tortuous, the edge being covered with pearly-white epidermis. Over the surface were large plaques of the same pearly-white epidermis, and between these plaques was a granulating surface. The discharge was very slight and the growth almost painless. I excised the entire mass and closed the gap by a shifting plastic operation, and recovery was complete. A microscopic examination of this case also revealed abundant giant cells, but careful search failed to detect tubercle bacilli. I have not heard of any recurrence of the disease.

The term "rodent" is objectionable, as it is associated with a tolerably definite form of ulceration in other parts of the body, which is obviously not identical with what Mr. Allingham describes as occurring at the anus or in the rectum; for, in the first place, rodent ulcer is essentially a disease of advanced life, while the cases given by Mr. Allingham vary in age from 17 years to 42. Secondly, he says, "Neither its edge nor its base is at all hard," while we know that a rodent ulcer of the face has always a layer of infiltrated tissue surrounding the ulcerated surface; and the fact stated by him that he has noticed repair taking place very rapidly, when suddenly all the granulations melt away, and the ulcer assumed its former character, is an occurrence familiar to surgeons in the case of lupus, but not at all consistent with the general progress of rodent ulcer, which is one of continued and chronic extension, without any effort at repair.

Lupus is a term which, in the present state of our knowledge, it is somewhat difficult to define. Possibly the discovery by Koch of the tubercle bacillus in the giant cells of lupus may enable us to limit more precisely in the future this disease. For the present, at any rate, it would appear that the recorded cases of the form of locally malignant ulcerations to which I have been referring are more nearly related to lupus than to other diseases.

Dysenteric ulceration.—As a sequel to true epidemic dysentery, loss of substance may result; but it is probable that many of the museum specimens which are stated to be instances of dysenteric ulceraation are incorrectly described. We must bear in mind that the symptoms of ulceration of the rectum, from whatever cause arising (*e.g.* continued diarrhœa, with the discharge of blood-stained mucus), can hardly be distinguished from the milder cases of dysentery ; and it is impossible, with our present knowledge, to draw a hard and fast line between simple catarrhal and true dysenteric inflammation.

The changes resulting from diphtheritic dysentery have, however, been fully investigated, and the process of ulceration is admirably described by Ziegler (Fig. 20).* In recent cases the mucous membrane is highly congested and swollen, and generally beset with minute extravasations of blood. The epithelial surface is overlaid with a glairy blood-streaked mucus. This presently becomes more slimy and blood-stained, and interspersed with flaky fibrinous shreds and films, which indicate the beginning of a superficial necrosis of the mucous membrane. Soon the necrosis is made sufficiently evident by the appearance of erosions and losses of substance.

In slighter cases the necrosis and loss of substance are, at first, merely superficial, but the deeper structures are successively attacked, and in severe cases the whole of the glandular layer of the mucous membrane at particular spots may perish. The necrotic tissue is reduced to a turbid mass, in which the structural elements, and the nuclei of the cells, soon cease to be recognisable. The parts which undergo necrosis seldom cover any great extent of surface, and are often confined to the prominent ridges and folds of the mucous membrane. These look dirty grey or black, while the intervening parts are still livid or dark red ; in other cases the necrotic tissue takes the

* "Special Pathological Anatomy," p. 288.

form of a flaky more or less adherent coating, or, more rarely, of broad, continuous sloughs. The underlying tissue is, in all cases, densely infiltrated with cells. The infiltration may extend through the entire thickness of the submucosa, and may at length invade the muscular layers. When the mucosa is removed, open ulcers are of course left behind. These may vary in their depth and extent. Sometimes over a great part of the bowel the mucous membrane remains only in narrow strips or islands.

The disease may become arrested at any of the various stages of its course, and repair then begins. And when the ulceration has been deep, atrophic cicatrices may result. During this process of cicatrisation a muco-purulent discharge takes place, constituting what has been called chronic dysentery, or cœliac flux; and it is in this stage that British surgeons most frequently meet with this disease in persons who have returned invalided from warmer countries. In these cases ulceration of the rectum can frequently be diagnosed by digital examination, or seen by means of a speculum, and the treatment will require much patience and care. For the ulcerations due to syphilis the reader is referred to chap. xii., and the irritable ulcer is a subject of such importance that I propose to devote a chapter to its special consideration.

CHAPTER VIII.

IRRITABLE ULCER, OR FISSURE OF THE ANUS.

IN the whole range of surgery there are but few diseases which, while of very limited extent, produce such extreme misery to the patient, and none in which surgical treatment is attended with more certain success than in the affection under consideration.

In order to elucidate intelligibly the **ætiology** of this remarkable disease, and before entering on the train of symptoms so characteristic of irritable anal ulcer, it will be necessary for us to review the more important anatomical features of the termination of the bowel. The outlet of the intestine is closed by two sphincter muscles, the external being subcutaneous, and consisting of an elliptical band of fibres closely surrounding the anal verge. The internal sphincter consists of the normal circular fibres of the rectum considerably increased in number, and averaging about two lines in thickness at the lower extremity, and gradually merging into the circular coat of the rectum above.

On the outer side, these muscles are separated by the attachment of the levator ani, the fibres of which are more intimately connected with the external sphincter; and on the inside these two sphincters are in close apposition, the line of demarcation corresponding accurately with the junction of the skin and mucous membrane. According to Hilton, this is indicated by a white line, which is generally to be seen, although in some cases it is very much better marked than in others. Hilton has also pointed out the important fact that the nerves, principally branches of

NERVE SUPPLY.

the pudic, which come down below the internal sphincter, pass out between the muscles at the junction to become superficial in this situation (Fig. 21). These nerves are very numerous, and he makes the ingenious suggestion that this is due to the peculiar physiological fact that the normal state of these

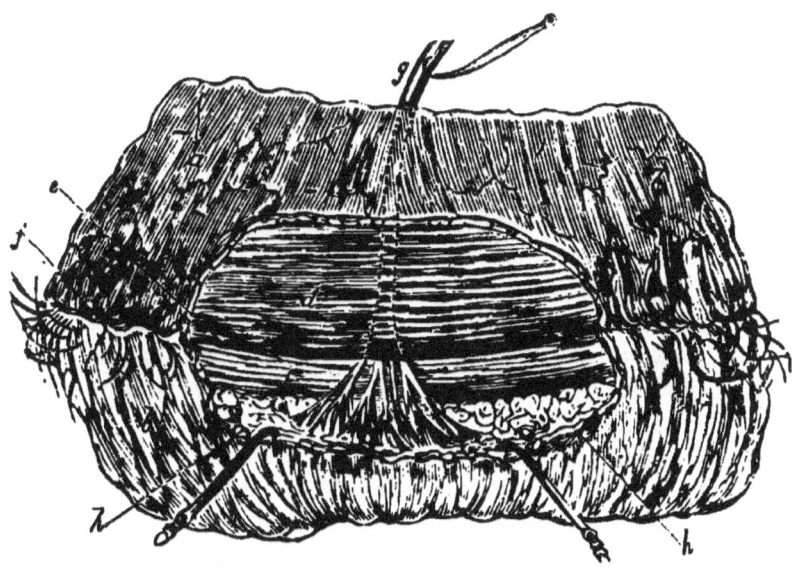

Fig. 21.—Nervous Supply of Anus. (Hilton.)

a, Mucous membrane of rectum; *b*, skin near the anus; *c*, external sphincter muscle; *d*, internal sphincter muscle; *e*, line of separation of the two sphincters; *f*, the overlying white line marking the junction of the two sphincters; *g*, nerve supplying the skin near the anus, which it reaches by passing first externally to the rectum and then through the interval between the two sphincters; *h*, flap of mucous membrane and skin reflected back.

muscles is a condition of contraction instead of relaxation. This theory would appear to receive support from the somewhat analogous arrangement in the bladder, as we know that the most sensitive portion of that viscus is at the neck. When one of these sensitive nerve twigs becomes exposed by the formation of a fissure, the extreme amount of pain so characteristic of the lesion can readily be understood.

As we have before seen, the sensibility of the interior of the rectum is but slight, so that ulcers situated inside the anus produce but little actual pain. When, however, the ulcer passes beyond "Hilton's white line," it is in the great majority of cases acutely painful, from implication of some of the small nerve filaments; but, on the other hand, we see occasionally a fissure-like ulcer extending over the anal verge which is not so acutely painful. Accordingly, we find Gosselin dividing these ulcers into two distinct varieties, the tolerant and intolerant; and Mollière suggesting the more suitable terms, tolerable and intolerable. Associated with painful fissure, there is always great spasm of the sphincter muscle: Boyer, indeed, considered that this spasm was antecedent to, and the cause of, the fissure. Van Buren speaks of this spasm as affecting the muscle in part only, what he terms "fascicular spasm"; * and he defines this term in a foot-note to be "the alternating contraction and relaxation of certain of its fasciculi, and not of the whole muscle, as the expression spasm of the sphincter would imply." I have never, however, been able to observe this condition, as it has always appeared to me that the muscular contraction involved the whole circumference of the sphincter. And in any case the distinction appears to me to involve a frivolous and practically worthless refinement.

The explanation of this condition of the muscle is best understood by a reference to the diagram (Fig. 22), taken from Hilton.

The sensory nerve filament exposed by the ulcer receives impressions, which are conveyed to that part of the spinal cord from which the lumbar, the ilio-

* Loc. cit., p. 221.

Chap. VIII.] PATHOLOGY OF IRRITABLE ULCER. 135

lumbar, sciatic, and pudic nerves, etc., spring, and we find, as a consequence, symptoms referable to reflected influences along these trunks: hence, pains in the back, down the leg, and in the genito-urinary organs are common accompaniments of irritable anal ulcer; and in the same way, reflex irritation of the nerve supplying the external sphincter produces spasmodic contractions of that muscle. A familiar analogous example of this reflected spasm is to be found in the retention of urine which so frequently follows any of the ordinary rectal operations.

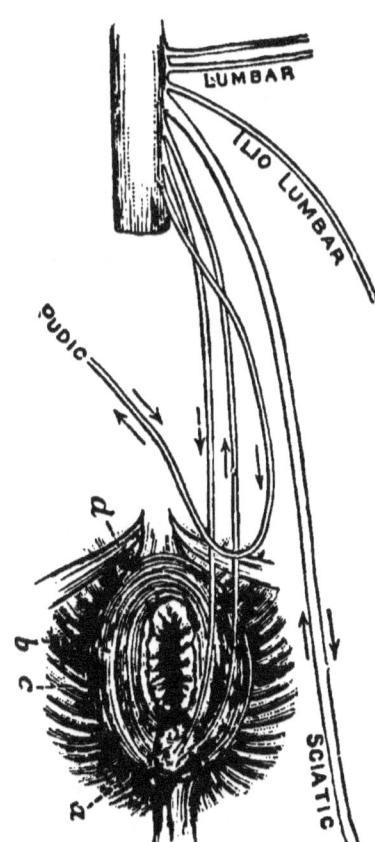

Fig. 22.—Diagram of the Nervous Relations of Irritable Ulcer of the Anus. (Hilton.)

a, Ulcer on sphincter ani; b, filaments of two nerves are exposed on the ulcer, the one a nerve of sensation, the other of motion, both attached to the spinal cord, thus constituting an excito-motor apparatus; c, levator ani; d, transversus perinæi.

To summarise, therefore, it has been suggested that the following sequence of events occurs as tending to produce the fully formed irritable ulcer. First: during the passage of a costive and large motion a rent in the mucous membrane is made; or an excoriation the result of syphilis, dirt, etc., exposes one of the delicate nervous twigs. As a result of the constant motion and distension, and by the lodgment of particles of fæces in the rent, continued irritation is set up, which, in turn, occasions spasm of the sphincter. The spasm once started, the irritation is increased, and so a vicious circle is established, and the result is that the ulcer is never allowed to heal.

Such, then, is the theory which up to the present time has been universally taught and readily accepted, the more so as the treatment based on this theory is so largely successful; but that it is (at any rate in a large majority of cases) erroneous and the treatment advocated unnecessary, I think I am in a position to show.

In order to arrive at a true understanding of the painful fissure it will be necessary to allude briefly to some further points in the anatomy and development of the termination of the bowel. As Symington* has pointed out by means of frozen sections, the anus is not merely an opening, but a canal nearly an inch long, which in the normal state remains closed; it is lined with the ordinary scaly epithelium of the skin, and at its upper limit joins the mucous membrane. At the point of junction there are several little mucocutaneous processes called anal valves, the mucous membrane being gathered up behind them so as to form pouches called rectal sinuses, resembling somewhat the sinuses of Valsalva behind the semilunar valves of the aorta and pulmonary arteries. These structures are subject to great variability; in some instances the pouches are so large that they constitute little false diverticula (the encysted rectum of Physick), while in others they are scarcely noticeable, and the anal valves rudimentary. These anal valves are apparently vestigial remains left in the development of the rectum. The mucous membrane is developed from the endoblast, and in early fœtal life

* *Journal of Anatomy*, vol. xxiii. p. 106; 1889.

is a closed cavity called the mesenteron. This is met and opened into by a process called the proctodæum, which commences as a depression in the epiblast, and eventually forms the anal canal. The failure of this junction causes the commonest form of imperforate anus, while the anal valves, which are so common that they may be considered normal, appear to be the vestigial remains of this coalescence.

Mr. Bland Sutton* has quite independently arrived at the conclusion that the hymen is a similar vestigial remnant of the development of the female organs of generation.

I am quite satisfied that the true explanation of the formation of the vast majority, if not all, painful fissures is as follows :—During the passage of a motion one of these little valves is caught by some projection in the fæcal mass and its lateral attachments torn ; at each subsequent motion the little sore thus made is reopened and possibly extended, the repeated interference with the attempts at healing ends in the production of an ulcer, and the torn-down valve becomes swollen and œdematous, constituting the so-called pile, or, as it sometimes has been called, the "sentinel pile" of the fissure. Most of us have experienced the little bits of skin torn down at the sides of the finger-nails, popularly called "torments," and how painful they are when dragged upon. Now the torn-down anal valve resembles closely this condition of the finger, except that in the former it is situated at the acutely sensitive anal margin, and subjected to the periodic strain of a passing motion; it is therefore not to be wondered at that the pain should be so excessive as seriously to affect the general health and render life miserable.

The first case which directed my attention to the theory above detailed occurred some years ago. An

* *Brit. Med. Journ.*, Dec. 10th, 1887.

otherwise healthy young woman consulted me, complaining of very severe pain after the bowels moved. So great was the pain that she postponed defæcation for several days, with the result that she was quite incapacitated for attending to her duties for several hours after the bowels moved. She was in no way a neurotic subject, and I was convinced that her suffering was extremely acute.

Upon superficial examination nothing was to be seen, but the left side of the anus was extremely tender to the touch, and any attempt to introduce the finger was attended with such pain and spasm that I did not persist in it, but prescribed an aperient, and the following day dilated the anus under ether, when a small ulcer rather triangular in shape was clearly seen; the base of the triangle was below, and was formed of an œdematous little tag of cutaneous tissue, which I recognised as an anal valve. Catching this in a forceps and drawing it down tended to enlarge the ulcer, and the resemblance of this to the "torments" on the finger at once suggested itself.

Since then I have seen thirty cases, and in all of them it was evident that a like cause existed for the ulcer. The torn-down anal valve could always be demonstrated, and although it was in some cases very small and apparently insignificant, in the vast majority it was much hypertrophied and œdematous. The accompanying illustration (Fig. 23), taken from a post-mortem case, which I accidentally obtained when examining the rectum for another purpose, well exemplifies the condition I have alluded to.

We may now approach the study of the **symptoms** of this disease with a more reasonable hope of our being able to comprehend their significance. The subjective phenomena of irritable anal ulcer present a train of symptoms which are eminently characteristic, and are frequently alone sufficient for diagnosis.

Chap. VIII.]   ANAL "TORMENTS."   139

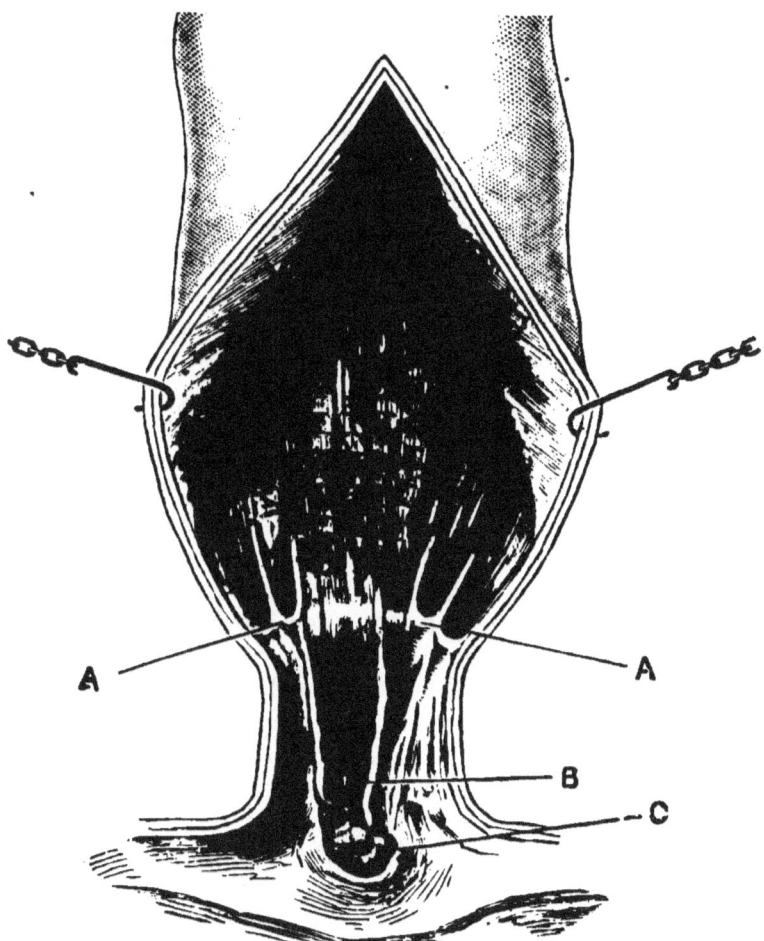

Fig. 23.—The illustration (taken from a drawing kindly made for me by my resident pupil, Mr. Kennan, from a specimen accidentally found in the hospital *post-mortem* room) well indicates the points to which I desire to call attention. The rectum and anus were removed from the body and opened by a vertical anterior incision. Where the contracted anal canal joins the rectum the white line described by Hilton is to be seen, and at this level four of the anal valves (A), with the pouches behind them, are distinctly observable. In this specimen they are much more pronounced than usual, so that it can be readily understood how liable they are to be caught and torn during defæcation. At C is to be seen the "sentinel pile," and at B the fissure which had cicatrised. This so-called pile is here obviously one of the posterior anal valves which was torn down and has become thickened and œdematous.

The pain is paroxysmal and always associated with the act of defæcation. During the actual passage of the motion, however, it is not usually severe, but shortly afterwards it comes on with great intensity; it is a dull, gnawing, and extremely distressing sensation, situated immediately within the anus, and not unfrequently associated with some of the reflected pains before alluded to. It lasts often for many hours, completely incapacitating the sufferer from following his occupation and necessitating the recumbent position while it lasts. It then subsides or entirely disappears, to be, however, reproduced in all its intensity when next the bowels move. The act of defæcation is therefore postponed as long as possible, with the result that when the evacuation does take place the pain is greatly increased. As a result of the constant pain, the constipation, and the frequent resort to narcotics, constitutional symptoms of a severe type become developed; the countenance becomes careworn and sallow, the appetite is bad, and there is considerable emaciation, a train of symptoms which, in many respects, resembles the cachexia of malignant disease. The fæces are passed in a narrow cylinder, or sometimes they are flattened and tape-like, due to the incomplete relaxation of the sphincter during defæcation, and not unfrequently a streak of bloody matter is to be seen on them. Serremone* considers that a contracted state of the anus is frequently congenital, and the cause, not the result, of the fissure, the narrow outlet being more liable to tears from overstretching. In a case that recently came under my notice this was certainly the fact. The patient, aged fifty years, had difficulty in getting his bowels relieved all his life, and had never passed a motion thicker than his little finger. The anus, although normal in

* "Inaugural Thesis." Strasburg, 1861.

other respects, was extremely narrow, and even under ether could not be stretched to a greater extent than would just admit the tip of the index finger without the epidermis cracking. He had frequently suffered from these superficial cracks when his bowels were constipated, but they never developed into the typical painful fissures. He was much benefited by forcible dilatation, followed by the use of conical bougies, which he has now learned to pass for himself. Among the more distant sympathetic affections to which fissure may give rise, Curschmann * has noted the frequent co-existence of spermatorrhœa and rectal disease, and more particularly fissure.

If, then, a patient comes to us complaining of severe pain, lasting for some time after defæcation, the presumption is strong that a fissure is present, no other rectal disease producing this characteristic distress. The disease is more common in females, especially those of neurotic tendency; and although more frequently met with in young adults, no age appears to be exempt; and it is sometimes met with amongst old people.

Upon making an examination, the first thing that attracts our notice often is a small "sentinel" pile. On passing the finger round the anus one part of the circumference is found to be tender, and any attempt to introduce the finger gives a great deal of pain, and is violently resisted. Upon separating the anal folds, the lower termination of the fissure can generally be seen, the surface being red or grey, and the edges somewhat indurated. The ulcer is usually somewhat triangular, the base being formed by the torn-down valve.

The position is, in the majority of cases, dorsal, or nearly so; although usually solitary, sometimes we find them multiple. This is most frequently the case when they are of syphilitic origin.

* Ziemssen's "Cyclopædia," vol. viii. p. 846.

Treatment.—In the more recent examples of this disease, and in those in which the origin is undoubtedly syphilitic, a cure can be generally accomplished by local applications; but in those which have existed for several months or longer, these means will, in all probability, prove ineffectual, and operative interference is called for. There is not in the whole range of surgery any operation in which the surgeon can speak so positively as to the certainty of cure and freedom from risk as in the trivial procedure necessary for the cure of this disease.

If the patient refuses to be operated on, or if the surgeon considers the case one suitable for the milder measures, he should prescribe a purgative of sufficient strength to ensure one soft and easy evacuation daily. Immediately after the evacuation the anus should be well washed with soap and water, and, if possible, defæcation should take place at night, immediately before the patient goes to bed.

For local treatment.—Touching the surface of the ulcer with a fine point of nitrate of silver will be found to give considerable relief; the layer of coagulated albumen, which is by this means formed on the surface, for a time effectually protecting the exposed nerve from the irritation of the fæces. With a similar object strong nitric acid has been recommended ; but it is more painful than, and not so efficient as, the nitrate of silver. Of late, a solution of chloral hydrate has been employed with good results as a local application.

If, however, these means prove ineffectual, or if the ulcer is of long standing, with great spasm and hypertrophy of the sphincter, operation should be undertaken.

Boyer first pointed out the fact that division of the sphincter was at once followed by a complete subsidence of the symptoms. He recommends that the incision should extend through the entire thickness of the sphincter. Indeed, he sometimes practised a double division, at different points, put in a large bougie, and plugged the rectum round this with charpie.*

Dupuytren further modified this operation by making the incision only through the superficial fibres of the muscle, and subsequent experience has abundantly proved that this more limited incision is amply sufficient. According to Curling, Copeland stated that he considered an incision through the mucous membrane alone sufficient, but, as Curling has pointed out, in the majority of instances the ulcer has already penetrated the mucous membrane, the fibres or the sphincter muscle being frequently visible in the floor of the ulcer. In these cases, of course, Copeland's suggestion is futile. The same may be said of the operation of Dumarquay,† which consists in a submucous division of the sphincter; he passes a knife up between the mucous membrane and muscle, and then divides the latter, much in the same way that the subcutaneous division of the tendo Achillis is effected.

Récamier, in 1829, recommended as a substitute for any cutting operation what he describes as "massage cadencé," and he performed the operation as follows: one or two fingers were introduced into the rectum, and then, with the thumb outside, the sphincter was pinched up and pressed so as to overcome its resistance. This was frequently repeated, and in a regular methodical way, so that no portion of the circumference of the anus was allowed to escape. Although he states that he had considerable success with this plan of treatment, it appears to have fallen into disuse; and later, Maisonneuve proposed to effect dilatation in a more rapid and thorough manner by introducing the fingers one by one, till finally his

* *Journal Complémentaire du Dictionnaire des Sciences Méd.*, Nov., 1818.
† *Arch. Gén. de Méd.*, 1846.

whole hand entered the rectum.* When this was accomplished, he closed his hand, and then withdrew it forcibly. This, no doubt, effectually stretched the sphincter, but as it was performed without anæsthesia, it must have been horribly painful. And although Maisonneuve afterwards modified his operation into a simple stretching of the anus with the two index fingers under chloroform, the method fell into disuse, and it is only comparatively recently that forcible dilatation of the anus has been practised on a large scale. Indeed, now it is used not alone for the cure of anal ulcer, but as a preliminary step to almost all rectal operations and explorations. This revival of the procedure was mainly due to the writings of Van Buren, who, in 1864, reported a number of cases in which this treatment had been successful,† and he, since, has insisted on the certainty with which a cure may be effected by this means.

The way in which this operation has hitherto been supposed to act is probably by producing a temporary paralysis, the result of hyperdistension. For this condition Van Buren uses the term *atony*, from the obvious analogy there is between the loss of power in the sphincter when over-distended by the pressure of the fingers and the inability of the bladder to expel its contents when retention of urine has passed certain limits. He further suggests that the stretching to which the sensory nerves are subjected so influences their function that they temporarily cease to convey impressions, the result of irritation, to the exposed fibres; in the same way that we find forcible stretching of the sciatic and sensory branches of the fifth nerve relieve neuralgia. This theory receives considerable support from the fact,

* "Clinique Chirurgicale," tome ii. Paris, 1864.
† "Transactions of the New York Academy of Medicine," vol. ii. p. 180.

which we frequently observe, that the first time the bowels move after hyperdistension there is complete immunity from the pain which before was so severe.

Until recently the operations above detailed have been universally employed, and undoubtedly the success attending them has been very great; but adopting the theory as to causation which I have above enunciated (page 137), it is easy to explain how these (as I now believe unnecessary procedures) have afforded such complete relief to symptoms. First as to dilatation. When the anus is widely stretched by the fingers beyond the limits of natural dilatation by the passage of a motion, the little valve is torn down beyond the extent to which it is torn in normal defæcation, and consequently rest is given to the ulcer, and healing ensues, the subsequent defæcation not reopening the sore. In like manner spontaneous cure occasionally ensues by the valve being torn down to its full extent during repeated acts of defæcation, then the fissure heals and the valve remains a little cutaneous tag, like many external piles; this had obviously occurred in the specimen from which the illustration was taken. In the "torments" at the side of the finger, if the little piece of skin is forcibly torn down, although the process is painful, a cure results, just in the same way that forcible dilatation cures the anal ulcer. Secondly, complete or partial division of the sphincter, if carried down through this tag, relaxes each side so that the subsequent motions do not catch in it and reopen the wound; but most surgeons of experience in this subject have met with cases in which both dilatation and incision of the sphincter have failed; and the advice given by many writers to supplement the operation by the removal of the so-called pile at the lower end of the ulcer is undoubtedly sound. I would, however, go a step farther, and say that the removal of this little tag is all that is necessary, and that either the forcible dilatation or incision of the sphincter is uncalled for.

What I would recommend in these cases where the classic symptoms of painful fissure are present is as follows :—Do not subject the patient to the pain of extended digital examination, much less occasion him the torture of passing a speculum; but, after having the bowels fully relieved by an efficient aperient, administer an anæsthetic, and dilate the anus with the fingers sufficiently to obtain a good view of the entire circumference of the muco-cutaneous junction; then, if the symptoms have been characteristic, a little ulcer will almost certainly be found, and at its lower extremity the torn-down anal valve sometimes greatly hypertrophied. All that is now necessary is to catch this in a forceps, and with a fine pair of scissors remove it by a V-shaped incision, with base towards the ulcer, so that nothing be left that can be caught by a passing fæcal mass. If there is any unhealthy granulation tissue in the ulcer this should be removed with the sharp spoon, and the surface well dusted with boracic acid; the cure will then be as immediate and certain as when the little "torment" at the side of the finger-nail is shaved off level with the skin. It is advisable after this little operation to examine carefully all the other anal valves, and if any of them are likely from their size and projection to be torn down and so form other fissures, to snip them off with the scissors.

I have now operated in this way upon thirty cases; in all the piece torn down by defæcation was found, although in some it was small. In two of the cases, in addition to the treatment above detailed, I slightly incised the sphincter, as it was very rigid and contracted. This I did in deference to old tradition, and I now believe it was inflicting a useless although trivial injury on the patient. One of these cases had been twice operated on by other surgeons—once by incision, and once, I am informed, by the truly barbarous method of Maisonneuve—namely, gradually introducing the whole hand into the rectum, closing the fist, and then forcibly withdrawing it. What this heroic treatment failed to accomplish the little operation of snipping away the torn-down anal valve surely effected. In another case, which had also been operated on before by incision, the bifid remains of the anal valve still caused an obstruction to the passing fæces, and so kept up the sore.

In conclusion, I would wish to point out that the cases alone suitable for this operation are the painful or intolerable fissures of the anus; they are not to be confounded with the superficial cracks of the anal skin common in eczema and some other conditions, nor with the superficial ulcers met with in syphilis; and they must also be distinguished from tolerable ulcers occurring wholly on the mucous membrane, which are frequently situated in the rectal sinuses, but which do not produce the characteristic symptoms.

CHAPTER IX.

NON-MALIGNANT STRICTURE OF THE RECTUM AND ANUS.

JUDGED from a clinical standpoint, **stricture of the rectum** would include all those pathological changes which result in a more or less complete retardation of the passage of the fæces through this tube, but it will facilitate the consideration of the subject if at present we confine the term to those changes in the wall of the bowel and anal outlet which produce a narrowing of the lumen of the gut, reserving for future consideration those neoplastic growths which are characterised by the clinical features of malignancy, and also those instances of obstruction due to the pressure of structures outside the rectum. Of the true strictures, the most important are the cicatricial stenoses, which, of course, necessarily pre-suppose the existence of loss of substance, the result of ulceration, or direct traumatism. What the exact nature of this preliminary destruction of tissue was, it may be quite impossible to determine. Even where we have to deal with an open ulcer the diagnosis is beset with difficulty; and where, as in the present instance. we may have only to deal with the cicatricial contraction subsequent to this process, it is obvious that the difficulty is much enhanced.

Whether **spasmodic stricture,** due to a spastic contraction of the circular unstriped muscular fibres, ever is present, has been freely discussed by most of the authors on rectal surgery, and considerable differences of opinion are to be found in their works on the subject, but the majority of recent writers refuse to admit muscular spasm as a cause of temporary stricture, much less of permanent stricture. Leichtenstern,* in his admirable article on constrictions, occlusions, and displacements of the intestines, says: "The existence of such an affection no longer calls for serious discussion." And Van Buren,† after an able criticism of the subject, expresses the opinion that "neither in imaginary nor in actual stricture of the rectum is muscular spasm an element of any practical importance." On the other hand, Dr. J. S. Bristowe‡ states that, "Not very unfrequently spasmodic contraction, with great hypertrophy of the muscular tissues, is met with as one of the troublesome sequelæ of dysenteric ulceration of the rectum." Mr. Harrison Cripps, also, in his book,§ brings forward very strong evidence in favour of muscular spasm being an important factor in the ætiology of rectal stricture. That a permanent condition of spastic contraction exists, he does not consider to be possible; but that long continued irritation, by exciting frequent intermittent spasm, may terminate in shortening of the muscular fibres, with increase of the connective tissue elements, he considers highly probable, and analogy furnishes us with numerous examples of similar changes in striped muscle, the result of long continued irritation. Of these, the most familiar is the condition of the knee joint, to which Mr. Barwell has applied the term "contracture," where, as a result of pathological changes in the articulation, spasm of the muscles is first produced. And, finally, such a continued shortening of the hamstring muscles results, that the knee is permanently flexed, and the head of the tibia

* Ziemssen's Cyclopædia, vol. vii. p. 484.
† *American Journal of Med. Science*, October, 1879.
‡ "Obstruction of the Bowels;" Reynolds' "Syst. of Medicine," vol. iii. p. 72. London, 1871.
§ "Diseases of Rectum and Anus," p. 205. London, 1884.

subluxated backwards. Many other examples could be quoted, if necessary. Quite recently I have seen a case in which aggravated talipes equinus resulted from a patch of lupus extending under the tendo Achillis, causing atrophic shortening of the muscles of the calf; and we have frequent opportunities of witnessing the considerable contraction, with increase of tissue in the external sphincter ani, which will result from the long-continued irritation of a small painful fissure. It may be argued that all these are examples of contraction of voluntary muscular fibre, and that, consequently, no deductions can be with safety drawn from them, justifying the conclusion that like changes can take place in unstriped muscular fibre; but we know that œsophagismus may last for a very long time from a small ulcer or excoriation of the œsophagus; although it must be admitted that hitherto it has not been demonstrated that spasmodic stricture of this tube has occasioned permanent contraction, recognised after death. In the bladder, however, the presence of a calculus, or of ulceration without direct obstruction, will produce such an amount of thickening and contraction of the wall of that viscus, that after death it may be found impossible to dilate it to its normal proportions. Cripps does not rely upon analogy alone for supporting his argument, but brings forward the following instance of rectal ulceration with some tendency to stricture. He says: "I was puzzled about the case, for upon the first examination I found ulceration in the posterior part of the bowel, with an annular stricture situated two inches from the anus, which would barely admit the tip of the finger. The examination was extremely painful. Upon examining the same patient a few days later under an anæsthetic, the ulceration was present as before, but, to my surprise, there was scarcely any stricture, for the finger would pass readily into the bowel, with only a sense of being slightly gripped at the spot which previously would not admit the finger-tip. I had this patient under observation some time, and soon learnt that by introducing the finger somewhat roughly into the bowel, the sense of stricture was immediately produced, but by keeping the finger gently in contact with the strictured part, a feeling of gradual giving way was experienced, so that the finger would lie comparatively easy in the narrow part, where upon any rough movement it could be felt to be palpably and immediately grasped." Two years afterwards Mr. Cripps saw this patient again; and on examination, in the place of the yielding and comparatively soft stricture encountered previously, there now existed a firm, hard, totally unyielding fibrous contraction, narrowing the bowel to the smallest circumference. Since the publication of Mr. Cripps's views on this subject I have had an opportunity of seeing a case which, to my mind, strongly supports his ingenious theory of the causation of stricture. The case was one of fistula complicated with stricture. Upon making an examination it was found that the fistula communicated with the bowel by a large ulcerated opening on the coccygeal aspect of the gut, and immediately above this ulcer, a tight annular stricture was to be felt, which barely admitted the tip of the finger; it gave the idea of being quite above, and unconnected with the ulcer. What the original cause of the ulcer was in this case I am not prepared to say, as it possessed at the time it first came under my observation no very distinctive features, but viewed in the light of Mr. Cripps's theory, I am of opinion that the following sequence of events occurred : the ulcer became established, and by the irritation thus set up a frequent peristaltic action of the muscular coat of the rectum was induced, tending to expel the source of irritation, and in time atrophic shortening of the overtaxed circular fibres of the muscular coat was produced: finally the ulcer by perforating the rectum gave origin to a suppurative periproctitis, hence the fistula. In support of this view of this interesting case, I may allude to the fact, with which most surgeons are familiar, that a not uncommon symptom complained of by patients suffering from rectal ulceration, is a forcing down and straining, as if there was something to be evacuated immediately after the bowel has been completely emptied, and at other times when it contains no fæces. This is undoubtedly due to the contractions of the tube *above* the source of irritation in their useless attempts to expel it.

In my opinion, the strongest support is lent to this theory (although this is not alluded to by Mr. Cripps) by the position of the internal opening of fistula when found in connection with stricture. Where a fistula forms in connection with a strictured urethra or other mucous tract, it is in nearly all instances found to open into the tube above the seat of stenosis, but in the rectum this is not invariably the case. In a large number of instances (probably the majority) the internal openings of such fistulæ are found below the stricture. That ulceration should occur leading to fistula in the dilated portion of gut above the narrowed part can be readily understood as being a direct result of the stricture; but how are we to account for the occurrence of fistulæ below? This has been a puzzle to many surgeons; and when reading Mr. Cripps's views on the pathology of stricture, it at once occurred to me that in it we had the true explanation of this fact, which may be thus briefly formulated: Ulcer of the rectum, by perforating, may produce a fistula, and at the same time may lead to a permanent contraction of the gut immediately above the seat of ulceration; so that the only relation existing between the fistula and the stricture is the fact that they have a common origin in the ulcer.

Granting, then, that some cases of stricture originate in long-continued spasm, the result of irritation, there remains a large number in which cicatricial contraction is the sole cause; and we may take as a type of this class those which originate in direct traumatism.

As will be seen in the chapter on syphilis, chancroids are of common occurrence at the anus, and are occasionally found in the rectum; and the question, what share they take in the ætiology of stricture, has been warmly discussed. It is quite impossible to come to any accurate conclusion from published statistics as to the relative importance of the various venereal diseases in the ætiology of stricture. Some surgeons, as Gosselin* and Mason,† would have us believe that it almost invariably results from chancroids; while, on the other hand, we find other authorities, such as Allingham and most English authors, discountenancing the views held by Gosselin and Mason, and not recognising chancroids at all as a cause of stricture. Again, it would appear that some surgeons, having it impressed upon their minds that the most common cause of stricture is syphilis, are too apt to accept a syphilitic history on insufficient, or even no *definite*, evidence of this disease. The published statements vary so much, according to the views of the surgeon describing them, that the only conclusion we can arrive at, from a study of them, is that in many instances the origin of the disease is rather a matter of conjecture than a scientifically ascertained fact; but sufficient evidence is, I think, forthcoming to establish

* "Des rétrécissements Syphilitiques du Rectum," Arch. Gén. de Méd., tome iv. p. 667.
† *Am. Journ. of Med. Science*, p. 22; 1873.

the fact that primary soft chancre, phagedænic ulceration, and the gummatous ulceration and cirrhotic changes of advanced constitutional syphilis, all have a share in the ætiology of stricture. It is quite impossible, in the great majority of cases, to determine what the starting point was, or even whether it was of a venereal nature at all, or due to some of the other numerous causes of stricture.

Dysentery has been credited with being the cause of a considerable number of the examples of stricture which come under our notice, but we must remember that the symptoms of rectal ulcer, from whatever cause arising, are in many respects similar to the milder cases of true dysentery. If every case in which there is a muco-purulent and sanious discharge is called dysentery, then, indeed, the ætiological influence of this disease must be considered great; but if the term is restricted to true diphtheritic dysentery, we must admit that the number of cases which can be traced to this disease is small indeed.

In that exhaustive treatise, "The Medical and Surgical History of the War of the Rebellion,"* Surgeon Woodward has entered very fully into this subject, and sums up the experience of the American army surgeons as follows: "No case of intestinal stenosis, resulting from the contraction of dysenteric ulcers, has been reported to the surgeon-general's office either during the war or since. The Army Medical Museum does not possess a single specimen, nor have I found in the American journals any case substantiated by post-mortem examination in which this condition is reported to have followed a flux contracted during the civil war." In view of the vast numbers of cases of diarrhœa and dysentery that occurred during the war, these facts would seem to

---

* Part II. ; med. vol. p. 504.

indicate that in America, at least, this complication is extremely infrequent.

The observations of Rokitansky on stricture of the colon have been frequently alluded to by authors on this subject, but the only case of which he gives details was one which terminated fatally *thirty years* after the attack of dysentery;* so we may reasonably question whether the connection between the two is put beyond doubt.

In Great Britain at the present time diphtheritic dysentery is such a rare disease, and the cases which are met with are of such a mild character, that but little evidence can be brought forward upon the subject.

In speaking of the stenoses of the intestine resulting from dysentery, Leichtenstern says: "The deep lesions of the mucous membrane and submucosa in the diphtheritic form of epidemic or sporadic diarrhœa lead during the often prolonged process of recovery to marked stenoses not unfrequently at several points. Unaffected islands of mucous membrane often persist between the diphtheritic losses of substance which are made to project into the lumen of the canal in the form of knobs and folds by the contraction of the adjoining cicatrices.

"By the contraction of unilateral cicatrices the intestine becomes bent. When the cicatrix is extensive, and on all sides, the intestine is drawn together in the direction of its longitudinal axis, and thus stiff, callous folds, bands, and sickle-shaped projections into the canal, lying one above the other, are produced. Constrictions of this kind are often increased by tough polypoid excrescences growing from the edges of the mucous membrane into the canal, which sometimes act like valves and increase the stenosis."

* "Ueber Stricturen des Darmkanals, Oest. Jahrb.," Bd. xviii. p. 37; 1839.

Tubercular ulceration is by some stated to be an occasional, though it must be admitted a rare, cause of stricture. Allingham states that he has met with cases of this kind, in which the diagnosis was confirmed by Sir James Paget;* and Cripps † gives one case as occurring in seventy cases of stricture at St. Bartholomew's Hospital; but until we have the presence of tubercle bacillus demonstrated in case of ulceration with stricture, it can scarcely be positively affirmed that stricture does actually result from tubercular disease.

The fact that a considerable proportion of cases of rectal stricture eventually die of pulmonary phthisis would, however, tend to show that in all probability a larger number of cases have really commenced in rectal tuberculosis than has hitherto been recognised.

Where **injuries** have been inflicted on the rectum, especially those which are attended with considerable loss of substance, or where extensive sloughing or long-continued suppuration has intervened, stricture is liable to occur; this is peculiarly the case where the entire thickness of the gut is destroyed throughout its *entire* circumference; on the contrary, where the mucous membrane *alone* is destroyed, or where the whole thickness of the bowel is destroyed in *part* of its circumference only, stricture will be a less probable result. But, on the other hand, in healthy individuals extensive wounds heal frequently without much contraction. This is notably the case in gunshot wounds. Curling gives a case of stricture following a wound from an enema pipe; and Van Buren‡ records the case of a man who had lacerated the rectum in his efforts to get rid of its contents: "He had been

* "Diseases of the Rectum," p. 250.
† "Diseases of the Rectum and Anus," p. 209.
‡ *Loc. cit.*, p. 264.

left in Texas in charge of cattle during the Civil War, and, cut off from communication, he was compelled to subsist on milk without any vegetable food. As a consequence of the unirritating qualities of this food, and the absence of cathartic medicine, his lower bowel became distended with fæces, to get rid of which he was forced in his extremity to use sticks and such rude means as he could command, and in this manner he caused injuries which led to a bad stricture at the usual seat."

Many other examples of stricture following accidental wounds could be adduced if necessary. As a result of operations for fistula, prolapse, and hæmorrhoids, occasional instances of stricture are observed. Where the old operation of widely excising a fistula was resorted to, this complication must have frequently supervened; but where the modern operation is adopted, there is but little risk of such an occurrence, unless from a debilitated state of the system, or from tuberculosis, extensive ulceration and suppuration complicate the case. Similarly in the removal of piles, unless the submucous tissue is much encroached upon, or too much of the skin at the anal border removed, there need be little fear of stricture supervening. Where the actual cautery has been very freely used, or where nitric acid, or other powerful escharotics, have been extensively applied in the treatment of rectal disease, stricture has been known to follow; but where ordinary caution is used in these useful plans of treatment, but little danger arises.

A considerable number of cases of stricture have been met with after parturition, which are apparently due to that process; in some, no doubt, the direct pressure of the fœtal head, by causing sloughing or inflammatory changes in the rectal wall, is the direct exciting cause; but probably in the majority of these cases the immediate cause is the contraction following a pelvic cellulitis, ensuing as a complication of child-birth. In a case recorded by Whitehead,* the long-continued wearing of a vaginal pessary appeared to be the exciting cause ; and similar cases have been noticed by other obstetric surgeons. It will be seen from the foregoing that a very great variety of injuries may eventuate in stenosis of the lower bowel.

From an analysis of 367 cases of non-malignant stricture which I have collected from various reliable sources, it would appear that 276 were females and 91 males, as nearly as possible a proportion of 3 to 1 ; whereas, it will be seen in the chapter on malignant disease that the male sex in this case is more commonly attacked. To what can this greater frequency of non-malignant stricture in the female be due? Various explanations have been given ; but, to state the case broadly, the true explanation lies in the anatomical relationships of the lower bowel to the organs of generation in the female ; in consequence of which secondary inoculation of the rectum from venereal disease of the genitals is more apt to take place than in the other sex, and also in consequence of which various displacements and diseases of the uterus are possibly competent to produce injurious effects, which of course are negatived in the male ; the traumatisms common in child-birth, too, no doubt tend to swell the number of cases.

In its **pathological anatomy,** stricture of the rectum must necessarily present numerous varieties of character, especially when we take into consideration the many diverse ætiological sources to which I have traced its possible occurrence. It may in the first place present differences in *situation,* thus : Stricture may be situated at the anus, when it owes its origin to congenital narrowing ; to too free a removal

* *Am. Journ. of Med. Science,* July, 1872 ; p. 114.

of external piles, or a liberal application of the cautery; and to the cicatricial contraction following the healing of chancroids or other forms of ulceration. In by far the majority of cases, however, the locality affected is the rectal pouch, the lower orifice of the stricture being within three inches of the anus. In rarer instances the position is higher up at the junction of the sigmoid flexure with the rectum, and a few cases have been recorded where a double stricture has been present, one at the upper portion of the rectum and the other in the pouch. The cause of these multiple stenoses is stated in most of the cases to have been dysentery.

There is considerable variety to be found in the *extent* of stricture, the amount of intestine involved in the stenosis varying greatly; in some the contraction is distributed uniformly around the entire circumference of the gut; but only a very small portion of the length of the tube is implicated. This form constitutes the so-called *annular* stricture of the rectum, and probably those cases which have arisen from the permanent contraction of the circular muscular fibres, as before alluded to, are of this nature. In such cases the intestine is sharply constricted, as if it had been included in a ligature, all the coats of the tube being contracted, and at the same time hypertrophied. In other instances we find that the contraction is due to puckering up, and protrusion into the lumen of the bowel, of one side more particularly of the intestinal tube: such cases are generally the result of cicatricial changes during the healing of an ulcer, and may be so sharply marked as to justify the term *valvular* strictures. We may recognise a third variety where a considerable length of the bowel is involved in the contraction, or where, by the increase in thickness of the rectal wall due to hypertrophic changes in the connective tissue elements, the lumen of the gut becomes narrowed, sometimes for a distance of several inches. These constitute the so-called *tubular* strictures.

Besides the coats of the rectum proper, the fascial structures of the pelvis may cause stenosis of the bowel by their contraction. Cripps * attributes to the levatores ani and their sheaths a considerable share in the ætiology of stricture. Speaking of the anterior fibres of these muscles, he says, "These fibres run from the inner surface of the pubis to the sides of the coccyx, crossing the rectum at an obtuse angle, about an inch and a half from the anus. Both the origin and insertion of these fibres being close to the middle line, when the muscles of opposite sides contract they act as constrictors of the rectum as it passes between them, and I believe that not a few cases of rectal stricture at this point are caused by the permanent atrophic shortening of the fibrous element of this muscular tissue."

That a permanent contraction of these fibres should constrict the rectum laterally is obvious, but could only do so at the sides, and the resulting stricture would therefore be slit-like, with the long axis in the antero-posterior direction. Mr. Cripps, however, does not bring forward any cases of this nature, and none such have ever come under my observation. In the cases which I have seen where the altered structure was confined to but a portion of the circumference, the anterior aspect of the rectum was that which was most implicated. This probably is to be explained by the very much closer fibrous attachments of this portion of the bowel to the vagina in the female, and the prostate and base of the bladder in the male, than are to be found in other parts of the circumference.

It is but seldom that the surgeon has an

* "Diseases of the Rectum and Anus," p. 201.

Chap. IX.] *PATHOLOGY OF STRICTURE.* 161

opportunity of seeing the post-mortem appearances of

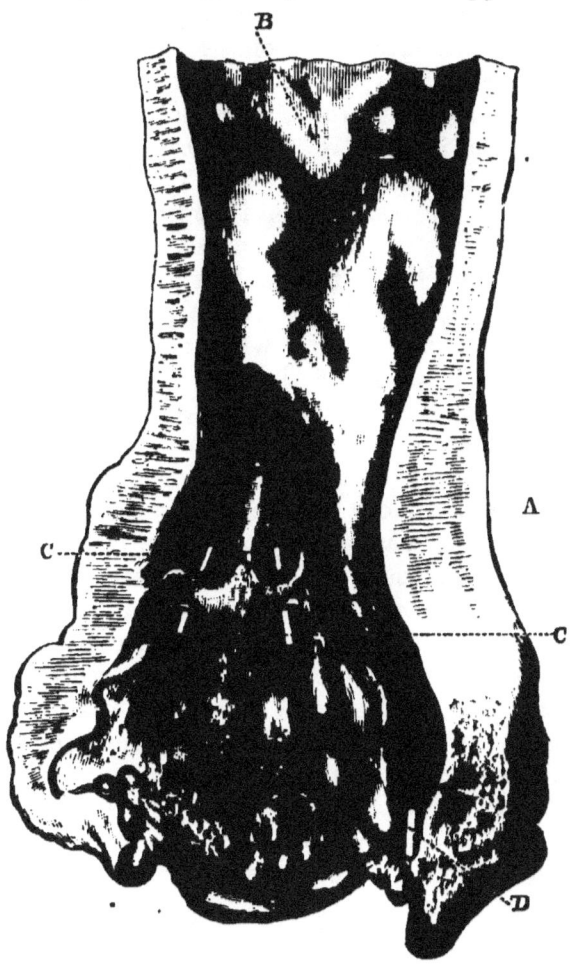

Fig. 24.—Non-Malignant Stricture of the Rectum.*

A, Greatly thickened wall of the rectum; B, termination of mucous membrane; below this point the entire thickness of the mucous membrane has been destroyed; C, bridles of cicatricial tissue; D, complete rectal fistula.

fibrous stricture of the rectum in its early stages. Indeed, the same may be said of the clinical phenomena,

* Museum of Richmond Hospital.

as it is only when some obstructive symptoms become developed that these patients usually seek medical aid.

Of the more extensive stenoses, however, the post-mortem appearances have been noted frequently, as this disease, although called non-malignant stricture, is one which frequently terminates fatally. From a review of the published reports of the morbid anatomy in such cases, it would appear that there is great thickening, as a rule, of all the coats of the bowel, the new tissue formed being extremely dense and hard (Fig. 24); hence the old term, "scirrho-contracted rectum," which was applied very loosely before the characters of fibrous, as contrasted with malignant, stricture became clearly differentiated.

In microscopic section it will be seen that there is great hypertrophy of the connective tissue elements in all the coats; more particularly is this to be noted in the muscular coats, the fibres of which are separated and compressed by new connective tissue formations; and a considerable amount of hard adipose tissue is always also present. In a case recorded by M. Malassez,* he found that the stricture, at its narrowest part, consisted of a material in all respects identical with granulation tissue, *i.e.* embryonic connective tissue; it was made up entirely of young elements, and broke down readily. In the lower portion of the stricture, where it was oldest, bundles of newly-formed fibrous tissue were found, surrounded by embryonic cells, as in cicatrices, and fasciculi of the muscular coat were isolated by these cells. According to Cripps,† the muscle fibres undergo marked atrophy, and in some instances disappear altogether, their place being occupied by a hypertrophy of the fibrous trabeculæ, normally present in the muscular coats.

* Cornil : "Leçons sur la Syphilis," p. 412. Paris, 1879.
† "Diseases of the Rectum and Anus," p. 202.

Both above and below the strictured point considerable alterations are to be observed in the intestine. Below, we sometimes find polypoid excrescences, occasionally of a considerable degree of density. The mucous membrane is, in the majority of cases, ulcerated, or replaced by cicatricial tissue. In other cases, however, the mucous membrane remains unaltered, and, when this is the case, it may be taken as evidence of the extrinsic origin of the stricture. The glandular structure of the lining membrane is atrophic, and the openings of fistulæ and internal rectal sinuses are not unfrequently met with. These gross pathological changes so frequently met with, are, in my opinion, a very strong confirmation of the opinion held by Mr. Cripps, as to the ætiology of stricture from spasm, which has already been discussed. By any other theory they appear to be quite inexplicable.

Of the changes which are to be observed in the bowel above the stricture, generally the most obvious is dilatation, and this may be present to a marked degree. The mucous membrane will frequently be found to have disappeared in patches, as a result of ulceration, and at the seat of these ulcers the wall of the rectum may be so thinned that rupture has taken place. If this should be above the peritonæal reflexion, extravasation of fæces and acute peritonitis will be the result; but if the perforation takes place into the areolar tissue of the pelvis, then stercoral abscess is the result. This abscess may open in various places, resulting either in a complete muco-cutaneous fistula, which will then present the characters described under the head of fistula of the superior pelvirectal space; while at other times a fistula bimucosa will result, and of these the most common, in the female, is a communication between the rectum and vagina, the vaginal orifice being situated generally high up, close to the attachment of the vagina to the cervix uteri. In the male the fistulous tract most frequently communicates with the bladder, causing then, in the majority of cases, extreme suffering, owing to the escape of fæces into the cavity of the bladder, a condition which is one of those most urgently demanding the performance of colotomy. In some rare cases, however, the bladder appears to be tolerant of its abnormal contents; and in one case which came under my notice, and to which I have already made allusion in another connection, the first condition which attracted attention was a peculiar deposit in the urine, the patient making no complaint of pain about the bladder or rectum. Upon microscopical examination, however, of the deposit, some particles of striped muscular fibre were to be seen, which were evidently portions of undigested food which had escaped from the bowel, and a rectal examination revealed the presence of a strictured rectum.

Although the greater number of cases of rectovesical fistula are to be found in the male subject, the female is not exempt, as will be seen by a reference to page 104.

One of the most remarkable instances of fistula bimucosa is that recorded by Quain* as having been found at the post-mortem examination of Talma, the eminent tragedian. The pelvis was filled with an enormous sac, formed by the upper part of the rectum largely dilated. When the sac was raised a circular narrowing of the gut was discovered, situated at a distance of six inches from the anus; this was proved upon closer examination to be wholly impervious. It was, in fact, a solid cord, but on the surface irregular, and having the appearance of a purse drawn tightly, and puckered with the strings tied around it.

* "Diseases of Rectum," p. 190. London, 1854.

The great dilatation of the bowel at its lower end dipped down below the level of the stricture in the form of a dependent sac, in which was an opening about an inch in diameter, from which fluid had been diffused into the abdominal cavity. The rectum below the stricture was no more than the size of a child's intestine, and upon it, close to the stricture, was an ulcerated surface, with a narrow opening, to which the edges of the aperture above the stricture had been adherent. A new communication, but an imperfect one, had thus been established between the two parts of the gut, severed one from the other by the stricture. But the connection had given way, doubtless in consequence of the violence of the expulsive efforts, and thus the contents of the bowel had escaped a short time before death. In this interesting case an effort had evidently been made by nature to overcome the obstruction. In a case recorded by Wagstaffe,[*] a somewhat similar condition was observed.

CHAPTER X.

SYMPTOMS OF NON-MALIGNANT STRICTURE OF THE RECTUM.

Symptoms of non-malignant stricture are generally in the earlier stages extremely vague. The most frequent are attacks of diarrhœa, alternating with constipation, and where these have persisted for some time the suspicions of the surgeon should be aroused, and a rectal examination instituted. The diarrhœa is generally slight, and is more noticeable in the morning: it is frequently associated with the discharge of

[*] Trans. Path. Soc. London, vol. xx. p. 176.

small quantities of bloody mucus, and brown matter resembling coffee-grounds. This diarrhœa is due to catarrhal inflammation, caused by the irritation of retained fæces above the strictured point, the mucous discharge softening down the fæcal accumulation, and so allowing it to pass the stricture. When the bowel has been evacuated, a period of constipation ensues, to be again followed by the fæcal accumulation and catarrhal discharge. As the case progresses the intervals of constipation become fewer, and the local irritation and discharge increase, unless, indeed, the case goes on to complete obstruction, which is not a very frequent occurrence, the patient generally becoming exhausted before this takes place. Associated with the diarrhœa there is a good deal of tenesmus in most instances; and pain, which is generally referred to a point above the symphysis pubis, or the middle of the sacrum, is common. Pain after food, and flatulence, are not so frequently complained of as when the constriction is situated higher up in the intestinal tube; and the same may be said of vomiting. This last is only to be observed as a very late symptom, and after long-continued obstruction, when it may occasionally become stercoraceous.

The symptoms, although at first mainly local, after a variable time produce a general impression on the system. The exhausting muco-purulent discharge, which is commonly derived from the altered surface of the dilated bowel above the stricture, more particularly when there is a large amount of ulceration, may produce hectic fever, or amyloid degeneration of internal organs, with which we are familiar as a result of protracted suppuration in other parts of the body; moreover, the possible occurrence of septic poisoning is always to be remembered. On the whole, the disease is of an essentially chronic character, and it may take many years to run its course. The most distressing symptom which a patient with well-developed stricture suffers from, is the constant desire to go to stool, attended with colicky pains, but the attempt to defæcate is frequently without result. This is caused by the accumulation of fæces above the strictured part, and is only temporarily relieved by the spontaneous diarrhœa referred to, or by the action of purgatives. As the constriction becomes narrower meteorism becomes developed; the greatly dilated and full colon may be felt through the abdominal wall; the feeling of doughy softness conveyed by fæcal accumulations on palpation of the abdomen may be made out; or the outline of the large intestine may be indicated by dulness on percussion. Owing to the increased efforts to obtain an evacuation the suffering becomes intensified, till finally death results from exhaustion, unless it is hastened by some of the complications, such as peritonitis from perforation, or the supervention of sudden and complete obstruction.

Much has been written on the shape of the stools as indicating stricture, but the idea taught in most text-books, that narrow, or tape-like fæces, are indications of the presence of a stricture, requires qualification. Such an appearance is often produced where no stricture is present, and, on the other hand, a well-formed motion may be passed by a person suffering from marked stenosis. In the case of stricture of the urethra, the twisted or forked stream of urine is not formed by the shape of the stricture, but by the collapsed meatus urinarius; the flow of urine not being sufficient to dilate fully this orifice. So also in the rectum, the margin of the anus gives the final form to the voided fæces. Where there is a contracted state of the anus, due to fissure or other cause, the fæces are passed in the form of narrow cylinders, or they may be flattened laterally. Where a stricture is situated at some distance from the anus, it is quite possible that the mass may be re-formed in that portion of the rectum below the stricture, and so be passed of normal calibre. More frequently, however, the fæces are found in little masses of a spherical or ovoid shape, reminding one of the appearance of sheep or rabbit droppings, and this is to be explained by the relaxed condition of the termination of the bowel and sphincters being unable to compress the mass sufficiently to render it again of uniform consistence.

There can be no doubt, however, that in some cases the fæcal mass is passed in narrow cylinders as a result of stricture, and Van Buren has given the true explanation of this condition.* When a stricture is situated low down in the rectum, it is, during the violent efforts accompanying defæcation, extruded through the anus far enough to give its final impress to solid matter passed under this extreme pressure; and Kelsey† records an interesting case in which he was able to observe the mechanism of an occurrence of this kind, which I give in his own words: "The woman suffered from a stricture one inch above the anus, which was of sufficient calibre to admit the ends of two fingers easily. She had never noticed any deformity of the fæces. While under the influence of ether, and after the sphincter had been very thoroughly dilated, an O'Beirne's tube was passed through the rectum, which was empty, into the sigmoid flexure, which was full. After resting there a few moments it provoked a movement of the bowels. The stricture was instantly crowded down into view, appearing at the anus, and taking the place of the anus, which, owing to the complete dilatation, ceased to have any action, and was merely a patulous ring. Through the stricture there came a long tape-like evacuation, the mould which gave it its peculiar form being the stricture

* "Diseases of the Rectum and Anus," p. 279.   † *Ibid.*, p. 189.

pressed to the surface of the perinæum, and greatly lessened in calibre by folds of mucous membrane, which were crowded into it from above. While remarking to those present on the peculiar mechanism of its production, the straining ceased, the stricture rose, the mucous membrane was relaxed, and a passage of natural formation was the result. This alternation was repeated several times. At each effort the stricture was forced down to the anus, the membrane above it was crowded into it, so as greatly to lessen its calibre, and a flat passage was the result. When the effort was less violent there was still a passage, but the stricture having risen to its place, and not being so tightly filled with the mucous membrane, the passage was natural. The lesson to my own mind was this: that a stricture of large calibre might, as a result of straining, cause a passage of small size, and that, to get this peculiar shape, the stricture must be crowded down so as actually to take the place of the external sphincter, and be the last contracted orifice through which the soft substance is expressed."

A very grave, but rather infrequent termination to stricture is **complete obstruction.** After a stricture has continued for a long time, possibly many years, without affecting the general health to any great extent, the bowels being relieved sufficiently by the process before alluded to, the symptoms of complete obstruction may supervene, and it is a remarkable fact that the onset of this condition is not uncommonly somewhat sudden: this abrupt complication may be due to one of two causes, either the impaction of a foreign body, or even an unusually hard mass of fæces in the narrowed gut; or, as a result of an inflammatory œdema of the mucosa and submucosa due to the chemical irritation of retained and decomposing fæces.

Of the former variety the following is a well-marked instance: I was called, in the year 1876, to see a man, aged thirty years, whom I found suffering from well-marked symptoms of obstruction of the rectum; the belly was tumid and tender, and he had vomiting, not, however, stercoraceous. There were frequent abortive attempts to defæcate, and I was informed that he had obtained no relief from the bowels for the past ten days. He stated that he had small-pox four years previously, which was followed by a discharge of matter from the rectum, and that since then he had suffered from alternating attacks of diarrhœa and constipation. Upon making a rectal examination, a stricture was at once detected within one and a half inches from the margin of the anus; projecting through the orifice of the stricture, which was very narrow, a hard substance could be felt, and with considerable difficulty I removed it with a forceps. It proved to be a plum stone. He assured me that it was over a year since he had swallowed it, as he had a distinct recollection of the fact of having done so. I treated the case with a limited incision and dilatation with bougies, and I afterwards heard that the patient remained in tolerably good health, although he still required the occasional introduction of a bougie.

It is a very remarkable fact, how frequently fruit stones have been found impacted in intestinal strictures, or collected in numbers in the pouch above them. A very interesting case of this kind is reported by Dr. Wickham Legg:* A woman, aged 28, had frequently before death vomited and voided by the rectum cherry stones, and during life a tumour composed of them could be felt through the parietes, giving to the hand a very peculiar sensation as they were rubbed together. At the post-mortem

* Path. Soc. Trans., vol. xxi. p. 171.

examination the ileo-cæcal valve was found strictured, and in the intestine above there was nearly an imperial pint of fruit stones. A number of similar instances are to be found recorded. In some cases, possibly, the mechanical irritation set up by the accumulation of stones may be looked upon as the *cause* not the effect of the stricture. Such, however, was evidently not the case in the instance I have brought forward. Besides cherry stones, many other hard and insoluble substances, as pieces of bone, apple core, etc., have been found lodged in a stricture, and so constituting the determining cause of complete obstruction.

Inflammatory swelling, as a cause of complete obstruction, is more often met with in cases of malignant disease of the rectum than in cases of the ordinary stricture; but even in the latter it is sometimes observable, more particularly after the injudicious use of bougies.

One of the most serious complications that may arise during the course of stricture of the rectum is **peritonitis.** In this case we find inflammation of the peritonæum occurring either as an acute and general manifestation, the result of rupture of the intestine and extravasation of fæces, or as a more chronic and limited disease; the former not being an infrequent termination to years of suffering from rectal stricture. It may be due to spontaneous rupture, during a violent effort at defæcation, of the attenuated and ulcerated wall of the bowel in the neighbourhood of the stricture; or, at other times, we have to admit that the treatment adopted by the surgeon must be held to be directly responsible for the fatal perforation. There is, unfortunately, no lack of cases in which the point of a bougie, an enema pipe, or even the index finger of the surgeon, has penetrated through a diseased intestinal wall, and permitted the extravasation of fæces to take place : this accident has occurred to the most accomplished surgeons, who, with a candour highly to be commended, have recorded their misfortunes, and so enabled us to learn a lesson that should ever be present with us when dealing with cases of this kind, to employ the utmost gentleness when examining or conducting the treatment of constriction of the rectum. When only the tip of the index finger can be insinuated into the aperture of a stricture, the temptation to force it through, so as to determine the length of the stricture, or to effect dilatation of it, is indeed strong, but it must be absolutely resisted, unless we wish to swell the already too long list of mishaps which have occurred. The old saying, "meddlesome surgery is bad," applies with more force to the disease under consideration, probably, than to any other surgical affection.

Chronic and limited peritonitis.—As a result of the inflammation and ulceration of the rectum associated with stricture, the pelvic peritonæum may become involved, and bands of adhesion be found in consequence, without any perforation having taken place. It is not at all infrequent to find this condition on post-mortem examination of old-standing cases of rectal stenosis, thickening of the peritonæum, limited effusion, and bands of lymph being the most common appearances met with. In some cases the adhesions produce by their contraction an increased narrowing of the lumen of the bowel, the most common seat of which is at the junction of the sigmoid flexure and the rectum, so that the obstruction thus formed may be considerable. The symptoms of this complication are, however, scarcely recognisable during the life of the patient, so that the treatment of it comes scarcely within the range of practical surgery. Where, however, it is suspected, it furnishes a strong additional argument for the relief of the irritation by colotomy.

Abscess and fistula complicate the case in a very large proportion of cases of rectal stricture, and they may be found in various situations. When occurring in the ischio-rectal fossa, the communication between the bowel and the suppurating cavity is generally to be found below the seat of stricture, and when so placed it is, as I have before endeavoured to show, a strong evidence that it has originated in an ulcer, which was also the exciting cause of the stricture; although it is quite possible that these ischio-rectal abscesses may occasionally form as the result of irritation, without any direct communication being established with the gut, in the same way that we find extra-articular abscesses occurring in the neighbourhood of a diseased joint. Where, however, they form as the result of perforation of the pouch above the stricture, they are then usually situated above the levator ani and recto-vesical fascia, in the superior pelvi-rectal space. From this position they may penetrate in various directions, as into the vagina or bladder, or into the peritonæum; or the first place at which they become superficial may be in the iliac region, as happened in one case under my care. If they come down to the perinæum, their most usual position is posterior to the anus, as the pelvic septum more readily allows the passage of matter there than at any other part of the circumference. These fistulæ have, however, been already considered in connection with the pathological anatomy of this disease, and their symptoms, when fully formed, are sufficiently obvious, so that the subject need not be further discussed.

Upon making an examination of a person affected with rectal stricture, it will generally be found that the anus is surrounded by hypertrophied flaps of skin, which no doubt owe their origin to the continued maceration and irritation of the parts by the acrid discharge.

Mr. Colles* considered these appearances as almost pathognomonic of stricture. He says: "We often observe at the orifice of the anus the following appearance, which is indeed almost always present when the disease is situated near to the external sphincter; namely, at each side of the anus, a small projection, which, on its external surface, appears as a mere elongation and thickening of the skin, but internally presents a moist surface not exactly like the lining membrane of the gut; nor yet can we say that it is ulcerated. These two projections lie close together below and divaricate above, presenting a resemblance to the mouth of an ewer. Whenever this external appearance exists, I feel almost certain of finding a stricture of the rectum before the finger is pushed as far as the second joint into the gut. In some cases, however, this external mark has not been present." In my experience, however, this condition has not been often present, and I have seen a similar appearance when no stricture could be detected.

Hæmorrhoids are of common occurrence as a complication, and this is not to be wondered at when we consider the pressure that must be exercised on the branches of the hæmorrhoidal veins in passing through a dense stricture.

It is also common to observe the openings of fistulæ, and these are frequently multiple; but of all conditions of the anus great relaxation is the most noticeable, the finger readily passing through without any difficulty.

This relaxed condition of the anus permits the involuntary escape of sanious muco-pus, which constitutes one of the most unpleasant subjective phenomena of stricture; and it also sometimes permits the extrusion of the stenosed portion of the bowel which has been already alluded to. True prolapse

* Colles' Works, by McDonnell, p. 370. New Sydenham Society, 1881.

of the rectum is, however, an extremely rare complication of rectal stricture. Upon introducing the finger into the rectum the stricture, if situated in its usual position in the rectal pouch, will be at once felt, if it has encroached much upon the lumen of the bowel; but if the amount of contraction is slight, so that two or more fingers can be passed readily into it, it may be more difficult of detection : in these cases a thickening and hardness, or ulceration of, or outgrowths from, the mucous membrane will excite the suspicions of the surgeon. In this preliminary examination the whole under surface of the stricture should be carefully felt, relative involvement of the various portions of the circumference of the gut made out, and the existence and extent of ulceration or outgrowths determined. Where the entire circumference of the gut is tolerably uniformly contracted, and where the amount of induration of the tissues is considerable, the sensation conveyed to the finger resembles closely the feel of the os uteri when the finger is in the vagina : this likeness is occasionally further increased by the fact that the finger can be passed round the stricture, which appears to project down into the bowel, and is due to a very limited intussusception following the violent expulsive efforts.

With the finger the lower aspect of the stricture can be completely examined; but, unless the opening is sufficiently large to admit the passage of the finger through the stricture easily, the surgeon is unable to form an opinion of the length of bowel involved. For this purpose bulb-ended bougies, as recommended by Bushe, are necessary. They are best made of ivory or ebonite bulbs fastened on to a whalebone or readily flexible metal rods; they should be of various sizes (Fig. 4, page 15). Having selected the largest that will easily pass through the stricture, the instrument should be introduced through the contraction until the end is felt to be free in the bowel above. By gradually withdrawing it now the surgeon will be able to recognise the moment it enters the superior opening of the stricture, and thus an estimate of the length of the stenosis can be arrived at. Where the stricture is situated beyond the reach of the finger great difficulties will be experienced in the diagnosis; indeed, it may be safely asserted that stricture of the upper portion of the rectum has been supposed to exist much more frequently than its existence has been demonstrated, even when the greatest care has been taken in the examination by accomplished surgeons. The diagnosis and treatment of non-existing strictures has been a favourite field of practice for charlatans; persons suffering from ordinary constipation being easily led to believe that their symptoms are due to mechanical obstruction. Kelsey* gives an amusing case of this kind. "A lady went to consult a rectologist, for some reason or other which is not stated, and a sound was introduced into her anus. Her husband, learning this, rushed to the house of the scoundrel in a violent rage, and armed with a whip. Half an hour later he returned, disconsolate. He had found out that, like his wife, he had a stricture of the rectum, and, like her, he had submitted to catheterisation!"

If the patient has symptoms which would lead us to suspect the presence of a stricture high up, such as diarrhœa alternating with constipation and paroxysmal colicky pain, great straining and pain while at stool, and the discharge of muco-pus or altered blood from the anus, and yet if under such circumstances no indication can be obtained by ordinary digital examination, the patient should be placed under ether, and an examination conducted in the lithotomy position. First of all the anus should be well stretched, and the bi-manual method adopted, with one hand pressing deeply down

* *Loc. cit.*, p. 182.

into the pelvis through the abdominal parietes, and the index finger of the other hand passed as high as possible up the rectum. By this means we may be enabled to make the diagnosis; failing this, an enema should be administered, and if this is at once returned without our being able to distend the colon, it is strong evidence of obstruction. A careful examination with a bougie should now be instituted, but the information obtained by its use is open to several fallacies. In the first place, the point may impinge against the promontory of the sacrum, and so its further progress may be arrested; or, it may be caught in some of the folds of mucous membrane. This may generally be obviated by having the bougie hollow and perforated at the point like an O'Beirne's long tube, and made so that an enema apparatus can be attached. When, now, the point becomes arrested, some warm water can be thrown up, and so the loose folds of mucous membrane lifted off the end of the instrument. Another source of error is the bending of the bougie upon itself, so that a considerable length may be passed although a mechanical obstruction exists. Whenever the progress of the instrument becomes arrested, the direction of the end of the bougie should be altered and a fresh attempt made. The utmost gentleness should be observed, the surgeon always bearing in mind that the coats of the bowel may easily be perforated. The only unequivocal indication that the instrument is really in a stricture is the feeling that it is grasped, a sensation with which we are quite familiar in the catheterisation of urethral stricture. Of course it is obvious, however, that before we can attach any importance to this symptom in rectal stricture, the sphincter ani must have been temporarily rendered paralytic by hyperdistension, unless indeed it is so relaxed as a result of the disease as to render this preliminary step unnecessary. The only means left at our disposal for the

M—23

further examination of the upper portions of the rectum is by introduction of the whole hand into the rectal pouch, and passing one finger up into the sigmoid flexure, a procedure the details and dangers of which have been elsewhere (page 16) fully entered into; but it can be but seldom that the requirements for diagnosis necessitate the undertaking of this operation, which has been shown to be attended with a considerable amount of risk. The importance of determining the position of a stricture, situated high up, when the symptoms of obstruction are urgent, is manifestly of the first importance, as the treatment to be adopted will in a great measure depend upon exact diagnosis; lumbar colotomy is only suitable to those cases which are situated so low down that they can be easily explored with the finger, and in which linear proctotomy is inadvisable; or to those cases of stricture higher up in which the position has been accurately localised. Where doubt exists, abdominal section is probably to be preferred, as will be shown in another place.

The only points of diagnosis which remain for our consideration are the differentiation of benign stricture from malignant neoplasms on the one hand, and extra-rectal disease producing pressure upon the other. It is generally easy to distinguish the obstruction due to the pressure of tumours, or by bands of adhesion from non-malignant stricture, the sensation conveyed by the finger readily estimating whether the obstruction is situated in the rectal wall or not; but the diagnosis between malignant and non-malignant disease, although in typical instances easy, is sometimes attended with very considerable difficulty, so that it may be impossible to arrive at a definite conclusion until the case has been kept under observation for some time, and its rate of progress carefully noted.

DIFFERENTIAL DIAGNOSIS.

The following table illustrates the more important points of difference :

NON-MALIGNANT STRICTURE.	MALIGNANT OBSTRUCTION.
Generally a disease of adult life.	Generally a disease of old age.
Essentially chronic, and not implicating the system for a long time.	Progress comparatively rapid and general cachexia soon produced.
The orifice of the stricture feels as a hard ridge in the tissues of the bowel. Polypoid growths, if present, are felt to be attached to the mucous membrane.	Masses of new growth are to be felt either as flat plates between the mucous membrane and muscular tunic, or as distinct tumours encroaching on the lumen of the bowel.
Ulceration of mucous membrane may be present, but without any great induration of the edges.	Ulceration, when present, is evidently the result of the breaking down of the neoplasm, and the edges are much thickened and infiltrated.
The entire circumference of the bowel constricted unless the stricture is valvular.	One portion of the circumference generally more obviously involved.
Pain throughout the whole course, in direct proportion to the fæcal obstruction, and only complained of during the efforts at defæcation.	In the advanced stages pain is frequently referred to the sensory distribution of some of the branches of the sacral plexus, due to direct implication of their trunks.
Glands not involved.	The sacral lymphatic glands can sometimes be felt through the rectum to be enlarged and hard.
There is usually evidence that ulceration has commenced at the anus and travelled upwards.	Usually commences well above the anus.

CHAPTER XI.

TREATMENT OF NON-MALIGNANT STRICTURE OF RECTUM.

THE various plans of **treatment** which have from time to time been advocated for rectal stenosis may be conveniently classed under the following heads : (1) Dietetic and medicinal; (2) dilatation, (*a*) gradual, (*b*) sudden ; (3) incision, (*a*) internal, (*b*) external ; (4) excision ; (5) colotomy ; (6) electrolysis.

By attention to the **diet** a considerable amount may be done in the way of making life more endurable, and it is obvious, only such food should be allowed as will leave a small fæcal residue. First in importance stands milk, which should form a large portion of the patient's food ; and strong soup, eggs, and meat, in moderation, also may be allowed. As most vegetables leave a considerable residue, they should be but sparingly used, and such articles of food as oatmeal, brown bread, etc., should not be permitted. One objection to giving too unstimulating a dietary is that the fæces which are formed produce such little excitation of peristalsis, that purgative medicines will, at the same time, be required in considerable quantity.

As to the **medicinal treatment** of stricture, it is obvious from the very nature of the case, that the use of purgative medicines must constitute an important element of our practice ; and that some discrimination must be experienced in our employment of such agents.

Of all aperients, the sulphates of soda and magnesia, as combined in some of the many mineral waters in the market, will be found to answer best. A sufficient dose should be taken early in the morning to ensure a free evacuation. Where the calls to stool are frequent, and where considerable straining exists, it is generally an indication of retention of fæces above the stricture, which are best dislodged by a copious enema of soap and water. The compound liquorice powder will often prove an efficient aperient; or, where patients can take it without nausea, castor oil. Frequently, however, it will be found well to change the medicine employed, care being taken not to use any of the more irritating drugs, such as aloes, colocynth, etc., as they only tend to increase the tenesmus. Belladonna is highly spoken of by many authorities, and in some cases certainly tends to relieve spasm; it is best given in the form of a suppository containing a grain of the extract. This may be with advantage combined with five grains of iodoform, especially if there is an open ulcer present. Where the bowel is very irritable, marked benefit will result from the use of small starch and opium enemas, and where the catarrhal discharge is considerable, injections containing liquor of bismuth, or tincture of rhatany, will probably diminish the secretion. Where the disease is of unquestionable syphilitic origin, mercury or iodide of potassium may be tried, but these remedies can only prove useful in those cases of recent origin where syphilitic deposit or ulceration is progressing. It is manifestly useless to expect that where atrophic shortening of the muscular fibres, with cicatricial or cirrhotic contractions, has taken place, that any good can possibly result. Indeed, it is probable that under these circumstances a positive injury will be inflicted by so-called specific treatment further lowering the already debilitated constitution.

Medicinal treatment, however, can only at best do little more than relieve symptoms; and for any permanent benefit we must look to some of the mechanical or surgical operative methods indicated in the above enumeration. Of these the first is **dilatation.** Where the obstruction is considerable, as is almost always the case when the patient comes under observation, an attempt may be made to dilate by means of the bougie; but I would have it to be borne in mind that this treatment is only suitable to those cases which are unattended with open ulceration, as where an ulcer exists, rupture through its floor is very likely to follow, so that in this case we should try and heal the ulcer by dietetic and local measures, failing which, I believe the treatment by incision to be decidedly preferable to any attempt at dilatation. In those cases, however, in which there is no open sore, the careful use of bougies is of the greatest service, and although it cannot be said with certainty that a permanent cure can be effected (by which is understood that no further treatment will be rendered necessary) still a considerable amount of good can be done, and the space necessary for a free motion can be maintained by the occasional passage of a bougie. When a stricture of the urethra has been relieved by gradual dilatation, no surgeon will admit that the case has been completely cured, but by subsequent occasional catheterisation the stricture can be prevented from contracting again to such a degree as to occasion serious symptoms. The same is true of a stricture of the rectum.

Much will depend on the form of bougie used. The so-called gum elastic instruments, which are made of platted cord covered with varnish, and which have been in general use for many years, are unsuitable, as they are too rigid, and they are made unnecessarily long. Kelsey recommends the use of soft rubber bougies similar to those occasionally used for the urethra, and these answer the purpose admirably, as it is almost impossible that any injury could be inflicted by their use: the only drawback to them is, that they are so soft that it may be found difficult to pass them into the orifice of the stricture. For general purposes those bougies made with an olive-shaped bulb, mounted on a flexible whalebone stem, answer all purposes admirably. In some cases it may be found advisable to leave a bougie in the stricture for some hours, thus adopting the principle of *vital* dilatation so authoritatively recommended in the treatment of urethral stricture. For this purpose the instrument should be short, about four or five inches only in length, and with a string attached, so that it can be passed entirely into the rectum, the string hanging out at the anus, and serving to withdraw the bougie when required; as where the anus is kept dilated for any considerable time great annoyance is given to the patient, and violent expulsive efforts induced. Cripps speaks favourably of conical bougies, by means of which, if gentle pressure is kept up, a gradual and continuous dilating effect is maintained.

A considerable amount of ingenuity has been expended in the construction of elastic hollow bougies, which can be inflated with air, or distended with water after they have been introduced into the stricture. This is a plan of treatment, however, which requires extreme caution, as the surgeon is unable to satisfactorily estimate the amount of force which he is using, especially if water is injected instead of air. In some cases, where there is a great deal of induration and perirectal thickening, this method may answer tolerably well. None of the special instruments which have been invented answer the purpose better than the "Barnes' bags" used by obstetricians for the purpose of dilating the os uteri, the fiddle-shape rendering them less likely to slip out of the stricture when once they have been introduced, and they can be introduced with the greatest facility while empty. I have treated some cases in this way with decidedly good results. Sudden dilatation by means of Todd's dilator, or other similar instruments, must be looked upon as a very hazardous proceeding, which cannot be recommended for any form of stricture.

If it be found that dilatation is impossible to any useful extent, or if, as before mentioned, the presence of ulceration render the attempt at dilatation inadvisable, recourse may be had to division of the contracted tissue, by either internal or external incision, by which is understood the simple superficial incision of the stricture alone; or the complete division of the rectal wall from a point above the stenosis through the contracted tissues, and also through the anus, external sphincter, and skin.

Internal incision is of but limited utility, and is not by any means devoid of danger. It is only applicable to those cases of very rigid stenosis in which there is a great deposit of indurated tissue, and then only as an aid to gradual dilatation by bougies. For this purpose a Cooper's hernia knife, or blunt-pointed bistoury, should be used, and several superficial incisions made round the circumference of the stricture. Great care should be taken that the division is not carried through the wall of the gut, else fæcal extravasation and stercoral abscess will be the inevitable result.

External incision, or linear proctotomy, may be adopted for those cases attended with ulceration of the mucous membrane below the stricture; and, indeed, in all those cases where gradual dilatation is ineffectual, or where the continued use of the bougie sets up such an amount of constitutional and local irritation that it is inexpedient to continue the treatment.

As it has frequently happened that fistulæ in connection with stricture have been operated upon without the stricture being diagnosed, and as the internal openings of these fistulæ are occasionally situated above the stricture, it is evident that complete division of the contraction has been thus sometimes unintentionally performed; but the first deliberate attempt to cure stricture in this way appears to have been made by Sir G. Humphry, of Cambridge,* who proposed this method in consequence of the "good results following longitudinal incisions in urethral stricture." And although he performed the operation on two occasions with excellent results, this method of treatment appears to have fallen into disuse until comparatively recently; and even still, especially in England, many surgeons prefer colotomy in those cases where gradual dilatation is either inefficient or inadvisable. To M. Verneuil is undoubtedly due the credit of having revived this operation, and of having brought it prominently before the profession. In a paper read at the Surgical Society of Paris,† he enters very fully into the subject, and enumerates ten cases of the operation. He has moreover extended the application of linear proctotomy to malignant disease, as a substitute for colotomy in the treatment of obstruction. Can we anticipate a complete and absolute cure by this treatment in cases of non-malignant stenosis? A more mature experience of this operation compels me to give a less favourable opinion as to ultimate results than was expressed in the former edition of this work. If the operation has been completely performed, that is, if the entire thickness of indurated structures has been divided, a cure will result in the majority of instances, and, as compared with colotomy, the advantages are obvious. After the recovery of the patient the power of controlling the evacuations is only partially recovered, so that the occurrence of a certain amount of incontinence must be

* *Association Medical Journal*, p. 21; 1856.
† *Bull. de la Soc. Chirury.*, October, 1872.

expected in the majority of cases, while the occasional passage of a bougie is afterwards necessary in order to avoid stenosis. The operation may be performed by different methods. Verneuil recommends the following plan :—The bowel having been well cleared out, and the patient placed in the lithotomy position, the finger is introduced through the stricture. If difficulty is experienced in doing this, a probe-pointed bistoury is first passed, and an incision sufficient to allow of the easy introduction of the finger is made in a direction directly backwards. A trocar and cannula is now entered at the tip of the coccyx, and pushed on till it enters the rectum well above the seat of stricture. A flexible bougie, or a piece of string, is by means of the cannula passed into the rectum, the end being hooked out at the anus with the finger. By means of this the chain of an écraseur is now passed, and the tissues thus surrounded are gradually divided. More recently Verneuil * has recommended the opening into the rectum to be made with a fine point of Paquelin's cautery, instead of the trocar and cannula. Van Buren is of opinion that the entire section is best made with a knife-shaped cautery, the charring of the edges protecting the wound from irritation of the fæces until granulation is established : he also avoids making it *directly* backwards, as he considers that the wound heals better when made a little to one side. Of all methods the knife is, I am satisfied, the best. With a probe-pointed bistoury the division can be more surely, expeditiously, and cleanly made than by any other means; and if it is confined to the middle line there need be no fear of bleeding. Any vessel that does spring can easily be ligatured; or, if there is general oozing, the wound may be plugged with an aseptic sponge, or, better still, with iodoform gauze, and the whole supported with a T-bandage. At first there

* International Medical Congress, Copenhagen, 1884.

will be a certain amount of incontinence, but as the wound heals the power of retaining fæces will gradually be regained. During the healing, frequent syringing with some antiseptic solution should be employed; or, until granulation is established, continuous irrigation with warm solution may be resorted to.

Excision of the stricture has been performed several times for non-malignant stricture, and in suitable cases has much to recommend it. It is preferable to linear proctotomy both as to the cure of stenosis and the ultimate control of the bowel, and general comfort of the patient. The first case of a simple stricture I excised was one in which there was a doubt between malignant growth and simple stricture; the result was extremely good. Microscopic examination showed that the case was one of benign stricture. I have since adopted excision in a case in which there was no question as to diagnosis, with equally good result. The method of operation differs in no respect from the perinæal excision of rectal cancer. (See page 360.) In cases of slight stricture attended with extensive ulceration, that have proved intractable to ordinary treatment, **colotomy** is indicated, even though the symptoms of obstruction are not very severe, the operation being mainly undertaken to afford physiological rest. If the ulceration can, under the new and more favourable conditions, be made to heal, and the stricture eventually dilated, the artificial anus may be finally closed. And again in cases of ano-rectal syphiloma where there is great infiltration and matting of the surrounding pelvic structures, the rectum feeling like a hole bored in a turnip, excision is obviously contra-indicated and colotomy the only course open.

Electrolysis has been recommended for the treatment of rectal stricture in the same way that it is used in stricture of the urethra. It is to me inconceivable that good results could possibly follow its use. If the current is strong enough to destroy tissue, injury may be done by destroying tissues not intended; while even if the stricture itself is necrosed the healing after separation of the eschars will only add to the contraction. The supporters of this treatment would have us believe that a current passed through the contractive material of a stricture has some elective

Fig. 25.—Diagram illustrating Dieffenbach's Procto-plastic Operati n.

and "alterative" effect on it; I am convinced that if the electrodes had been passed *without any current* with the same assiduity they would by their mechanical pressure have produced as good results.

Stricture of the anus is most frequently the result of some operative interference, and can generally be treated by dilatation sufficiently to obviate any inconvenient obstruction, unless there is a great deal of dense cicatricial structure, in which case the procto-plastic operation of Dieffenbach (Fig. 25) may be had recourse to. Vertical incisions are made through the contracted tissues, one anterior and the other posterior, and from the outer ends of these, two radiating incisions are carried through the skin, thus forming an angular flap of integuments. This flap is dissected up, and its apex brought up to the inner extremity of the vertical incision, where it is retained by sutures. The incisions are in the shape of the letter Y, and the resulting cicatrix in the shape of the letter V. Cases requiring this operation are, however, of rare occurrence.

CHAPTER XII.

SYPHILIS OF THE RECTUM AND ANUS.

FROM the fact that syphilis is credited with causing a large number of the cases of stricture met with in practice, considerable attention has been directed to the subject; and at all stages of this protean disease the anus and rectum may be the locality affected. Considerable looseness of description is, however, to be found in the accounts given by various authors, particularly in reference to the primary lesions met with.

There can, however, be no doubt that both the chancroid and true chancre are met with not only at the anus, but in the interior of the rectum.

At the anus, especially in the female, the **chancroid** is of common occurrence; and in this sex it may be the result of auto-inoculation from similar disease in the vulva, the discharge trickling down over the perinæum, and so infecting any excoriations of the part that may be present; or it is quite possible that the accidental contact of the penis during coition may be the means of conveying infection. In the male, however, primary soft sore in the neighbourhood of the anus is exceedingly rare, and when present furnishes strong presumptive evidence of sodomy. According to Péan and Malassez,* nearly one-half of the superficial anal ulcerations observed in females at the Lourcine, in 1868, were due to soft chancre; and according to Fournier,† one-ninth of the cases of chancroids in the female are situated at the anus; whereas in the male he met with only one case in four hundred and forty-five.

* "Étude clinique sur les ulcérations anales." Paris, 1872.
† "Dict. de Méd. et Chirurg. Pract.," art. Chancre, p. 72.

In position these ulcers may be found on the skin in the immediate neighbourhood of the anus, or between the folds of the outlet, and extending over the border of the sphincter. They are mostly multiple, with sharply-cut edges; in fact, they in no way differ in appearance from the same form of ulceration met with in other parts of the body. When the ulcer extends over the margin, the pain is considerable, especially after defæcation, and bleeding is not uncommon; in rare instances extensive phagedæna may supervene and occasion considerable destruction of the parts; when this has been the case, or, indeed, when the chancres have been numerous and large, an anal stricture may be the result, but generally these sores heal without difficulty.

That chancroids may extend up into the cavity of the rectum has been put beyond doubt by the observations of Bumstead and Taylor,* Van Buren,† and others; and there can be but little doubt that the much greater relative frequency of non-malignant stricture in the female is in a great measure due to this fact, although in all probability other causes are concerned in the production of the same result; for a further consideration of which the reader is referred to the chapter upon stricture.

That a primary soft sore may be found in the interior of the rectum without involvement of the anus has been denied by many authorities. The following case recorded by Neumann,‡ from the very full and elaborate way in which it has been investigated, appears to set the matter at rest:

"Upon examination, a sharply cut sore, having the characters of a soft chancre, was found on the posterior wall of the rectum, about 4 cm. above the

* "Venereal Diseases." Philadelphia, 1879.
† "Diseases of the Rectum." London, 1881.
‡ *Allgem. Wien. med. Zeitung*, No. 49; 1881.

sphincter; the anus and genital organs were healthy. Inoculation of the discharge on the patient's arm produced characteristic soft sores. The patient's husband was then examined, and was found to have a soft sore on the margin of the prepuce. He admitted that he might have infected his wife directly. Subsequently two chancres appeared among the anal folds, presumably from secondary inoculation."

The first and most important indication in the treatment of soft sore in this region is absolute cleanliness. The bowels should be kept somewhat free by means of a saline aperient; and iodoform, or black-wash, used as a local dressing. Where the ulcer presents a spreading edge, cauterisation with nitric acid is indicated; and, should the ulcer become chronic and assume the characters of irritable fissure, division of the sphincter or forcible dilatation may be required.

True chancre at the anus is very rarely met with. This fact is accounted for by Péan and Malassez by the very slight disturbance to which the disease gives rise, so that the sufferers do not usually seek advice. True chancres here, as elsewhere, have a hard and raised outline, with indurated base, and might be mistaken for fissure, from which, however, they may be distinguished by the freedom from pain. In doubtful cases the diagnosis should be suspended, pending the appearance of secondary symptoms. Primary hard chancre within the rectum is even rarer still; and it is scarcely possible that it can occur in this situation except as a result of unnatural connection. Cases of it have, however, been put on record by Ricord, Fournier, and others.

Of all syphilitic diseases in the neighbourhood of the anus, **condylomata,** mucous patches, or moist papules, are the most frequent. According to the statistics of Bassereau, condylomata were present in this situation in 110 out of 130 cases in the male; and in the female, if we except the vulva, the skin immediately surrounding the anal outlet is found to be the locality most commonly affected.

It is not likely that typical condylomata could be mistaken for anything else, their appearance being so characteristic (Plate III., Fig. 2); the raised flattened surface, pearly-grey colour, and abominably fœtid discharge, generally rendering the diagnosis easy. They are due to an inflammatory change in the epidermis and corium, especially the papillæ, which swell up, owing to the infiltration with exudation cells and fluids; the epidermal covering becoming macerated and softened. Their growth is much fostered by inattention to cleanliness, and they have a great tendency to relapse. They sometimes undergo considerable increase in size by branching, and proliferation of the papillary structure. When this is extensive they present a cauliflower-like appearance, or even a distinctly pedunculated wart may result. On the other hand, ulceration may take place, and irregular ulcers, called by the French writers "rhagades," may be found. These are situated in the anal folds, and may be diagnosed from the ordinary fissures in this locality by being multiple, by having one or both edges *elevated*, and by being considerably less painful. After these ulcers are healed the elevated edges may persist, as folds of hypertrophied skin, sometimes with a markedly crenated border like a cock's comb. They are of a pale pink colour, soft, glistening, and moist. In speaking of these, Sir James Paget [*] says: "I will not venture to assert that these cutaneous growths are never found except in syphilitic disease of the rectum, but they are very common in association with it, and so rare without it that I have not seen a case in which they existed either alone or with any other disease than syphilis."

[*] *Med. Times and Gazette*, p. 280; 1865.

PLATE III.
Fig. 1.—Eczema of Anus.
Fig. 2.—Condylomata of Anus.

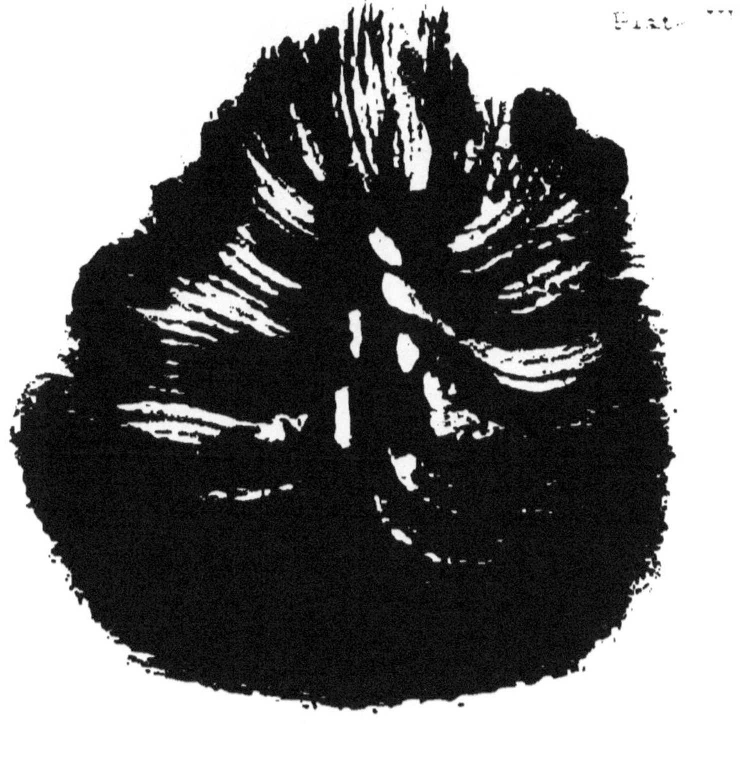

E. Burgess del. lith.

There is no doubt, however, that growths which are not to be distinguished from these do result from other causes, such as piles, carcinoma, and other internal rectal diseases. Kelsey* and other American authorities confine the term "condylomata" to the non-syphilitic hypertrophic folds of skin found round the anus resulting from piles.

Gummata.—There is no part of the body, in which connective tissue is present, in which the gummy deposit so characteristic of the later stages of syphilis may not be found; and we find that the lower bowel and anus prove to be no exception to this rule.

Cases of localised gumma in the rectum have been put on record by Leisol,† Mollière,‡ Verneuil,§ and Barduzzi.‖ The most interesting case of this kind is, however, described by Zappula.¶ The patient, a man of thirty-six years of age, had gonorrhœa, and an ulcer on the glans penis fifteen years before. Mercurial treatment was at once adopted, and no lesions of syphilis subsequently appeared. Fifteen years later he began to suffer from pains situated to the right side of the anus, and in the right tuberosity of the ischium; and afterwards very soon the symptoms of rectal stricture became developed; and so severe was the obstruction, that a large fæcal accumulation, which could be felt through the abdominal wall, formed. Upon digital examination, smooth, elastic elevations were recognised, which appeared to be enlarged folds of mucous membrane. At a distance of 4 cm. from the anus there was found a painless swelling, the size of a hazel nut, globular, smooth, and elastic; it was apparently situated under the mucous membrane, to

---
* "Diseases of Rectum and Anus," p. 146; 1883.
† Archiv f. Dermatol. u. Syph. Wien, 1876.
‡ *Op. cit.*
§ Quoted by Fournier, *op. cit.*
‖ See Bumstead and Taylor, "Venereal Diseases," p. 607.
¶ Arch. f. Dermat. u. Syph. Prag, pp. 62—90; 1871.

which it was not attached. The diagnosis lay between syphilis and cancer; and as a complete cure resulted from the exhibition of iodide of potassium, the former diagnosis was established.

Ano-rectal syphiloma.— Under this name Fournier has described* a remarkable specific infiltration of the rectal wall, which, as he states, begins in the submucous layers, the mucous membrane being only secondarily affected and at first free from ulceration. This disease, he has noted, is more common in females in the proportion of eight to one, and its usual situation is the rectal pouch, but the anus may be involved. The tendency of this infiltration is to undergo ulceration, or sometimes to end in cirrhotic contraction without any ulceration. What the exact pathology of this condition is does not appear to be settled. Fournier speaks of it as a hyperplastic proctitis, passing at a later stage into a fibro-sclerous condition, and it appears that it is more closely related to the diffused sclerotic changes which take place in the spinal cord, liver, and other organs during the later periods of syphilis, than that it is, as Van Buren describes it, an infiltrated form of gumma. It is to be diagnosed by the stiff, lumpy feel of the intestine, usually free from ulceration. Van Buren states that he has seen it entirely disappear under a mercurial course, but it is obvious that treatment, to be at all effectual, must be undertaken before the stage of cirrhotic contraction has set in.

In **congenital syphilis** the only common manifestation at the anus is the mucous patch, sometimes associated with radiating fissure-like ulcers. I have, however, met with one case of stricture in a child, aged ten years, who had well-marked "Hutchinson's teeth" and interstitial keratitis; and, from the great infiltration and firm feel of the coats of the bowel, I

* "Lésions Tertiaires de l'Anus et du Rectum." Paris, 1875.

have but little doubt that it was the result of a similar cirrhotic change to that described by Fournier, as resulting from the later stages of the acquired disease. Bodenhamer * alludes to inherited syphilis as an occasional cause of congenital stricture.

Trelat† speaks of small superficial fistulæ, perforating the tabs of skin usually found in syphilitic disease. He says that they are all healed and dry within, like the holes for ear-rings, and are characteristic of syphilis. They have sharply-cut orifices, and are found in cases of ano-rectal syphiloma.

During the later stages of syphilis a form of ulceration is not uncommon in the rectum, which may assume extensive proportions, and finally, by the contraction which takes place during its cicatrisation, occasion stricture of the bowel. A very instructive case of this kind is given by Sir J. Paget, in a clinical lecture delivered at St. Bartholomew's Hospital.‡ The following is an abridged account of this interesting case:

The patient was twenty-eight years old, and stated that she had suffered from syphilitic sores seven years previously, shortly followed by a cutaneous scaly eruption. About a year subsequently she became subject to an itching about the anus, and a growth of skin appeared reaching a short distance into the rectum. Two years after this a large ulcer formed in the neighbourhood of the anus, and she was received into University College Hospital. The ulcer was destroyed by the application of some corrosive fluid, and the growth before mentioned was removed. Rectal bougies were passed for stricture, which was already in process of formation. At the end of a fortnight, being much relieved, and her general

---

* "The Congenital Malformations of the Rectum and Anus," p. 63. New York.
† *Progrès Médical*, p. 473; June 22, 1878.
‡ *Med. Times and Gaz.*, vol. i. p. 279; 1865.

health having much improved, she was made an outpatient, but soon becoming pregnant, she ceased to attend. The child she afterwards gave birth to was born dead. A year subsequently she was admitted into St. George's Hospital, having in addition to the previous disease of the rectum a recto-vaginal fistula; the sphincter ani was divided; bougies smeared with unguentum hydrargyri were frequently passed, and she was placed under the influence of mercury by means of the calomel vapour bath. Under this treatment she improved rapidly, and was soon discharged. After the lapse of another year, having in the interval borne another child, she applied at King's College on account of a relapse of her previous condition, and having received relief from the same kind of treatment as that before employed, she soon left the hospital. She subsequently became a patient at St. George's Hospital. The canal of the rectum was so much narrowed that only a catheter could be passed through the stricture. Her general health, which up to this period had been tolerably good, began to fail, and suffering from sickness and diarrhœa for some days, she lost flesh rapidly. She was finally admitted into St. Bartholomew's. At this time she was in a state of extreme emaciation and misery, and evidently suffering from pulmonary phthisis, so that any expectation of affording her permanent relief seemed hopeless. She shortly afterwards died. At the post-mortem, the points of chief interest were to be found in the rectum and colon. The anus of this patient did not present more than remnants and traces of the cutaneous growths, which are generally significant of syphilis. In the rectum were found the results of widespread ulceration. The whole mucous membrane was destroyed, except one small patch, which was thickened and opaque; the exposed submucous surface was lowly tuberculated, undulating, and uneven, and was thickened by infiltration. In the early stages the tissue was soft, as if from recent inflammation, effusion, or œdema; but as the infiltration organised, it became callous, with fusion of the mucous and submucous coats, and then contracting, thus brought about a state of stricture. On the mucous membrane of all parts of the colon there were ulcers of regular (round or oval) shape from a sixth of an inch to about two-thirds of an inch in diameter, with clean, sharp-cut, scarcely thickened edges, surrounded by healthy or only too vascular mucous membrane. Their bases were, for the most part, level, or with low granulations resting on the submucous tissue, nowhere penetrating to the muscular coat, with no marked subjacent thickening or hardening. On some of them were ramifying blood-vessels; on some few there was at the centre of the base a small island of mucous membrane, giving to the ulcer an evident likeness to the annular syphilitic ulcers of the skin. At some places two or more of these ulcers coalesced into a large ulcer of irregular shape, and rather deeper than the smaller ones, but in all general characters similar to them. By such coalescence, some of the ulcers in the lower part of the colon were continuous with the ulcerated surface of the rectum, making it probable that, at first, similar forms of ulcers may have existed in the rectum, though now superadded thickening and partial scarring had destroyed nearly all traces of any primary shapes of ulcer. The ulcers of the colon were placed without plan or grouping, except that they decreased in number and closeness, and, on the whole, in size also, from the rectum to the cæcum. In the cæcum there were none; in the ileum only one, very small, and of rather doubtful character.

Extensive **amyloid degeneration** of the rectum, in common with that of the rest of the intestinal tract, is frequent in old-standing cases of extensive syphilis; but this condition is not of much practical importance, as it gives rise to no symptoms of significance except diarrhœa, which is due to the extensive changes of the whole intestinal tube rather than to any local disease of the lower bowel.

The **treatment** of syphilis of the rectum presents no peculiar features; the various manifestations demanding similar treatment to that employed for the corresponding lesions in other parts of the body.

CHAPTER XIII.

PROLAPSUS RECTI.

By the term prolapsus, or procidentia recti, is understood the protrusion of portion of the rectal wall through the anus. The old term "prolapsus ani," which is to be found in many text-books, is so obviously erroneous that it is best discontinued. Of prolapsus, we can recognise three distinct varieties: 1. Where the mucous membrane alone protrudes (partial prolapse); 2. Where the entire thickness of the intestinal wall is included in the protrusion; and 3. Where there is invagination as well as prolapse; or, in other words, the external appearance of an intussusception.

1. **Partial prolapse.**—When the extruded mass consists of mucous membrane only, the muscular coats of the intestine remaining *in situ*, the condition is spoken of as partial prolapse. This is of somewhat common occurrence, and is a very much less serious affection than either of the other varieties.

A slight protrusion of the mucosa can be produced voluntarily; and normally occurs during and immediately after defæcation. In some animals this is more especially noticeable than in the human subject, the horse being a familiar example. Horner has described a special arrangement of muscular fibres tending to produce this physiological prolapse. He states that a portion of the external longitudinal muscular coat of the bowel terminates by passing between the sphincters, and then turning directly upwards is inserted into the mucosa.

We may conveniently group the causes of pathological prolapse of the mucosa under three heads : (1) that due to the effusion of inflammatory products in the lax tissue of the submucosa ; (2) where the mucous membrane is dragged down by piles, polypi, or other neoplasms attached to it ; and (3) where the folds of prolapsed membrane have been protruded by peristalsis, the muscular structures of the anus and perinæum being relaxed.

Mollière* has shown, by a simple experiment, the mechanism of the first of these causes. On the dead body of a young girl he inserted the point of a blow-pipe beneath the mucous membrane of the lower end of the rectum. Upon injecting air into the submucosa, the mucous membrane bulged out at the anus, and the same procedure at another portion of the circumference was followed by a like result. In this case the anus was not at all unduly relaxed. In the decomposing bodies of animals, the gas generated by the putrefactive changes in the loose submucous tissue frequently produces a like result ; this being more especially liable to occur on account of the great mobility of the mucous coat, and the very loose areolar connections that exist between the middle and internal tunics of the lower bowel.

In cases of catarrhal proctitis and dysentery, inflammatory exudations frequently cause the protrusion of bright-red folds of mucous membrane from the anus, a condition which has by Roser not inaptly been compared to the inflammatory ectropion of

* "Maladies du Rectum," p. 199.

the ocular conjunctiva. Prolapse occasioned in this way is generally, in the first instance, of limited dimensions; but a prolapse once started has a tendency to increase in the same way that intussusceptions of other portions of the intestine are progressive. So that a prolapsus recti which has attained large dimensions may have been, in the first instance, due to a trivial inflammatory exudation in the rectal submucosa.

The protrusion of folds of mucous membrane associated with the prolapse of internal piles is exceedingly common, but it seldom occurs to any great extent. This is probably due to the fact that the presence of the hæmorrhoids has produced a certain amount of inflammatory thickening in the submucosa, which renders the extensive separation of the mucosa improbable, and the prolapse will certainly disappear entirely if the piles are subjected to efficient surgical treatment.

The form of prolapse produced by the adenoid polypus is peculiar, and deserves some notice. In this form, instead of a broad fold of mucous membrane being protruded, a narrow funnel-shaped portion is drawn out, sometimes of upwards of two inches in length, and constituting the pedicle of the polypus. This likewise gives rise to no trouble after the growth has been removed.

The most important cause in the production of this variety of prolapse is the occurrence of violent and long-continued expulsive efforts, especially if associated with a relaxed condition of the muscles around the anus; consequently we find prolapsus recti a common accompaniment of vesical calculus and phimosis in the child; and of enlarged prostate and urethral stricture in the aged and adult. Other cases are apparently due to the irritation of intestinal parasites, or of diarrhœa; while the custom, which is common amongst nurses, of leaving young children to sit on the chamber utensil for a long time after defæcation is completed, undoubtedly tends to favour the production of prolapsus recti.

This is a disease which is very much more common at the extremes of life, by far the majority of cases being met with in young children and old people, its primary occurrence between the ages of fifteen and fifty years being quite unusual. The greater liability of children to be affected is probably due to their much greater susceptibility to reflex irritation, and to the more numerous sources of direct irritation to which they are subject. Most of the authors who have written on rectal disease enumerate as one of the causes of the greater prevalence of prolapse amongst children, the want of support to the lower bowel in consequence of the greater straightness of the sacrum. This, however, I think, can not be considered an important ætiological factor. In old age the general relaxation and want of tone favour the formation of rectal prolapse.

Symptoms of partial prolapse.—The diagnosis of this disease is easy, the protrusions of mucous membrane appearing as bright red folds, arranged with sulci between

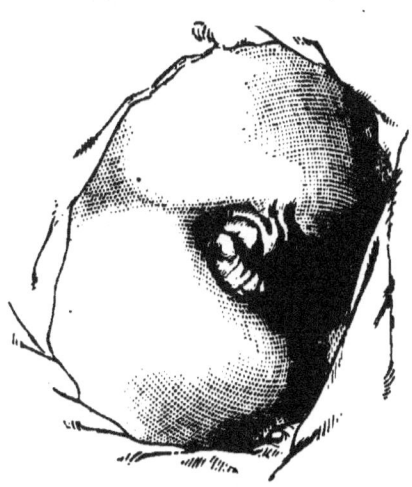

Fig. 26.—Partial Prolapsus Recti (Bryant).

them, which radiate from the aperture (Fig. 26); whereas the sulci in complete prolapse are principally parallel to the anal margin of the bowel

(Fig. 27); and, again, in partial prolapse the size of the tumour is usually of much more limited dimensions. The principal masses in partial prolapse are placed laterally, and on the surface of the prolapsed intestine superficial catarrhal ulcerations are frequently to be seen. At first the protrusion

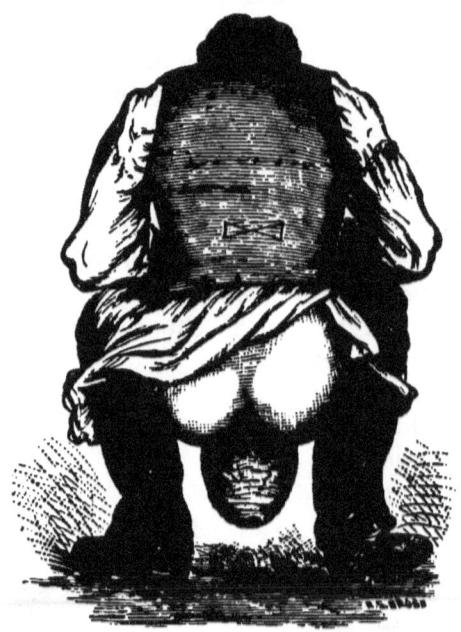

Fig. 27.—Complete Prolapsus Recti (Van Buren).

only occurs after defæcation, and is easily returnable; in more chronic cases, however, it becomes more difficult to replace, and may reappear independently of defæcation. The mucous membrane also becomes thickened, and the submucosa infiltrated. A muco-purulent discharge is common, and anal bleeding to a slight extent often occurs. As has been elsewhere stated, the protrusion of internal piles is frequently associated with more or less prolapse of the mucous membrane, but this is a condition which ought always to be readily distinguished from the disease under consideration. Prolapsed hæmorrhoids are more isolated tumours, and firmer to the feel than a mere flap of mucous membrane; and when of the venous variety, the livid colour will serve to establish the diagnosis.

Of the **complications** of partial prolapse, the most important is **inflammatory gangrene**, which is an occasional termination, by means of which nature sometimes effects a spontaneous cure of this disease, the gangrene being due to inflammatory stagnation, and not to the strangulation of the prolapse by the sphincters; indeed, in these cases relaxation of the sphincters precedes usually the disease, and in cases of gangrene the finger can be readily passed into the rectum, showing that there is no strangulation by the muscles.

Spontaneous recovery may also sometimes result without the actual occurrence of gangrene; the prolapsed mucous membrane being injured either by the passage of a hard fæcal mass or external agencies, to such an extent that inflammation sufficient to effect a permanent cure is set up.

2. **Complete prolapse.**—After partial prolapse has existed for some time, it is apt to merge into the more serious form where all the tunics of the bowel are involved; or sometimes the complete prolapse comes on suddenly, the entire thickness of the rectal wall being protruded by one expulsive effort. True comprehension of the **symptoms** and **diagnosis** of complete prolapse readily follows from the understanding of its nature and mode of production. When the protrusion reaches any considerable dimensions, it is obvious that the serous coat of the intestine will be involved, and, owing to the fact that the peritonæal pouch descends much lower on the anterior than on other

aspects of the rectum, the first appearance of a sac is to be looked for in front (Fig. 28). If, however, the prolapsus continues to increase in length, so that the upper portion of the rectum and sigmoid flexure

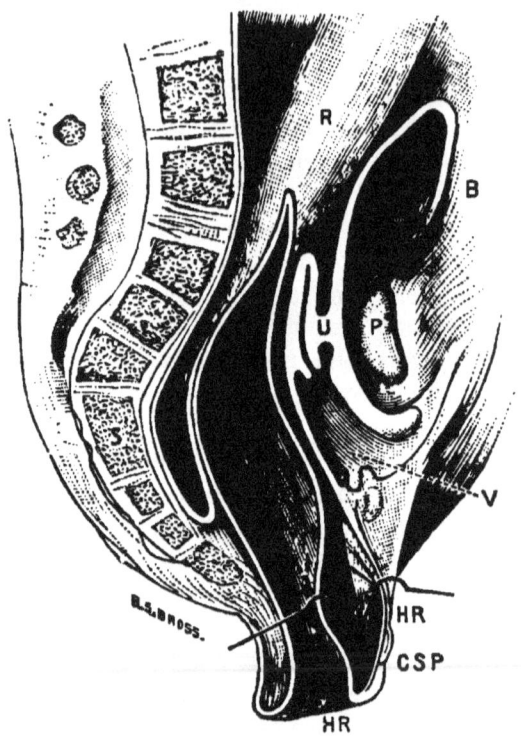

Fig. 28.—Complete Prolapsus in a Child.

R, Rectum; B, bladder; S, sacrum; P, pubes; U, uterus; V, vagina; H R, Prolapse; C S P, peritonæal pouch (Cruveilhier).

become protruded, a peritonæal sac will be found surrounding the tumour, except where the mesorectum is attached. As a rule, where the protrusion measures more than three inches in length it is generally curved, the concavity looking towards the coccyx, and in very extreme cases it may be arranged in a spiral manner. The diagnosis between this variety and prolapse with invagination is easily made, as in the latter the outer layer retires again within the anus, leaving a sulcus between the margin of the anus and the tumour; along the entire circumference of which a probe, or the finger, can be passed; while in the latter this sulcus does not exist, the outer layer of the prolapse being directly continuous with the anus. Complete prolapse may assume very great proportions; in rare instances the greater part, or even the entire, of the colon being protruded. In one case which came under my notice the whole large intestine was thus extruded. The patient was a child, aged four months. The prolapse was arranged in a spiral manner, making two and a half revolutions. The apex was formed by the cæcum, the ileo-cæcal valve being distinctly discernible, while the other end was continuous with the anus. The mucous membrane was in several places superficially ulcerated. The patient died of peritonitis. From the history I received of this case, I have no doubt that it commenced as an ordinary prolapsus recti, which gradually increased in dimensions by successive portions of the colon becoming involved, until the whole protruded. A similar result has been known to follow from an ileo-cæcal intussusception, the large intestine becoming invaginated from above downwards. In this latter case the first part to appear through the anus will be the ileo-cæcal valve, while in the former it will make its appearance last. In cases of colic or rectal intussusception, which have secondarily involved the entire lower portion of great intestine down to the anus, it may be impossible to arrive at a conclusion as to whether the disease has originated as an ordinary prolapse, or as an intussusception higher up. The mass may also assume large proportions, even to the extent of hanging down between the thighs. (*See* Fig. 27.) This may be the case to such a degree that all attempts to return it within the abdominal cavity will prove abortive.

Prolapsus recti is generally unattended with any very severe pain, the mucous membrane, from a point immediately within the anal margin, being of low sensibility; but the constant mucous discharge, and the incontinence of fæces, which in severe cases is always more or less present, render the patient very miserable.

The **complications** of complete prolapse are both important and serious; and they owe their chief gravity to the involvement of a peritonæal pouch brought down with the prolapse. In intussusception, the invaginated peritonæum is usually soon the seat of a localised adhesive peritonitis, which obliterates the sac; but this is not the case in prolapsus recti, the protruded portion of peritonæum retaining its continuity with the general cavity. Into this peritonæal pouch a portion of intestine is apt to fall, constituting a hernia of a peculiar variety, which may be appropriately described under the term *prolapsal hernia*. It is usually the small intestine that is thus implicated; but the ovaries, bladder, and other viscera have occasionally been found similarly involved. In the first instance the hernial tumour is to be looked for on the anterior aspect of the prolapse, in consequence of the peritonæum descending lower in front than behind. In aggravated cases, however, the small intestine may descend behind as well as in front. A hernia thus formed is to be recognised by the same signs that we are familiar with in the more usual seats of this disease: the tumour is usually tympanitic on percussion; gurgling can be felt when the swelling is manipulated; and the sac can generally be emptied by pressure, the enclosed portion of intestine being reduced into the abdominal cavity. As Allingham has pointed out,* the

* "Diseases of the Rectum," p. 168. Fourth edition.

position of the orifice of the prolapse is so altered when a hernia is present as to become a diagnostic sign. He says: " Directly the bowel is protruded, you can tell that there is a hernia also present, by the fact that the opening of the gut is turned towards the sacrum. When the hernia is reduced, the orifice is immediately restored to its normal condition." The same changes which are observed in hernia generally have also been noted in this form. It may become irreducible by the formation of adhesions between the sac and the contained viscera; or strangulation may result from the constriction of the neck by inflammatory swelling. According to Esmarch,* the formation of what he terms an "anus preternaturalis in ano" has been known to occur as a result of this strangulation.

A very rare complication of complete prolapse is the spontaneous rupture of the entire rectal tunics, and the protrusion through the rent of a portion of the small intestine. In a very interesting paper by M. Quénu,† the whole subject of spontaneous rupture of the rectum is fully discussed, and the recorded cases of this rare injury collected. In all, nine cases have been noted; and of these, five are described tolerably completely, and in four of the five an old and extensive prolapsus recti preceded the rupture. The immediate exciting cause in these cases was some violent muscular effort, as defæcation, vomiting, or lifting a heavy weight; and in all the protrusion of small intestine through the anus was followed by a reduction of the prolapsus recti. All the cases occurred in adults, and but one was noted in the male subject, the others occurring in women. In none of the cases did the structures appear to present any macroscopic change, such as ulceration or inflammatory

* *Loc. cit.*
† "Des ruptures spontanées du rectum," *Revue de Chirurgie*, Mar. 10, 1882.

softening preceding the accident; but in the case which came under the personal observation of M. Quénu, there was a considerable amount of inflammatory exudation to be seen under the microscope, although to the unaided eye the tunics appeared to be absolutely normal.

The position at which the rupture occurred varied from one to five inches from the anal margin; and the direction of the rent was transverse in some and longitudinal in others. The diagnosis is readily made, the escape of small intestine at the anus, with its smooth peritonæal covering and attached mesentery, being pathognomonic of the lesion. The prognosis is very bad, none of the recorded cases having recovered. The treatment adopted by Adelman* is, however, based on sound surgical principles. Having failed to reduce the hernial protrusion by direct pressure, this surgeon performed laparotomy, and drew back the small intestine into the peritonæal cavity; and then, by pulling down, and so reproducing, the prolapsus recti, he brought the rupture, which was situated at two and a half inches from the anus, into view, and carefully stitched it up. With the improved methods of dealing with peritonæal wounds now at our disposal, should the case be seen early enough, it is possible that success might attend a similar procedure.

In forming a **prognosis,** in any case of prolapse, it is well to bear in mind that, when the mucosa is alone implicated in young children, there is a strong tendency to spontaneous cure, and but mild measures will be necessary to assist nature; but where the disease exists to any extent in the aged, no such result can reasonably be looked for. And at any age, if the prolapse is complete, and attended with much inflammatory thickening, and if it is of long duration, relief can scarcely be expected by any means short of surgical

* *Journal für Chirurgie und Augenheilkunde,* 1845.

operation. In cases of extensive protrusion, with formation of a peritonæal sac, the danger to the patient's life is considerable, peritonitis not unfrequently supervening as a result of sloughing, ulceration, or the inflammation or strangulation of a hernia.

Treatment.—The first step which is usually required in the treatment of this disease is to effect reduction. Indeed, in many cases, it will be found that the unaided efforts of the patient will be sufficient to effect this when the prolapse is recent; or, in the case of young children, the mother will have returned the protrusion within the anus before the surgeon is called in. But, where the disease has become chronic, or where the submucosa has become distended with inflammatory effusion, considerable difficulty may be experienced in effecting reduction: and in a few very rare cases the hypertrophic changes may be so extensive that it may be quite impracticable to return the mass into the abdominal cavity. In order to effect reduction in the child, the little patient should be laid across the knees, and gentle pressure of the whole mass of the tumour should be for some moments exercised, so as to reduce its bulk by the squeezing out of the contents of the bowel prolapsed, and of any fluid effusion in its tunics. After this reduction should be proceeded with, efforts being made to return the more central parts first, and in the great majority of cases but little difficulty will be experienced. It will often be found that the prolapse immediately reappears after the removal of the finger. In order to obviate this Sir C. Bell has suggested an ingenious manœuvre. He advises a small piece of paper to be twisted into the form of a hollow cone, in the way that paper bags are made. This is to be well greased on the *inside*, and the apex of the cone placed in the aperture of the prolapse. The index finger is then to be placed in the cone, and steady pressure kept up. As the paper

o—23

advances it draws with it those portions of the prolapse which last came down, till finally the whole is reduced. The finger is now withdrawn, and the paper left *in situ*. In the case of the adult similar means will usually be found sufficient, the most convenient position being with the patient resting on his hands and knees. Should any difficulty be experienced in effecting reduction, it is better at once to administer an anæsthetic ; and to be prepared, if the case is severe and somewhat chronic, to operate for the radical cure at the same time. Reduction having been effected, it is necessary to apply some retentive apparatus, and the best evidence we can have of the difficulty of effectually keeping the prolapse up is the vast number of appliances which have been invented for the purpose. The best temporary means is to apply a pad of tenax, absorbent cotton, or some similar material to the anus, and then to strap the nates together with strips of strong rubber plaister.

In order to prevent the recurrence of the prolapse, great attention must be paid to regulation of the action of the bowels, and the usual sitting position during their evacuation should absolutely be interdicted, defæcation being effected either on a bed-pan, lying on the side, or in the erect position. It is also a useful plan for the patient to accustom himself to have the bowels moved the last thing before going to bed, so that he may at once lie down after the act. This is a good rule in many other rectal diseases, and with a little practice the habit can readily be acquired without discomfort. The anus should be well washed with cold water, solution of alum, or decoction of oak-bark ; but I am convinced it is a bad practice to use astringent injections, as, if they are strong enough to be of any practical utility, they are apt to produce tenesmus and straining, which will be productive of much more harm than any good they can accomplish. But a small enema of cold water will often prove of service; or suppositories containing ergotin, nux vomica, etc., will sometimes be found useful.

It is obviously a matter of the greatest importance to arrive at a diagnosis as to the cause of the prolapse, and, when this is practicable, to direct treatment to the removal of that state, whatever it may be; but, unfortunately, in a number of cases it is impossible to arrive at any conclusion as to the cause, so that our treatment must be somewhat empirical. In the case of children, however, we should examine carefully for rectal polypus, oxyurides in the rectum, for phimosis, or symptoms of vesical calculus. In certain cases it would appear that the exciting cause is an intestinal catarrh. Should any of these conditions be made out, it will usually be found that efficient treatment for their removal will prove sufficient to effect a cure of the prolapse. Where constipation is present the bowels should be regulated by change of diet, and, if necessary, plain enemata, rather than by the administration of purgatives, which cannot fail to prove harmful. In the adult, one of the most common causes for prolapse of moderate degree is the presence of internal piles, and the surgeon can promise the patient an absolute cure of the prolapse by efficient surgical treatment of the piles, the cicatrisation following the removal of the hæmorrhoids by any of the usual methods, detailed in the chapter on that subject, tending to produce an adhesion between the mucosa and the deeper layers of the rectal wall, and so preventing that free movement of the mucous coat which is essential to the formation of the prolapse in its earlier stages; and, again, the denser tissue of the cicatrix tends to prop up and stiffen the lower portions of the rectum. Where there are, in addition to the piles, many redundant folds of mucous membrane, they may be subjected to operation at the same time that the haemorrhoids are treated, portions being pinched up with a forceps and crushed, or cauterised; or if the ligature is preferred, the circle of mucous membrane may be divided into four or five segments with scissors and each portion separately ligatured. Such diseases of the urinary organs as stricture or enlarged prostate should receive appropriate treatment; and where constipation is present, the bowels should be relieved in such a way as to prevent straining. Sufficient has, however, been said as to the importance of recognising, and as far as possible removing, the cause of prolapsus recti; and in children these measures, in addition to the mild local treatment already indicated, will usually prove sufficient, the disease having a decided tendency to disappear as the patient approaches the age of puberty. In a few cases, however, in childhood, and in the majority of cases in the adult, something more will have to be done. **Pessaries** of various kinds may be tried, but are usually of but limited utility. The best form that I am acquainted with is shown in Fig. 29. It was devised by a patient of Dr. R. McDonnell's for his (the patient's) own use. It consists of an oval knob of vulcanite, with a very slender curved shank, which is perforated at the extremity for the reception of a piece of twine, so that if the instrument should happen to slip within the rectum it can readily be withdrawn. Instruments of this kind can be obtained of various sizes, and are used in the following way: the prolapse having been sponged and replaced, the knob is introduced into the rectum, the slender curved shank lying between the nates, and the mechanical stimulus afforded by the foreign body tends to brace up the rectum and anus, and keep the prolapse from

Fig. 29.-Pessary for Prolapsus Recti.

protruding; the very slender shank allows the sphincter to contract nearly to its full extent, and also affords a healthy stimulus to this muscle. Any instrument which keeps the anus distended only tends to weaken the sphincter, and so favour the production of prolapse. In several cases, even in adults, I have procured a complete cure by means of this simple pessary.

Subcutaneous injection of various fluids into the ischio-rectal fossa has from time to time been recommended for the relief of prolapse. Amongst the agents employed for this purpose may be enumerated: ergotin, nux vomica, and carbolic acid. Of these the first alone appears to me likely to be attended with favourable results. Nux vomica has proved dangerous in use, at least one fatal result having been recorded: and although strongly recommended by Kelsey,[*] I am unable to see how the injection of a strong solution of carbolic acid could possibly be of service. Vidal[†] records three cases of cure by the subcutaneous injection of ergotin. The first case was that of a man aged thirty-nine years, who had suffered from prolapse for eight years. Injections of a solution of ergotin were practised every two days, and after the fifth injection the mucous membrane scarcely protruded at all. After the eleventh injection it came down only during defæcation, and returned spontaneously. Twenty-two injections in all were used, and the patient was seen and found to be well four years after treatment. The second patient, a female aged sixty-four years, was cured after twenty-four days' treatment, and remained well two and a half years afterwards. The third patient, a female, aged forty-five, was cured by six injections in a period of twenty-four days. The solution recommended by Vidal was composed of fifteen grains of Bonjean's

[*] "Diseases of the Rectum," p. 116.
[†] *Paris Médical*, Aug. 28th, 1879.

ergotin, dissolved in seventy-five minims of cherry-laurel water, and of this solution a dose varying from fifteen to twenty-five minims was employed. The injection was made deep into the ischio-rectal fossa at a distance of one-fifth of an inch from the anus. Severe pain usually followed this treatment, but in no case was it followed by local inflammation or abscess. Spasm of the sphincter lasting several hours was induced; and the local action of the ergotin in causing contraction of the involuntary muscles in the neighbourhood was further exemplified by the occurrence of spasm of the neck of the bladder, with retention of urine on several occasions. Dr. Ferrand* has also recorded a case of a lady who had suffered a great deal with a prolapse of considerable size, and who was very much improved by a similar course of treatment. In my own experience, however, I have been disappointed with the hypodermic injection of ergotin, in the few cases in which I have tried it. Considering the safety and certainty of linear cauterisation, I am inclined strongly to give the preference to this method of treatment, which, moreover, has the advantage of being decidedly less painful than the often repeated hypodermic injection.

Various methods of **cauterisation** have been recommended, especially strong nitric acid; acid nitrate of mercury; butter of antimony; and other chemical agents. The actual cautery has also been advocated in the shape of the hot iron, galvano-caustic, etc., but the benzoline thermo-cautery of Paquelin (Fig. 30) is now generally preferred by most surgeons.

The application of strong nitric acid has been strongly recommended by Allingham.† He directs that the acid should be applied on the whole surface of the prolapse after it has been carefully dried, care being

* *Gaz. Hebdom.*, Jan. 2nd, 1880.
† "Diseases of the Rectum," p. 165.

taken not to touch the verge of the anus or the skin. The part is to be then well oiled and returned, and the rectum stuffed thoroughly with wool; a pad must, after this, be applied outside the anus, and kept firmly in position by strapping plaister, the buttocks being kept closely together by the same means. Although

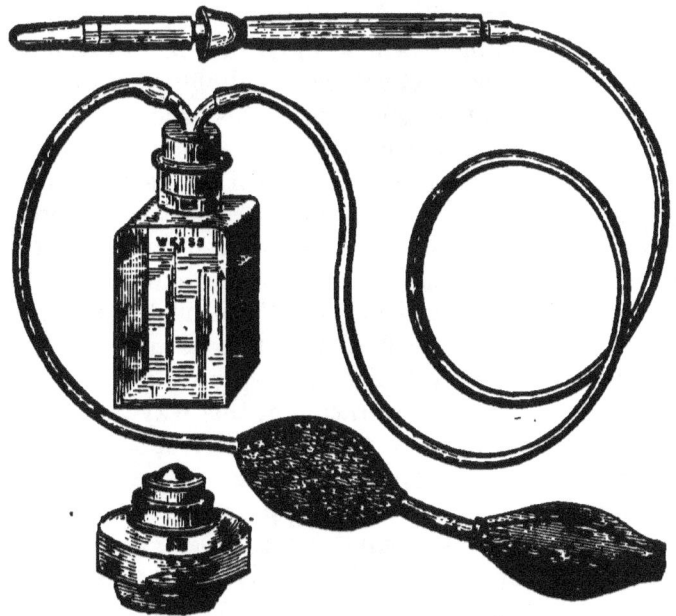

Fig. 30.—Paquelin's Thermo-cautery.

good results almost invariably follow this treatment in the case of children, it is not by any means so suitable for the treatment of the disease when occurring in adults, and is not without danger in old and debilitated persons. Extensive sloughing may be produced, and during the separation of the sloughs, hæmorrhage to an alarming extent may be induced. A case of this kind has come under my own observation. An elderly man was treated for extensive prolapse by the application of strong nitric acid. Seven days afterwards hæmorrhage of a very severe character came on, and although the bleeding was arrested by plugging the rectum, he died exhausted shortly afterwards. Inflammation of the rectum of a very severe type sometimes follows the application of the chemical caustics, especially when applied to an extensive surface of prolapsed intestine.

Another danger which must not be lost sight of is the occurrence of stricture. Allingham states that he has on several occasions seen this occur.

The actual cautery may be applied as recommended by Mr. Smith, with the assistance of his clamp. A portion of the mucous membrane is caught in a forceps and tightly clamped. The compressed portion is then burned away with the hot iron, or Paquelin's benzoline cautery. In using either of these instruments the heat should not exceed a dull red, as otherwise the tissues will not be sufficiently deeply charred to arrest hæmorrhage; but in operating in this way the surgeon must ever bear in mind the possibility of a peritonæal pouch being present in the prolapse. And should the tumour from its size and appearance suggest the possibility of this complication, the greatest care should be taken to include only the mucous membrane.

Mr. Smith has recorded[*] an interesting case which, although it terminated favourably, should serve as a caution in using the clamp and cautery: "The patient was an elderly man who had a prolapse as big as a cocoa-nut, always coming down and rendering his life a burden; he had already been operated on twice by a hospital surgeon, but in vain. The patient was then sent to me, and formidable as the case looked, I determined to undertake it. I applied the clamp deeply in three different directions. There

* *Lancet*, March 15, 1880.

was a great deal of bleeding, and I had to apply the cautery over and over again before I could stop it. And then, just as I was finishing the operation a most untoward event occurred, severe vomiting as the result of the anæsthetic took place, the prolapsus was forced still farther down, and before I and my assistants could return the parts, the violent action of the abdominal muscles was such that the weakened coat of the bowel gave way, and a knuckle of small intestine actually protruded through the rent thus made. I carefully returned this as soon as the vomiting ceased, and anxiously waited the result. Our house-surgeon, Mr. Newmarch, watched the patient with the greatest care, and treated him with great skill, keeping him constantly under the influence of opium, and locking up his bowels for several days. The result was not a single bad symptom of any kind. On the first action of the bowels there was no protrusion nor afterwards, and as soon as the man was fairly recovered, I removed three longitudinal folds of skin from the anus, so as further to tighten the parts. The man was completely cured. Now, the lesson this case teaches is this: not to employ an agent which could cause vomiting, because, of course, in such a terribly severe case as this, it is absolutely necessary to clamp deeply and thus weaken the bowel. It was a most unlooked-for accident, not likely to occur again; in fact, it is hardly reasonable to expect to meet with another such case for operation. I have, however, been called to cases as bad or worse, but where no operation could be recommended."

It is well to remember in connection with this case that forcible attempts to reduce a large prolapse have been attended with a fatal rupture of the peritonæal pouch,* and also, that the peritonæal cavity has been opened in an operation undertaken for the excision

* Roché, *Revue Méd. Chir.*, 1853.

of a complete prolapse, on the supposition that it consisted of mucous membrane alone.*

The **galvanic cautery** is not at all so suitable for the removal of prolapse as the other forms of cautery, as it does not produce sufficient charring of the structures to completely arrest bleeding.

The treatment, first strongly advocated by Van Buren, is that which is now generally adopted by surgeons, and is the one which I believe to be far the best in those cases in which the milder measures before-mentioned have not been successful. It is best described in Van Buren's own words: " Having etherised the patient, elevated the hips as in Sims's position, reduced the prolapse, and introduced a speculum of his pattern of the largest size, proceed to draw a line upon the mucous membrane with the cautery at a dull red heat, parallel with the axis of the gut, and repeat this four or more times at equal distances, carrying the cautery each time from a point three inches or more above the anus slowly down through its orifice, and terminating the line of eschar externally, where the delicate integument covering the sphincter joins the true skin; you will thus have a series of parallel vertical stripes of cauterised tissue, the lower extremities of which will appear as rays diverging from the anus. The lines of eschar may be made more numerous, deeper, and broader, according to the volume and duration of the prolapse. In a child, or where the protrusion is not voluminous, nor of very long duration, I would use a delicate cautery, perhaps not thicker than an ordinary probe, but for a larger tumour in an adult a more bulky iron; but in any case it should be bent nearly at a right angle, a short distance from the button at the extremity, so that it may reach all points of the concavity of the rectal surface. By operating in this

* Van Buren, "Diseases of the Rectum," p. 60. Second edition.

manner I believe you would get the full effect of the cautery in producing retractile cicatrices with the least amount of danger of subsequent stricture. Where after cauterisation a cicatrix is left which encircles the whole circumference of the bowel, constriction in some degree must follow. In a very bad case an operation of this kind might be repeated, new lines of eschar being made in the intervals between the old ones. This I did in the case of a young girl of thirteen with defective intelligence, who had an enormous prolapse, which had existed from infancy. In this case I added to the linear eschars small scattered points made with a slender probe-pointed cautery; the effect of the latter when applied over the sphincter was remarkable in arousing its contractility."

The after-treatment recommended is to keep the patient confined to the horizontal position for a week, a bed-pan being employed when the bowels act for at least double this period, to diminish the possibility of a relapse; enemata of tepid water being given when required to assist defæcation. The rationale of this simple operation is easy to comprehend. Any traumatism sufficient to produce inflammatory adhesion between the mucous and deeper coats of the rectum during the process of repair, would be sufficient to effect the desired result, but by no method can this be so effectually carried out as by the cautery applied in the way described. It will be observed that Van Buren directs that the lines of cautery be carried through the margin of the anus for the purpose of stimulating its sphincter muscle to contract, and also with a view of producing a certain amount of cicatricial narrowing of this abnormally dilated outlet. Great relaxation, frequently associated with fatty atrophy of the external sphincter, is a common accompaniment of prolapsus recti. It should, however, be dealt with only in connection with efficient surgical treatment of the protruded mucous membrane. That this muscle can be stimulated to contraction by local irritation is abundantly proved by the spasm and hypertrophic thickening produced by the presence of an irritable anal fissure; and it is with a view of, to a certain extent, imitating this pathological process that the cauterisation of the anus itself is resorted to. Allingham* recommends the following modification of Van Buren's operation. If the prolapse is not already protruded, he draws it down with a vulsellum; he then makes four or more longitudinal linear eschars from the base to the apex of the protruded mass, taking care to make the cauterisation deeper at the base than at the apex, where peritonæum may be close. Care should be taken to avoid any large vessels which may be seen. The prolapse is then reduced, the sphincter partly cut through with the cautery on either side, and a little oiled lint placed in the rectum. The recumbent position is absolutely enjoined for a month or longer if the wounds are not by that time perfectly healed. I do not think that this plan of treatment is an improvement upon that recommended by Van Buren, as if the cauterisation is made through a speculum when the parts have been returned to their normal position there will be a close correspondence between the deeper and more superficial portions of the wound, whereas if the cauterisation is effected before reduction of the prolapse, considerable displacement of the wound will necessarily take place, and firm adhesion between the mucous coat and deeper structures is less likely to be ensured.

The establishment of a traumatic anal stricture for the purpose of keeping up a prolapse has been occasionally recommended. Thus Dieffenbach reduces the prolapse and stuffs the rectum with charpie to make the anus bulge; he, then, with a cautery,

* *Loc. cit.*, p. 169.

deeply chars the entire margin of the anus, so that when cicatrisation is complete, a stricture will be formed, which will support the weakened bowel. And although a somewhat similar free use of the cautery receives the sanction of such a distinguished modern surgeon as Esmarch,* it cannot be, in my opinion, compared with the operation of linear cauterisation, for in the first place the protruded portion is only mechanically retained by forming an artificial stricture, whereas by the operation of linear cauterisation the coats of the bowel are so strengthened that the tendency to protrude is obviated; and any surgeon who has witnessed the suffering and annoyance occasioned to the patient by the presence of an anal stricture, will agree with me in thinking that it is very doubtful whether the cure thus produced is not worse than the original disease.

Dupuytren has recommended the removal of elliptical folds of skin and mucous membrane from the margin of the anus, and a somewhat similar practice has been sanctioned by Robert, Dieffenbach, and Mott; but with the improvements in the use of the cautery, I do not anticipate that this operation is likely to be again resorted to, except under some very special circumstances.

In some rare instances it may be found that the prolapse is so voluminous that its reduction is impracticable, in the same way that we sometimes find that the abdominal cavity is incapable of receiving back the contents of a large hernia; and it becomes a question as to whether any operative measures should be entertained in such a case. Also in those cases in which linear cauterisation has failed, it is obvious that any of the operations which have hitherto been enumerated are unsuitable, and the only procedure that can be contemplated

* *Loc. cit.*, p. 152.

is the **excision** of the mass ; and I believe that this operation is amply justified when we consider the miserable state to which the patient is reduced, and the much greater success with which peritonæal wounds are dealt with now than formerly. One of the earliest operations of this kind is exemplified in a case recorded by Dr. Kleberg,* so I give it in full: "On the previous day a dose of castor-oil was given, and on the morning before the operation an enema of lukewarm water was administered high up the bowel. Immediately before, a glass of wine and one grain of opium were given. After the patient had pressed down the gut as far as he could he was placed in the lateral position, with the pelvis raised and shoulders turned downwards. Chloroform was then administered." (In two cases Kleberg has operated without chloroform, because the patients were in such a miserable condition that he was afraid to narcotise them thoroughly.) "I carefully examine about the rectum at the junction of the skin and mucous membrane, in order to discover the sphincter ani, a procedure that was more difficult than one would think, because it had become so stretched and atrophied that I could only make it out by feeling under the fingers the coarser fibres running across the longitudinal axis of the bowel ; of anything like the normal muscle there was nothing to be discovered. An assistant at this point surrounded with all the fingers the prolapse from above, the points of the fingers being directed towards the free end of the prolapse, and pressed as hard as possible into the gut at a point perhaps half an inch below the supposed sphincter. Immediately in front of the ends of the assistant's fingers I then placed a good fresh unfenestrated drainage tube of rubber, one and one-half lines in diameter, around the prolapsus, and drew it only as

* Quoted by Kelsey, p. 123.

tight as seemed necessary to stop circulation. The elastic ligature was kept at the necessary tension by means of an easily untied slip-knot of silk thrown under it. The assistant now had both hands free, and from this time on the operation was performed under the carbolic spray. A few lines beneath the ligature I now made a longitudinal incision, two inches long, through the prolapsed gut, and in this way opened the sac formed by the drawing down of the peritonæum. Then I seized the elastic ligature with the forceps and fixed it firmly. It was thus an easy matter to push back into the peritonæal cavity a protruding loop of intestine without the slightest bleeding taking place into the wound, or any air entering the peritonæal cavity, because the elastic pressure follows so rapidly all the movements that no opening can exist anywhere. After I had convinced myself that the peritonæal sac was empty, and that only that part of the gut which was to be removed lay in front of the ligature, I thrust the largest size of Luer's pocket trocars through the prolapsus immediately below the elastic ligature from before backwards, and passed through the cannula two elastic drainage tubes of one and one-half lines in diameter, and after removing the cannula tied them as tightly as possible, one on the right side, the other on the left. These knots were secured against slipping by means of a knot of silk. The first provision against hæmorrhage, the elastic ligature applied after Esmarch's plan, was then removed, and the prolapsus cut off with the scissors one inch in front of the permanent ligatures. After a few minutes' time, during which I kneaded the parts which still remained and lay above the ligatures thoroughly, and as far as possible removed the fluids from them, I covered the parts around the stump with cotton, and soaked that part of the prolapse which still remained above the ligature with a solution of chloride of zinc, dried it, squeezed the soft parts once more thoroughly, applied the chloride of zinc again, then covered the whole with dry cotton batting, giving the patient instructions to remove this as soon as it became moist, and to replace it with dry, and to give the air all possible access to the parts. No fever followed the operation, and the pain was bearable with the aid of an occasional opiate. On the next day the parts were so shrunk as to leave a concavity at the anus, where before there had been a bulging. There was no bleeding, no peritonæal irritation, and only slight tenesmus. On the fourth day the first ligature cut out, the second on the fifth. The rectum was irrigated twice a day with water and permanganate of potash; and on the seventh day a dose of castor oil was followed by a large evacuation while the patient was on his back, without pain or hæmorrhage; the passage was, however, involuntary. On the fourteenth day the wound was healed, the general condition of the patient excellent, and the evacuations regular but still involuntary. The sphincter at this time began to be appreciable, and there was no protrusion of the bowel, the patient going about and wearing a bandage. One month later he had control of solid fæces, but there was still a slight discharge of mucus, and after another month he was entirely well."

In this case the prolapse was about a foot long, and six inches in diameter; the mucous membrane was excoriated and ulcerated; the patient had been sick for two years, had been bed-ridden for two months, and was waxy pale.

Another case recorded by the same surgeon and treated by the same method ended fatally; but in this instance the case can hardly be considered a fair test of the dangers of the operation, owing to the bad condition of the patient.

More recently Mikulicz has recorded six successful cases of circular resection of prolapse, and Billroth and Nicoladoni have also recorded cases.* I have operated in this way on a patient with very extensive prolapse similar to that shown in Fig. 27, which had existed for eighteen years. After thorough evacuation and antiseptic cleansing of the bowel, the patient, a female aged forty years, was retained in the lithotomy position by Clover's crutch, and the pelvis was well raised, which caused the prolapsed hernia of small intestine to slip back into the abdomen. The inner intestinal tube of the prolapse was grasped with catch forceps and drawn down to the fullest extent. An incision was now made anteriorly, close to the junction of skin and mucous membrane, and carried through the entire thickness of the outer intestinal tube, and opening the peritonæum. A finger was now introduced into the hernial sac in order to make sure that the entire small intestine had been returned to the abdominal cavity. Finding the sac empty, the inner intestinal tube was cut through at the same level with curved scissors. Both portions of the tube were divided exactly on the same level, but in small sections at a time; the two edges of serous membrane were closed by fine catgut, and the muscular and mucous coats by silk. The first portion, about one-sixth of the entire circumference, was thoroughly sutured before the next portion was cut. No ligatures were required, the bleeding, as in Whitehead's operation for piles, being completely arrested by sutures. When in this way the entire circumference had been resected it was found that the anal orifice was much too large. Accordingly the external sphincter was separated from the bowel posteriorly, the intestine kept at the anterior portion of the incision, and the exposed surfaces of the sphincter brought together posteriorly, leaving

* "Annals of Surgery," vol. i., 1889, p. 65.

a racquet-shaped scar. In this way the anus was reduced to its proper dimensions. Rapid recovery took place. I saw her recently, one year after operation, without the slightest evidence of recurrence, and with very good control over the bowel, except when she has a bad attack of diarrhœa.

It is probable that excision performed in this way will prove a much safer operation than an excision of stricture, either malignant or benign; for when the operation is performed for prolapse the edges to be sutured together are already in apposition, so that they can be brought together without the slightest disturbance of surrounding parts; while in the other case they are widely separated and can only be brought together with difficulty. That excision is the best treatment for a very considerable prolapse is, I think, assured, and the success that has attended Whitehead's operation of excision of piles leads one to believe that possibly the minor cases of prolapse may be better treated by excision where the milder and less certain operations may have been unsuccessful.

It is, of course, a very different matter operating as above, with the full intention of opening up the peritonæal cavity, and with all the necessary precautions against protrusion of the small intestines through the wound, and operating on a prolapse under the impression that it consists of mucous membrane only, the first indication of involvement of the peritonæum being the escape of coils of small intestine, some of which may have been injured.

3. **The third variety of prolapse.**—This variety, where there appears in the rectum, or through the anus, a portion of the upper sections of the intestinal tract that has become invaginated, can only be briefly alluded to here, and that only so far as the diagnosis is concerned. For further particulars of this

interesting subject the reader is referred more particularly to the admirable book in this series on "Intestinal Obstruction," by Mr. Treves, and to the article on the same subject in Ziemssen's "Cyclopædia," by Leichtenstern. The important point in the diagnosis is the presence of a sulcus between the anal margin and the prolapse. If this exists it is obvious that the case must be one of intussusception, and if with a finger or ordinary probe the bottom of the sulcus can be reached, it will indicate that the intussusception has taken place in the rectum. If, on the contrary, the starting-point of the invagination cannot be felt in this way, the case will be one of colic, or the more common form of ileo-cæcal intussusception. Leichtenstern, whose statistics on this subject are the most comprehensive that have been yet published, states that out of a total of 220 cases of intussusception, in forty-one the tumour projected from the anus, while in thirty-one other cases it was felt in the rectum.

These tumours in the rectum have been not unfrequently the source of errors in diagnosis, and they have in consequence been removed, the intussusception having been mistaken for a polypus or malignant growth; so that the possibility of a rectal tumour being due to intussusception should be carefully remembered by the surgeon when trying to arrive at a diagnosis.

CHAPTER XIV.

THE ÆTIOLOGY AND PATHOLOGY OF PILES.

IT would appear from the earliest historical accounts, and from the most ancient medical writings, that hæmorrhoids, or piles, were recognised as constituting a common disease at a very remote period, and at the present day it may be safely said that the majority of persons who have reached middle life have suffered in some degree from some manifestation of it; in fact, there is scarcely a disease of more common occurrence. The word hæmorrhoids was used in a much broader sense formerly than it is now. It was a term applied to a bleeding from the anus from any cause whatever; and, indeed, by some of the ancient writers it was applied to any hæmorrhage, whatever its site might be. By Hippocrates the term was held to include varices of the extremities of the hæmorrhoidal veins, as will appear by the following passage, as translated by Bodenhamer:*
"A defluxion of bile, or of pituitous matter, to the veins of the anus inflames the blood which those veins contain. The veins themselves being inflamed attract the blood of the others near them, and being filled with it, raise and swell the internal parts of the rectum. The little heads of the veins are then conspicuous, and partly from the pressure of the fæces, and partly from their own fulness, are liable to break and emit blood, particularly at the time of dejections."

At the present day the term is so universally applied to vascular tumours of the rectum that it would be inconvenient to make any change. At the same time the word, as implying a symptom not

* "On the Hæmorrhoidal Disease," p. 8. New York, 1884.

always present, is objectionable. The word piles (from *pila*, a ball or swelling) is, as implying no theory or symptom, perhaps, on the whole, more suitable; and I would define the term as a *tumour originating in a diseased condition of the blood-vessels of the lower end of the rectum, the vessels having undergone dilatation and proliferation.*

Ætiology.—When we come to inquire into the ætiology of this disease, we find that the causes of piles which have from time to time been given are indeed numerous; and that hæmorrhoids are met with under almost all conditions of life (in the male and in the female, amongst the opulent and poor, the sedentary and active), so that we must look for one common cause as the most essential in the production of the same disease in such diverse circumstances. That common predisposing cause is, I feel sure, an anatomical one; and that the erect position occupied by man is the most essential element is, I think, a plausible theory. There is no similar disease, so far as I am aware, known amongst quadrupeds.

In order to estimate fully the importance of this question, it will be necessary for us briefly to review the more important anatomical facts in connection with the vascular supply of the lower bowel. The arterial supply of the rectum is derived from three sources; the superior hæmorrhoidal branch from the inferior mesenteric, the middle either directly or indirectly from the internal iliac, and the inferior hæmorrhoidal from the pudic. Of these the most important is the superior, which differs from the others by being a single vessel, and not symmetrical. The superior hæmorrhoidal artery passes down between the layers of the meso-rectum, and then divides into two branches, each of which forms a loop at either side of the bowel, the convexity looking downwards. From these two loops a variable number of vessels pass directly downwards, perforating the muscular coat, and then, lying immediately below the mucous membrane, they descend, one in each of the columnæ recti, to the anal verge, inosculating freely with one another, and at the lower end of the rectum communicating by transverse branches of some considerable size. Thus it will be seen that the entire mucous membrane, down to the very anal termination, is supplied by branches of the superior hæmorrhoidal artery, the middle and inferior hæmorrhoidal vessels being distributed to the tissues on the outside of the lower end of the bowel and the skin immediately surrounding the anus. The distribution of the veins is somewhat similar: the blood from the lowest portion of the mucous membrane of the rectum is first collected in a number (generally about twelve) of little dilated pouches, arranged in a circular manner, situated immediately within the anal verge; from these, small venous radicles ascend, inosculating to form larger branches, and, at a distance of about three inches from the anal margin, the veins, generally about five or six in number, perforate the muscular coats of the rectum by passing through oval slits in the muscular structure. Verneuil has laid great stress on this arrangement in the ætiology of piles, for as these apertures are unprotected by any fibrous canal, the veins must be subjected to considerable pressure; and retardation of the flow of blood through them during the contraction of the muscle must result. The veins having thus emerged from the rectum become tributary to the inferior mesenteric vein, and finally discharge their blood into the vena portæ. Like all the veins going to form the vena portæ, these vessels are destitute of valves, consequently the walls of the hæmorrhoidal vein have to withstand the pressure of a considerable column of blood, the tension of which is at all times liable to be increased by hepatic obstruction and intra-abdominal pressure. Veins also accompany the branches of the inferior and middle hæmorrhoidal arteries, but, unlike the vessels just described, they discharge their blood into the general or caval circulation and consequently are unaffected by changes in the portal system. It is usually asserted in the anatomical text-books that a very free anastomosis takes place between these two systems of vessels: that, however, as I have several times demonstrated by injection, is proved not to be the case. If a fine-size injection be forced through the superior hæmorrhoidal artery, or, preferably, through the accompanying vein (for, as there are no valves, this can readily be accomplished), it will be found that the capillaries of the mucous membrane can be with facility fully injected right down to the point where the mucous membrane joins the skin, but that the skin itself, and the radicles of the middle and inferior hæmorrhoidal veins, will not be injected. This is what we might expect from a consideration of the development of the lower bowel. The rectal pouch is at first a cul-de-sac situated at some distance from the perinæum, and as it descends it carries with it its own proper blood supply, which is similar in its source and arrangement to that of the rest of the intestinal tube; the anal depression, with its cutaneous appendages, which finally unites with the rectum, being supplied with vessels analogous to those of other cutaneous surfaces. I may, while dealing with this point, mention, in passing, that this want of communication between these two systems demonstrates the futility of applying leeches around the anus with a view of relieving portal congestion, a method of treatment which, nevertheless, is still frequently had recourse to by physicians. It is obvious that, applied to the skin in the vicinity of the anus, they can be of no more use than if they were applied to any other part of the surface of the body.

Another anatomical point which must not be lost sight of is the extreme mobility and dilatability of the structures constituting the termination of the bowel, as favouring the free growth of tumours. To sum up, then: small dilatations occur normally at the commencement of the rectal veins; the blood returning through them has to pass up against gravity while the body is in the erect or sitting posture; and, as the vessels are unprovided with valves, the whole weight of the column of blood presses continuously upon their radicles; the veins are subjected to pressure while passing through the muscular wall; and, again, at the liver obstruction is common.

The act of defæcation has much to say to the ætiology of piles. During the passage of a solid mass of fæces along the great intestine considerable pressure is exercised on the blood-vessels. In the colon and in the first stage of the rectum this is rather salutary than otherwise, as the vessels are arranged, for the most part, at right angles to the axis of the bowel, so that the contraction of the tube empties the veins into the larger channels; but when the lower portion of the rectum is reached, this is not the case. As previously mentioned, the vessels of the lower bowel are arranged parallel to the direction of the intestine, consequently the passage of the fæcal bolus forces the blood in the opposite direction to that in which it should flow in the veins. An illustration will make this more clear. If an elastic band is passed round the arm, and gradually rolled down towards the hand, the superficial veins below the point of constriction are rendered full and prominent. Now, a strictly analogous thing occurs during defæcation, with the exception that the veins are to be found in the outer and compressing tunic, and it is the solid mass of fæces in the interior that moves on. The passage, therefore, of every hard and constipated motion subjects the veins of the hæmorrhoidal plexus to a very considerable dilating strain, and thus furnishes one of the most important, if not one of the principal, factors in the ætiology of piles. When the abdominal muscles are called into forcible action, a considerable amount of pressure is brought to bear on the entire portal system of veins, and in ordinary circumstances the sphincters and levatores ani, acting simultaneously, equalise the external pressure on the hæmorrhoidal plexus, so that no dilating strain can be experienced; but when defæcation is taking place, the latter muscles are relaxed, while the abdominal muscles are contracting forcibly; consequently there must be a strong tendency to regurgitation of blood into the rectal veins. All prolonged efforts of straining at stool are, therefore, to be avoided by persons with any tendency to hæmorrhoids; and at the same time the bowels should not be allowed to become too constipated; and food calculated to leave a hard fæcal residue is to be avoided.

Sex.—Much has been written on the relative frequency of piles in the male and female sex, and very opposite opinions are to be found in the works of different authors on the subject. In my own practice I find the ratio is about equal; but statistics are at all times apt to be misleading, and especially is this the case in the subject under consideration. Females, from natural delicacy, are apt to postpone consulting their medical attendant for a disease which occasions such slight annoyance as ordinary internal piles, unless they are strangulated or inflamed; and, again, they are so accustomed to observe the menstrual flow that they attribute but slight importance to a bloody discharge from a neighbouring organ. It very frequently happens that we find anæmic women suffering from bleeding piles, who have either not noticed or paid little heed to the bloody discharge from the rectum until their attention has been directed to it by their medical attendant. In men, on the contrary, the discharge of blood generally at once attracts their attention, and they forthwith consult their doctor. There are several reasons why we should à *priori* expect piles to be more frequent in women than in men. In the first place, the pressure of a gravid uterus tends to produce dilatation of the veins of the hæmorrhoidal plexus in the same way that it produces varicosities of the labial veins, and those of the lower extremity generally; and, as a matter of fact, piles are extremely common, both during pregnancy and immediately after parturition. Another cause that must not be lost sight of is the habit which women undoubtedly have more than men of permitting their bowels to become habitually constipated, although this may be to a certain extent counterbalanced by the greater pelvic capacity which they possess. In women, also, at the menopause, a discharge of blood from the rectum, frequently attended with the presence of hæmorrhoidal tumour, is of extremely common occurrence; and the pressure of a retroflected uterus, or of tumours connected with the ovaries or other pelvic viscera, may sometimes be admitted as a cause of this disease. On the other hand, men are, as a rule, more exposed to the deleterious influences of excess in eating and drinking.

Age.—Piles are essentially a disease of the middle period of life, their occurrence under the age of puberty being extremely uncommon; and when they do occur in children they are generally found to be formed of simply dilated veins. I have seen two cases (one in a child aged six years, and another in a child aged eight) in which there were single external piles manifestly due to varix, and similar cases have been noticed by other observers. In a case recorded by Mr. F. Ogston, jun.,* the disease appeared to be congenital. The earliest age at which an internal

* *Lancet*, May 12, 1886.

pile has been noticed, so far as my knowledge goes, is that recorded by Allingham,* in which in a child aged three years he found three well-marked venous hæmorrhoids. On the other hand, piles but seldom originate in old people, except as a result of paralysis of the sphincter muscle, either associated with general paralysis, or from intrinsic relaxation. When this takes place, owing to the pressure of the portal system, from contraction of the abdominal muscles being unopposed by the relaxed sphincter, varicosities in the hæmorrhoidal plexus of veins are apt to occur.

Heredity.—It is almost impossible to form an opinion as to the influence of heredity in a disease so widespread and common as piles. No doubt the habits of life that help to the formation of hæmorrhoids are frequently inherited, and to this extent heredity may be admitted as a cause; but I do not think there is any evidence to justify us in attributing any great importance to this subject.

Excessive eating must be admitted as an important element in the formation of this disease, as, in addition to the general plethora which is induced, Niemeyer † has in detail pointed out that a general engorgement of the portal system of veins especially takes place during digestion, as evidenced by the temporary enlargement of the spleen during that time. This engorgement, when frequently repeated and carried to an abnormal extent, will produce a more permanent dilatation of the tributaries of the vena portæ, and so the formation of hæmorrhoids may be originated. The increased bulk and frequently irritating nature of the fæcal residue in those who eat inordinately, contributes to this result.

A catarrhal condition of the mucous membrane, by the congestion which is produced, tends to the

* " Diseases of the Rectum," p. 92, *note*. Fourth edition.
† " Pract. Med.," vol. i. p. 586. 1871.

formation of hæmorrhoids. This catarrhal condition is often, to a great extent at any rate, caused by the habitual use of strong purgative medicines. Many persons are morbidly sensitive about the action of their bowels, and consume enormous quantities of medicine, with the result of keeping up a constant irritation of the mucous membrane, which is quite uncalled for. If the habit of getting the bowels to move once a day at a certain hour is cultivated, it can generally be acquired without the use of cathartics; and for persons with hæmorrhoidal tendencies, the best time is immediately before going to bed, as the subsequent rest in the recumbent position tends to relieve the congested mucous membrane.

Pathology.—Various terms have been used to designate the forms of hæmorrhoids. Now although these varieties possess points that render them clinically important, it cannot be maintained that they are distinguished by marked pathological characteristics. When taken broadly, the pathology of all is practically identical — namely, increase and dilatation of the blood-vessels, more particularly the veins, with proliferation of the connective tissue. In the case of external piles the covering is the scaly epithelium of the anal canal, while internal piles are covered with mucous membrane; but their internal structure is alike.

Contrary to the teaching of many authors on this subject, I am convinced that extravasation of blood is not an important factor in the pathology of piles. In the simple venous external pile, which appears apparently suddenly, what really happens is this: A varicose condition of one or more of the inferior hæmorrhoidal veins has previously existed without attracting the attention of the patient; it is only when thrombosis and perivascular inflammation supervene that the tumour is noticed, and the patient has what is known as "an attack of piles." That

STRUCTURE OF PILES.

this is indeed an intravascular, and not extravascular, coagulation of blood can easily be demonstrated. If an incision be made into a simple inflamed external pile a blood-clot can be turned out, leaving a smoothly-lined cavity in which an endothelial lining can be observed. Moreover, as a result of stretching of the

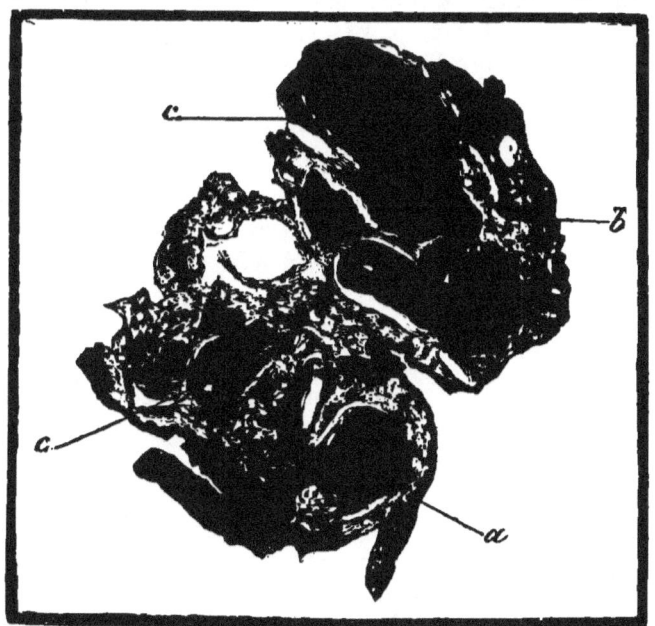

Fig. 31.—Section of Intero-External Pile. (× 3 diameters.)
a, Skin of anal canal; *b*, mucous membrane; *c*, laminated blood-clots in the interior of dilated veins.

anus—now the prelude to almost all rectal operations —extravasation of blood is of very common occurrence; yet we invariably see it as a widely diffused ecchymosis, which is rapidly absorbed, and not as a localised blood-clot similar to what we know as an external pile.

That, except in external covering, there is no essential difference in the structure of an internal and an external pile the specimen figured (Fig. 31) will show. This illustration is from a microphotograph kindly made for me by Professor J. A. Scott.

Here it may be seen that above and below the muco-cutaneous juncture (which is marked by a sulcus) large thrombosed veins (*c*) are filled with laminated coagula, and between these there is considerable proliferation of the connective tissue, in which moderately large arteries ramify. The mucous membrane (*b*) covering the internal portion has lost much of its character from frequent prolapse, but is still quite recognisable.

External piles that have been frequently thrombosed eventually become converted into but little else than connective tissue covered with scaly epithelium, the large dilated veins disappearing, the whole structure being not much more vascular than the surrounding skin. This is called the cutaneous or fleshy pile. A precisely similar change may take place in internal piles, resulting in the production of what is known as fibrous polypus of the rectum. This change, however, is not so common as in the case of external piles, no doubt owing to the more frequent irritation to which the latter are liable. From an examination of a somewhat large series of pile sections, it can be demonstrated that in cases of tolerably long standing internal hæmorrhoids the character of the section illustrated in Fig. 31, which was taken from a case of recent origin, become somewhat modified, the dilated and thrombosed veins are less remarkable, and the connective tissue more abundant and fully developed. This is a change that can be readily recognised in the living body. The soft and brighter red internal hæmorrhoid becomes firmer to the touch and whiter in colour—indeed, in extreme cases the covering is almost the same colour as the adjoining skin, instead of the bright red colour of normal mucous membrane. Once this change has occurred, the altered structure tends to become more and more pedunculated, until the fibrous polypus, attached by a more or less narrow pedicle, is produced.

The determining cause of hæmorrhoidal phlebitis is, according to Hartmann and Lieffring,* due to the presence in the interior of the veins of the bacterium coli commune. This opinion is based upon the following grounds:—In sections of thrombi from inflamed piles this micro-organism was invariably found, and it abounded in cultures obtained from the blood-clot; while on the other hand blood taken from uninflamed piles was found quite free from this bacterium, and remained sterile when cultivation was attempted.

CHAPTER XV.
EXTERNAL PILES.

THE classification of piles into external and internal is time-honoured; and as these varieties present many distinct clinical features, the terms are best maintained, although it is by no means uncommon to find both varieties associated in a given case; and sometimes it may even be difficult to make out the line of demarcation between them. By external piles we understand those which are covered by the skin of the anal canal, while those which are situated above the anal canal and consequently have a covering of mucous membrane, are called internal; and where an external pile is directly continuous with an internal one across "Hilton's white line," without any sulcus existing between the two, it is called intero-external.

External piles.—Of external piles several varieties have been described, but for practical purposes the division into *venous*, *cutaneous*, and *compound* will suffice. The condition described as œdematous pile by Mr. Cripps can scarcely be classed as a variety,

* *Bull. de la Soc. Anat. de Paris*, Fascic. No. 3, 1893.

but is more properly considered as one of the complications that may be grafted on the other forms.

The essential element in the production of a **venous pile** is a varicose condition of the external hæmorrhoidal veins, which is predisposed to by any of the causes before enumerated. As long as the vein remains pervious and free from phlebitis, the patient is free from discomfort, and in all probability ignorant of the existence of the disease. When, however, thrombosis takes place, or the swelling becomes acutely inflamed, the pain is severe, and attended with considerable constitutional disturbance : the tongue becomes furred, the febrile condition being quite out of proportion to the local cause ; the skin in the neighbourhood becomes inflamed and swollen; spasmodic contractions of the levator ani and sphincters add much to the patient's suffering, this being peculiarly annoying just when he is going asleep, the violent contractions and acute pain completely waking him up; there is a sensation as if there was a foreign body in the rectum, producing tenesmus and painful strainings ; the symptoms are all aggravated by walking or by any sudden contraction of the diaphragm, as in coughing, sneezing, etc. ; constipation is usually present, and when the bowels do move the pain at the time and for some hours afterwards is much increased. Such is the train of symptoms that characterise what is known as an acute "attack of piles." If an examination be made of a patient in this condition, one or more livid tumours, varying in size from that of a pea to a filbert, will be seen in the neighbourhood of the anus. These are acutely sensitive, and have a tense glistening surface ; pressure fails to empty the blood out of the tumour ; at least, this has been invariably my experience, and I believe that thrombosis of the varicose veins is the first element in the production of an attack of external piles.

At the anus of the majority of adults small varicosities are to be found in the plexus of veins. These can usually be emptied by pressure, but prior to thrombosis they give rise to no annoyance to the patient, and are so common as to be scarcely considered pathological.

The second variety or **cutaneous pile** is not unfrequently a sequence of the first, the inflammation surrounding the thrombosed vein producing hyperplasia of the connective tissue and skin, constituting a tumour, usually of small size and pale colour. These may become somewhat pedunculated, being attached by a narrow neck; while more commonly they are found with a broad base of support, the hypertrophy of the skin being confined to the radiating folds that normally surround the anus. These little cutaneous excrescences are called by some of the American authors "condylomata," and no little confusion has arisen in consequence, the latter term being in England only applied to the soft mucous patch of unquestionable syphilitic origin.

Cutaneous piles are apt from local inflammatory causes to become œdematous; they then become much increased in size, smooth and shiny on the surface, and acutely painful. Amongst the direct exciting causes of œdematous piles may be enumerated fissures or other breaches of the muco-cutaneous surface, or the eczematous inflammation that is not uncommon in this region. This inflammatory swelling usually subsides in a few days, leaving the pile somewhat permanently increased in size; or suppuration may result.

A form of external pile, to which the term **compound external pile** might suitably be applied, is not uncommon. It is usually of considerable size, about as large as a filbert, with smooth surface, and very prone to inflammation; if incised in the usual manner it is found to consist principally of connective tissue, and contains *several* thrombosed veins of considerable size, instead of one central cavity as in the common variety of venous pile. It is found in persons who have suffered from repeated attacks of inflamed external hæmorrhoids; and is almost always placed laterally, the long axis being antero-posterior.

While free from inflammation, external piles give rise to but trivial annoyance, which is caused by the mechanical inconvenience due to their size; but when inflammation supervenes, the pain becomes extremely severe, so that the strongest man may be thereby quite incapacitated for any active employment.

Hæmorrhage from external piles is an unusual occurrence, and when present is not generally severe. I have, however, several times seen it, and in each case its source could readily be determined, the blood being seen to flow from minute orifices in the pile. Reflex pains are commonly complained of, and the bladder participates in this reflex irritability, as evidenced by frequent micturition, or sometimes by retention.

Of the more important conditions with which external piles are found associated, painful fissure deserves prominent notice. It is but seldom that fissure has existed for any length of time without external piles being also present, and the character of the pile which exists at the lower termination of the fissure may be quite pathognomonic of this disease. This, which has sometimes received the name of "sentinel pile," is crescentic in shape, the fissure terminating in the concavity of the crescent. Such piles have a totally different origin; they are due to anal valves which have been torn down by a passing motion, and afterwards become swollen and œdematous, this tearing down of an anal valve being the essential element in the production of painful fissure (page 145).

Suppuration is a very common termination to an attack of inflamed external piles, and when it occurs tends to the production of a complete spontaneous cure. In rare cases, however, the abscess cavity opening in a second place, a small marginal fistula forms (page 71), which may require the intervention of the surgeon.

I have already (page 174) quoted the description of a condition of the anus given by Colles, and considered by him to be pathognomonic of stricture of the rectum; and it appears clear to me that the projections at the anal margin to which he refers are constituted by some form of hæmorrhoidal tumour: but while admitting that a group of small external piles nearly always surrounds the anus in cases of stricture, together with a certain amount of relaxation of the sphincter, in my experience these growths have not assumed the characteristic appearance described by Colles; so constant, however, is the association between these two diseases that, in every case of external piles, the rectum should be thoroughly explored with the finger for stricture.

A rare complication of piles is caused by the calcification of the enclosed thrombus, constituting the so-called phlebolith. Concretions so produced are to be recognised by their hardness to the touch, feeling like grains of shot or peas inside the dilated vein. The occurrence of phleboliths in piles was first pointed out by Bodenhamer,* who states that he has met with six cases. In structure they correspond with similar concretions found more commonly in some of the larger varicose veins in other parts of the body. According to Frankland † these are formed of concentric layers, which consist of proteid matters, and phosphates: the former, consisting of about 20 per cent. of the calculi, are nearly all albuminous or fibrinous; the latter, though mainly phosphate of lime, are associated with a

* "Hæmorrhoidal Diseases," p. 109. 1884.
† Holmes' "System of Surgery," vol. iii. p. 373. Second edition.

little sulphate of potash and sulphate of lime: that is to say, the phleboliths consist, as might be expected, of the coagulated proteid compounds of the blood together with the less soluble salts. When situated in the hæmorrhoidal veins they may give rise to irritation and the establishment of an anal fistula, as happened in three of the cases recorded by Bodenhamer.

The **treatment** of external piles is usually sufficiently simple. It may be divided into the palliative and radical; the latter of these is in nearly all cases preferable. If, however, the patient will not submit to the trivial operation necessary recourse must be had to local application during the period of acute inflammation. Of these **palliative treatments,** the best, in my opinion, is the application of a mixture of extract of belladonna and glycerine smeared over the part, and followed by a warm stupe. At the same time the bowels should be freely cleared, and a light, unstimulating diet, with rest in bed, prescribed. The inflammation will then usually subside in a few days; but it leaves behind a thickened projection of skin ready at any time to again inflame on the slightest provocation; or, if suppuration occurs, the cure may be radical, but only at the cost of much unnecessary suffering.

The **radical cure** may be accomplished either by incision or excision. When the pile consists of a simple thrombosed varix, treatment by incision and turning out the clot may succeed. The question whether these procedures should be carried out while there is inflammation present has been frequently discussed, many surgeons preferring to wait until the acute symptoms have subsided. This I believe to be quite unnecessary, only subjecting the patient to prolonged pain. It is but seldom that the surgeon is consulted about external piles, except when they are inflamed, and the most certain and rapid way of giving relief is immediate operation. I have never seen the slightest ill effect follow operation in these circumstances.

Simple incision may be applied when the pile consists of a single dilated and thrombosed vein. It is only necessary to incise the tumour freely with a sharp bistoury, and turn out the little contained clot. As the vein has already been occluded by the inflammation, bleeding need not be apprehended; the cavity should be dusted with iodoform, and the patient kept quiet for a day or two. The relief is usually immediate and complete; the sides of the cavity shrivel away, and the cure of the individual pile incised, at any rate, is permanent. As the incising is extremely painful, the parts may be first painted with a 4 per cent. solution of cocain, which acts fairly well in a number of cases; or freezing with ether spray may with advantage be employed.

If the pile be of the variety which I have described as compound external hæmorrhoid, simple incision will not give relief. The clots cannot be turned out, and the tumour will not collapse. For this form, therefore, and also for the cutaneous piles, excision is the proper remedy.

Excision undoubtedly is the form of treatment of most general applicability. If there are several tumours to be removed, it is better to give the patient a general anæsthetic, as the pain is considerable; and in performing the operation, the surgeon must be careful not to cut away the folds of skin about the anus too freely, else an anal stricture may be the unpleasant result. Only the distinct tumours should be dealt with, and of these only about two-thirds of each should be removed. The bases will then shrink, and all danger of stricture will be obviated. Simple œdematous folds of skin need not be interfered with, as they will quite subside when the source of irritation is removed. For performing this operation, by far the most useful instrument is Richardson's toothed scissors, its chief advantage being that it never slips, but cuts through the tumour exactly where it has been applied by the surgeon. I have not, however, found that there was less bleeding after its use than after the common scissors or knife. Hæmorrhage after this operation, however performed, is usually trivial, and readily arrested by the firm pressure of a sponge. Occasionally, however, when a large and fleshy pile has been removed, a small arterial branch may require a ligature. In the treatment of broad-based piles I have often found it a good plan to bring the cutaneous edges of the wound together with a few points of catgut suture. By doing so, healing will usually be much more rapid than if the surface is left to granulate. After the operation is completed, the surface of the anus should be well dusted with iodoform, and a sanitary towel tolerably firmly applied, which will check any tendency to bleeding, and also diminish the painful spasm of the levatores ani, so troublesome after many rectal operations. The bowels should be kept confined for three days, when a mild aperient, followed if necessary by an enema, should be prescribed. It is better to confine the patient to bed until the wounds are closed, as the congestion caused by the erect position tends to interfere much with the healing process. Sometimes there is some difficulty in getting the wounds to cicatrise well, unhealthy little ulcers forming. Indeed, I think that there is more difficulty in this way after the operation for external piles than there is after the removal of internal piles, although of greater size. Should the wound become sluggish, it may be painted with balsam of Peru, or the compound tincture of benzoin; and in other cases lightly touching the ulcer with sulphate of copper will frequently stimulate a healthy action.

CHAPTER XVI.

SYMPTOMS OF INTERNAL PILES.

MANY varieties of internal hæmorrhoids have received special descriptions from various authors, but the most simple, and at the same time practical, classification is that given by E. Hamilton.* It consists of the *venous*, the *columnar*, and the *nævoid*. It must not, however, be inferred that the characters which distinguish these varieties are always definitely marked ; but although they merge one into another more or less, the essential characters of typical examples are such as to justify the division.

The **venous pile**, when situated within the sphincter, resembles much the external pile already described, with this important difference, that the latter is covered with skin and has not much tendency to bleed ; whereas the former, being covered only by the thin mucous membrane, and more exposed to injury by the passage of hard fæces, is exceedingly prone to hæmorrhage. If these venous internal hæmorrhoids have existed for a long time and been prolapsed, the covering may become much thickened, and paler in colour, so that it more nearly resembles skin ; but even under these circumstances they can be distinguished from external piles by the presence of a more or less deep notch at the position of the muco-cutaneous junction.

The **columnar pile** is thus described by Hamilton : " The second variety, for which I would suggest the term 'columnar' pile, to denote its pathology,

---

* " Clinical Lectures on Diseases of the Lower Bowel," p. 32. Dublin, 1883.

consists essentially of hypertrophy of the folds of mucous membrane surrounding the anal opening, the pillars of Glisson. They have a red, almost vermilion, colour, an elongated form, and contain within them one of the descending parallel branches of the superior hæmorrhoidal artery."

This variety is, I believe, much the most common form of internal pile, and at the same time it is the most important, owing to its great vascularity. The arteries leading to piles of this kind become much enlarged, and can frequently be felt pulsating forcibly, and apparently quite as large as a radial artery, at the upper portion of the tumour.

If a section of one of these piles be examined microscopically, it is found to consist of hypertrophied submucous tissue, with numbers of arteries and veins, and, if there has been any inflammation, many of the latter will be found thrombosed. It can therefore be readily understood how, when ulcerated or abraded, these tumours will bleed copiously.

The third form, or the **nævoid pile,** has also been described under the name of "vascular tumour of the rectum," and very closely resembles capillary nævus. It may be present with the other variety, causing the mucous surface to assume a bright red, spongy, or villous appearance, not inaptly compared by Mr. Houston to a strawberry; or this change in the mucous membrane may occur without any other manifestation of disease, in patches as big as a sixpenny piece. This is the form of pile in which the bleeding is most constant and continuous, although the amount lost at any one time may not be so copious as in the second variety.

Of the **symptoms** of internal piles, the most important, and frequently the first, symptom to attract attention is bleeding. So constant is this symptom that the terms "bleeding" and "internal" piles are practically synonymous. It is very rarely that the disease has existed for any time without this symptom being marked. It is at first only at stool that this loss of blood is noticed, the tender and highly vascular mucous membrane being bruised and lacerated by the passage of a hard fæcal mass, the blood continuing to drip from the anus for some time after the rectum has been evacuated. As the disease progresses the bleeding becomes more frequent, till it occurs daily after each evacuation of the bowels; indeed, in extreme cases, it is not confined to the act of defæcation, but comes on at irregular times, without any apparent exciting cause. In this way a condition of the most profound anæmia may be induced, attended with pallid features, dizziness, and palpitation of the heart; indeed, it often happens, more especially in women, that these symptoms of anæmia have directed the attention of the medical attendant to a daily loss of blood from the rectum, which the patient has either not noticed or not heeded.

The amount of blood which may be lost in this way is occasionally very great, and in certain rare cases a fatal syncope has undoubtedly been produced. Bodenhamer has collected a number of cases in which it is recorded that excessive quantities of blood have been lost; but the majority of them are probably much exaggerated, a little blood making a great show when distributed over a patient's clothing, or mixed with urine and fæces.

It is obviously necessary to distinguish between hæmorrhage resulting from piles, and bleeding coming from some other part of the intestinal tract, such, for instance, as the stomach or small intestine. When not very excessive the latter is altered in colour by the digestive action of the intestinal contents, being of a dark, or sometimes tar-like, appearance; and it is, moreover, intimately mixed with the fæces. The blood from piles, on the contrary, is not mixed with the fæces, but has evidently escaped subsequently to the evacuation of the rectum; and if it has been exposed to the air for any time is of a bright red colour. The fact that it is usually seen by the surgeon some time after it has been passed, has given rise to the belief that it is invariably of arterial origin; but venous blood, if exposed to the air, will become bright in colour, even after it has escaped from the body. And again, when a patient is examined with his piles prolapsed, the blood may be seen escaping in jets, leading to the belief that it is coming from a lacerated artery. That this, however, is not invariably the case is contended for by Cripps.* He says: "It is a matter of some interest to consider the source of this bleeding. The fact of the blood escaping in jets has led many high authorities to regard it as arising from some arterial twig. With due deference to such eminent authorities as Brodie and Van Buren, I am of opinion that they are mistaken, and I do not believe that it ever comes from the arteries, but that the jet is caused by its being forced as a regurgitant stream through a small rupture in a vein by the powerful pressure of the abdominal muscles. If it really came from an artery, why should the jet only appear when the abdominal muscles act?"

Cohnheim† considers that intestinal hæmorrhage does not necessarily presuppose a breach of mucous surface, but that, when inflammation is present, the discharge may be produced by a diapedesis of red blood corpuscles.

Mucous discharge is commonly present, either alternately with bleeding, or replacing it altogether.

* "Diseases of the Rectum and Anus," p. 69.
† *Virchow's Archiv*, Band 41, p. 221.

This has been specially described by Richet,* under the name "hémorrhoïdes blanches," which, however, I cannot help thinking a singularly inappropriate term to designate what is nothing more than a catarrhal discharge, resulting from the long-continued irritation of the rectal mucous membrane. (See Proctitis, page 49.)

Except when strangulated or acutely inflamed, pain is not a prominent symptom of internal piles; nevertheless, a certain amount of uneasiness or discomfort is generally present, with a sense of fulness and weight in the pelvis; and reflex pains at a distance from the anus are not uncommonly complained of. These are most frequently situated in the back, groins, or genital organs. In one patient under my care there was severe pain located immediately above the symphysis pubis, which was completely removed by crushing some small piles.

CHAPTER XVII.

THE CLINICAL RELATIONS AND COMPLICATIONS OF INTERNAL PILES.

FROM what has been already said on the ætiology of piles, it will be obvious that they are not unfrequently but symptoms of some remote and more important visceral disease, and the surgeon who looks at these cases with the eye of a specialist, and directs his treatment solely to the rectum, will be sure to be disappointed; while in some instances positive injury will be inflicted on his patient. Of the more important organic diseases with which piles may be associated, those of the heart and liver occupy the most

* *Irish Hospital Gazette*, July 12, 1874.

prominent position, and it is incumbent on the surgeon to carefully examine every patient with piles for any symptom of these affections. I know of no case in which the surgeon may have greater difficulty, than in deciding on what treatment to adopt for rectal hæmorrhage from internal piles when they are associated with grave visceral lesions. As illustrating this, the following case may prove instructive : In 1885, in consultation with Dr. James Little, I saw a gentleman who had suffered from aortic patency for several years, and who also, for many months, had lost a very considerable quantity of blood from internal piles, each time the bowels moved. He was markedly anæmic, and it was obvious that his failing heart could not very long continue to maintain the circulation. The daily loss of blood was, by diminishing the arterial tension, still further taxing the weakened heart, and it was obviously imperative to make some attempt to arrest the bleeding, although the undertaking of any operation in the patient's wretched state of health was by no means promising. I ligatured five large piles, and the bleeding was from that time completely stopped. Although he died six months subsequently from heart failure, the operation unquestionably prolonged his life, and rendered the remainder of it much more endurable. No surgeon would hesitate to operate in a case of strangulated hernia, no matter what condition the patient was in ; yet in some of these cases of combined heart disease and bleeding piles, the continued loss of blood may be as surely, though perhaps not so rapidly, killing the patient.

The connection between disease of the liver and piles is a matter of common observation, so much so that the latter are looked upon as one of the usual symptoms where there is portal obstruction, as is commonly the case in cirrhosis of the liver ; but I think it will be within the experience of most physicians that when rectal hæmorrhage from internal piles is copious, another prominent symptom of cirrhosis, namely, ascites, is either entirely absent òr very slight; and I have twice seen persons who had been operated on for bleeding piles rapidly develop ascites after the hæmorrhage had been stopped. Consequently it becomes a very important matter to decide whether an operation should be undertaken for the relief of bleeding piles when associated with cirrhosis of the liver. If the hæmorrhage is not very great in amount and the person well nourished, it may be well to allow the discharge to continue, the slight loss being sufficient to relieve the obstructed portal system, and so prevent the transudation of serum into the peritonæal cavity. If, of course, the hæmorrhage should be excessive and the patient much run down, operation is unequivocally called for; but it is well to bear in mind the fact that when the piles are dependent upon obstructed vena portæ, recurrence is very likely to follow operation, and the wounds are slow in healing.

The discussion of this subject brings us to the oft debated question : Is the bleeding from piles ever salutary, and if so, under what circumstances should it be allowed to continue unchecked? If we refer to the ancient writers we find that they considered that a hæmorrhoidal flux served as an emunctory, by means of which bile and other acrimonious humours were excreted from the turgid extremities of hæmorrhoidal veins. Hippocrates plainly teaches that hæmorrhoids perform the function of evacuating "the black bile of melancholic humour"; consequently he recommends, when piles are operated on, that one should be left in order to obviate the danger of dropsy or phthisis. Bleeding from the rectum was thus rather considered as a physiological function than a pathological condition, and when suppressed, many methods were resorted to for the purpose of restoring the hæmorrhoidal flux. This teaching of the ancients has been, to a certain extent, continued down until quite recently; and even still we frequently see patients suffering from bleeding piles, who have been told by their medical attendant that if the bleeding was stopped there would be danger of other more dangerous diseases supervening. Trousseau,* writing on this subject, says: "The physicians of past ages have, perhaps, too much exaggerated the importance of hæmorrhoids in the scale of pathological phenomena, while those of our own time are fallen into the contrary extreme. It cannot be denied that the suppression of the hæmorrhoidal flux, when habitual, may be productive of general disorders among men almost as serious as the suppression of menses in women. Moreover, it is generally admitted that in certain persons who have not only regularly, but at indeterminate periods, a draining or hæmorrhoidal flux, the existence of this pathological condition is attended with a state of general good health, although it may remain for a long time uncertain and variable, provided the hæmorrhoids do not manifest themselves as soon as usual. Observation shows also that persons who have had hæmorrhoids for a long time suffer generally if this flux entirely ceases, and it often happens that there is a call for its restoration."

In the same article, Trousseau recommends the application of cupping-glasses to the anus, or the use of suppositories containing a quantity of tartarated antimony, sufficient to produce a considerable inflammation of the lower end of the rectum, with the object of reproducing the suppressed piles; and many of the old writers recommended the attempt to produce

* *Journal des Connaissances Médico-Chirurgicales*, p. 101; Sept., 1836, Paris.

hæmorrhoidal flux, *de novo*, merely as a therapeutic expedient.

It is not to be wondered at that, during an age when the periodical loss of blood was considered essential even for the healthy, this spontaneous bleeding from the rectum was considered not to be in any sense prejudicial, but rather on the contrary a thing to be encouraged, and even, if possible, in some cases, initiated. Now, however, most sensible surgeons will agree that the loss of blood, such as takes place from internal piles, is a pathological condition which it is well to free the patient from as soon as may be, the only possible exception that I know being some few cases of cirrhosis of the liver such as I have before mentioned, in which an occasional rectal hæmorrhage tends to prevent the occurrence of ascites; and some middle-aged and very plethoric people, who, although they frequently have slight bleeding, appear to remain otherwise in perfect health, and not to suffer in any way from symptoms of anæmia. In these cases the surgeon may well await the onset of indications that the loss of blood is ceasing to be well borne, before recommending any operation for radical cure, unless, indeed, the inconvenience (other than bleeding) is so great that operation may be demanded. Each case, therefore, must be judged on its own merits, and not treated according to any fixed rules.

The association of hæmorrhoids with uterine disease is not uncommon, and such cases are occasionally still further complicated by the presence of irritable bladder. According to Allingham, these cases are extremely unfavourable for treatment, and he insists on the necessity of curing the uterine flexion, or other disease, before attempting to treat the rectum. In dealing with them we must always bear in mind the fact that uterine disease will frequently give rise to reflex pain in the rectum, when no indication of disease is to be found in that organ, and, conversely, we not uncommonly see women who are being treated for supposed uterine disease, when the symptoms may be all referable to a small anal fissure or some other rectal disease.

Strangulation and gangrene of internal piles may arise from one of two quite distinct conditions. Either the piles may be extruded from the anus, and caught by the sphincter; or inflammation may be started from some trivial abrasion, and, owing to the extreme vascularity of the part, gangrene may rapidly ensue. Now it becomes a matter of importance to determine which of these conditions is present, as in some respects the treatment varies with the cause. The diagnosis is not difficult. If the congestion is produced by constriction of the sphincter, the history given will be that the patient had prolapse of the piles after defæcation, which probably occurred on frequent previous occasions, when he was able to replace them himself. He, however, fails at last, on this occasion, to reduce them, or they descend again immediately after they are replaced. They soon become swollen, and the pain becomes extreme, with considerable fever and other constitutional disturbances. Upon inspection (Plate IV., Fig. 1), there will be found protruding from the anus one, or more frequently several, livid, tense tumours. Round the margin of the anus there is some slight inflammatory œdema. Any attempt to touch these piles produces the greatest pain; in fact, there are but few diseases in which the pain is of such a severe character as in the one under consideration. If the finger be passed into the rectum great resistance is experienced, and the forcibly contracting sphincter can be felt grasping the finger tightly. If left unreduced, these intensely congested piles rapidly pass into a state of gangrene, and are thrown off,

PLATE IV.

Fig. 1.—Prolapsed and Gangrenous Internal Piles
Fig. 2.—Prolapsed Internal Piles.

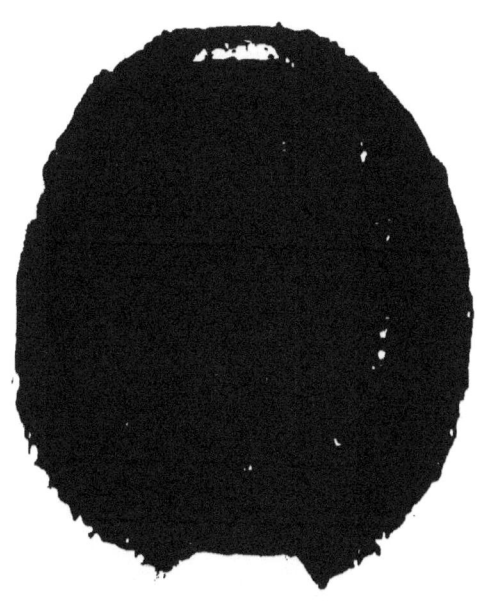

eventually producing a spontaneous cure, which, however, is frequently incomplete, as the entire pile seldom dies, and the ragged portion which remains is likely to give trouble at some future time. Bleeding during the separation of the sloughing pile is not of uncommon occurrence.

The treatment to be adopted may be either temporary, or radical. In the first instance reduction of the piles within the sphincter should be attempted: and if the case is seen early, and is not complicated by much external inflammation, reduction can usually be readily effected. In order to accomplish this, it is best to try and replace the most central portion first by passing the finger into the rectum, and as it is withdrawn to force up the remainder of the prolapsed portion. It is, however, better practice to obtain the permission of the patient to perform the radical operation at once; and, owing to the very severe pain usually attendant on this condition, there is no difficulty about this. I on one occasion had under care a patient who had previously obstinately refused to have any operation performed on his piles, but who became loud in his entreaties to have them removed once strangulation supervened.

Should piles be operated on during inflammation? is a question upon which considerable difference of opinion has been expressed. Those surgeons who oppose operation appear to me to base their objections purely upon theoretical grounds: no bad results, as far as I know, have been recorded; and, on the other hand, those surgeons who have made a practice of operating express themselves favourably to it. My own practice, when called to a case of strangulated piles, is to recommend the immediate administration of an anæsthetic, dilatation of the sphincters, and the complete removal of all piles that can be seen, just as if the case was one uncomplicated by strangulation.

The history of a case of inflammatory strangulation of piles differs from that of one of acute strangulation, the result of nipping by the sphincter; and it is important to make the diagnosis. In the former, the patient may be conscious of having abraded the anus during defæcation, and a day or two afterwards pain and throbbing in the rectum gradually come on; rigors and febrile disturbance supervene; and there is great swelling and tumefaction about the anus. Upon examination the internal piles will be seen protruded, together with œdematous folds of mucous membrane; the external piles, if present, will also be tumid from inflammation, and swollen folds of skin will radiate from the anus. Often, as a result of inflammation, internal and external piles run into one another, the line of junction being only marked by a slight sulcus. If the finger be passed into the rectum, the sphincter will not be felt unduly tense, the contrast with the condition of things in the previous case being most marked. If an attempt is made to reduce these inflamed piles within the anus, the most acute suffering will be induced, and the result will be quite futile. I have several times seen cases in which continued attempts had been made to reduce what really were inflamed external piles and œdematous folds of skin, or internal piles, which from the inflammation of the surrounding structures were quite irreducible, under the impression that the case was one of strangulation by the sphincter. The writings published on rectal disease do not, in my mind, attach sufficient importance to this subject, but I think that the truth of what I have above stated will be evident to any surgeon of extended experience.

Inflammation of the structures surrounding the

anus of a severe character may sometimes co-exist with strangulation by the sphincter, but this I consider unusual; therefore the following case, which I saw along with Dr. Roe, may prove of interest, especially as it illustrates also an unusual, and, in this case, eminently satisfactory plan of treatment. The patient was an officer who had seen a good deal of hard service in the Soudan. While there he commenced to suffer from bleeding piles, which continued to trouble him at intervals for twelve months. On a certain occasion, after passing a hard motion, which he felt had severely scraped his anus, the piles gradually became swollen and painful. When seen by Dr. Roe and myself, he had suffered extreme pain for four days, during which time several attempts had been made both with and without the aid of anæsthetics to reduce the swollen mass within the rectum, but without success. Upon examination a mass of internal piles of the size of a hen's egg, was to be seen, obviously gangrenous; and surrounding these, and continuous with them, were a number of very large and partly gangrenous external hæmorrhoids. It was quite evident that the tumours were absolutely irreducible, and also that the gangrene was the result, not of strangulation, but of inflammation, probably of septic origin. Upon introducing the finger into the rectum, however, it was very tightly grasped by the sphincter, and the continued spasm of this muscle was causing the patient very great pain. As the gangrene had done its work so completely, no operation for the removal of the piles was required, it being quite apparent that they would, in this case, entirely slough away, but as the continued spasm was keeping up irritation, and causing very great pain, we determined to set it at rest. Owing to the gangrenous condition of the surrounding structures, we deemed forcible dilatation inadmissible, so I introduced a tenotome beneath the mucous membrane, and divided the sphincter muscle freely. The operation was followed by immediate relief to pain. The anus and its surrounding skin were saturated with a solution of iodoform in ether, and charcoal poultices applied till the sloughs separated. The patient made a rapid and good recovery, and although there was destruction of the entire circumference of mucous membrane and anal skin, no stricture resulted, and he was able to go on foreign service again.

Fissure of the anus is not an uncommon complication of internal piles, and when a patient complains of pain *during* defæcation, this may always be suspected, as internal piles produce only a sense of discomfort rather than real pain, except when inflamed. Since the introduction of preliminary forcible dilatation of the sphincter when operating for internal piles, the practical importance of diagnosing the presence of fissure has been much diminished, as the dilatation, or at any rate the removal of the "sentinel pile," will generally effect a complete cure; indeed, I have sometimes first been made aware of the presence of a small fissure, situated far in, when dilatation had been accomplished.

In cases of old-standing piles small **polypi** are sometimes met with; they are usually of the fibrous and not of the adenomatous variety, and are due to the connective-tissue changes elsewhere alluded to (page 238). They can with ease be removed at the same time that the piles are being dealt with.

Simple stricture and **malignant neoplasms** are sometimes met with co-existing with piles, and it is necessary, of course, to determine whether this is the case by careful digital examination. Indeed, the complete rectal examination of every case of piles that presents itself should be the invariable rule.

The only other local affection likely to complicate piles is **fistula**, and usually there will not be much trouble about the diagnosis of this. The principles of diagnosis laid down in chapter v. apply equally here.

CHAPTER XVIII.

PALLIATIVE TREATMENT OF INTERNAL PILES.

THE treatment of internal piles may be classified into the palliative and radical. Unquestionably, in some cases, the palliative treatment will be followed by complete immunity from further trouble; yet these cases are quite the exception, and where the piles are large, and attended with the formation of much new tissue, it can hardly be expected that anything short of some surgical operative procedure will effect a cure. There are, then, two classes of cases in which the surgeon will confine himself to the medical treatment of the case: one that in which he considers there is a fair chance of cure by such measures; and the other in which operation is declined by the patient, or is considered inexpedient on account of some complication, and in which, therefore, the surgeon has to confine himself to mere palliation. I think the recommendations of a surgeon to have the trivial operation necessary for complete cure, are more often neglected by the patient suffering from internal piles than any other surgical malady. This is partly due to the fact that the public are still imbued with the teaching of former days, that piles are beneficial, and believe that many ills and disabilities follow this trivial operation. Again, the pain suffered is not very considerable unless inflammation has supervened. Yet there is no class of cases in which, with such slight risk to life, so much good can be done by operation; and patients who have suffered years of trouble and annoyance with piles are most grateful for the relief afforded by surgical treatment. What, then, are the indications or contra-indications for resorting to or not resorting to operation? I can point out some of them. If the case is an uncomplicated one of long standing, with large piles prolapsing after defæcation and with considerable hæmorrhage, radical treatment is clearly indicated; and, on the other hand, if the piles are associated with grave visceral disease, more especially diabetes or disease of the kidneys, with albuminuria, operation should not be undertaken, unless the bleeding is so copious as immediately to threaten life, as in the case before recorded. Should piles be operated on during pregnancy? is a question that often arises, in consequence of the very common connection between the two, many pregnant women suffering great inconvenience and copious bleeding from this cause. Now, in order to answer this question, we must remember that the cause of the piles in this case is frequently due to the mechanical pressure of the gravid uterus, and that they will subside as soon as delivery takes place, and that, owing to the close nervous relationship which exists between the uterus and rectum, any operation must occasion some risk of premature delivery. In this as in many other cases the indication for operation rests with the answer to the question whether the loss of blood is making itself constitutionally felt to any great extent or not: if it is, the piles should certainly be operated on. I have on one or two occasions operated in these circumstances with the best results, and without curtailing the duration of pregnancy.

The most important indication to be fulfilled in the treatment by medical measures is the regulation of the bowels. If allowed to become costive, the piles are liable to be excoriated during defæcation, and increase of bleeding and possible subsequent inflammation is the necessary result; whereas, if an easy and soft evacuation is secured each morning, a state of comparative comfort can be maintained.

Numerous purgative medicines have been especially recommended for persons suffering from hæmorrhoids. In a work of the present kind of course it would be impossible to enter upon a discussion of the comparative merits of all the numerous purgative drugs which have been advised in cases of the kind of which I am treating. Such disquisitions are suitable to, and should be left to, special works on therapeutics. I can only refer to, and indicate, those practices and principles in medicinal treatment which strike me as being of more especial practical importance; or which demand notice from their habitual and customary employment. Of purgative medicines which it has become the fashion to recommend, the most recent appears to be the cascara sagrada. I have tried the fluid extract frequently, and think it is decidedly inferior to many of the older and well-established medicines. It is unpleasant to taste, it frequently causes pain, and its action is inconstant. The use of aloes is by many writers supposed to be contra-indicated in rectal disease; for what reason I do not know: on the contrary I think a pill composed as follows the best for ordinary purposes:

℞ Ext. aloes socotrinæ . . . . gr. 1¼
Ext. nucis vomicæ . . . . „ ¾
Ext. belladonnæ . . . . „ ⅛

One or two for a dose as required.

If these pills are taken occasionally, and a dose of some of the well-known ordinary purgative mineral waters in the morning, motions of the proper consistency will be ensured. If the pills alone are used, the dose will have to be augmented; and if the mineral water only is employed, it is apt to produce only a small fluid motion, after a time leaving some larger masses in the great intestine. When this occurs, it is, I think, the indication for the employment of pills containing aloes. Far better, however, than any purgative medicine is the use of an enema, the employment of which is too much neglected in England, most people, especially ladies, preferring to take any quantity of medicine to the use of this simple expedient.

I cannot do better than quote the plain terms in which Mr. Mitchell Banks speaks of the disinclination of English people to use enemata.* "Why do not English lecturers and text-books tell students more about the value of enemata? A Frenchman looks upon his enema very much in the light in which an Englishman regards his tub; and a most excellent and cleanly thing it is. But our silly false modesty has induced us to attach something almost of indecency to this very innocent operation. I find some eminently respectable persons who are quite shocked at being asked to employ so un-English a remedy. And, again, the public have got an idea in their heads (fostered chiefly by medical men) that the frequent use of enemata produces distension and subsequent paralysis of the gut, and is consequently a very dangerous thing. I suppose if a man were to throw a quart of water into his rectum, and try to walk about with it there all day, that he would do himself mischief, but that the occasional use of six or eight ounces of tepid water can do anybody any harm, however long the habit may be kept up, I do not believe. A large number of persons are always purging themselves for constipation, in whom, I believe, the fæculent mass is lying in the rectum quite ready to move on if only the bowels could

* *Liverpool Medico-Chirurgical Journal*, p. 293; July, 1886.

effect its expulsion. The introduction of a little water at once lubricates the canal, and gives the gut the stimulus to a smart contraction, which is all it wants."

After the bowels have been moved, the anus should be carefully washed with cold water. This is one of the most important conditions to fulfil in order to make the hæmorrhoidal patient comfortable; and it is one of the most powerful means of checking the tendency to hæmorrhage that we possess. Banks tells us* of a method which a patient of his found out for himself, and it appears to be at once the most effectual, and has the advantage of being simple and always applicable. " So soon as the bowel was evacuated, he remained perfectly still, emptied the pan of the water closet, and, keeping the rush of water on with his right hand, he threw it up with his left on to the everted mucous membrane. By this means he thoroughly cleansed his piles, and, by the direct application of cold water to them, ensured contraction of their blood-vessels, and their complete retraction. Carefully drying the parts with a soft towel (and not using paper), he then washed his hands, and was comfortable for the rest of the day."

The diet should be regulated. All food tending to leave large fæcal residue should be avoided; but the error most common is eating too great a quantity of food rather than the quality, and all persons suffering from piles who are great eaters will derive great benefit from restricting their diet. Stimulants are better avoided, except under special conditions.

The internal administration of glycerine has been recommended, and when taken in large quantities it acts as a mild laxative, and so far is beneficial; but beyond this it does not appear to have any decided action. Another remedy which has for many years had a reputation when administered internally is

* *Loc. cit.*

"Ward's paste," or confection of black pepper. What it was supposed to do I never was able to find out, and I certainly have not been able to see any good results from its use.

Numerous topical applications have been from time to time recommended. Amongst these, the compound gall and opium ointment of the British Pharmacopœia is a general favourite, and it certainly, in many cases, succeeds well both as a local astringent and means of relieving pain. The following ointment has, however, the advantage of being more cleanly and of more definite strength:

℞ Morphiæ hydrochlor. . . . . . gr. x.
   Ext. belladonnæ . . . . . ʒ j.
   Acid. tannici . . . . . . ʒ j.
   Vaselin. .
   Lanolin. . . . . . . āā. ʒ j.
        Misce; ft. unguent.

A small piece to be rubbed over the piles when prolapsed.

Many of the American writers speak in the highest terms of the use of ferrous sulphate in powder, to be applied either dry or in the form of ointment or suppository; and where there is much bleeding it is a most admirable astringent, but it has the disadvantage of sometimes causing very considerable pain. The application which I prefer for ordinary cases is a dry powder composed of equal parts of oxide of zinc and subnitrate of bismuth; this should be dusted over the tumours after they are washed subsequent to defæcation, and will be found very comfortable and quite sufficiently astringent for most cases. Where there is much mucoid discharge, this powder will be found eminently suitable.

We find sometimes with piles an eczematous excoriation of the skin surrounding the anus. When this is the case, the following application acts very well :

℞ Liq. carbonis detergentis . . . ʒj.
Liniment. calcis . . . . ad. ʒvj.
Misce.

To be applied to the piles with a piece of soft sponge.

When the piles are inflamed, the application of a piece of absorbent cotton moistened with dilute lead lotion warmed, will prove very comfortable, especially if it alternates with light poultices. The application of an ice-bag is sometimes very comforting, but its use requires careful watching, as, if too long continued, the tendency to gangrene will be fostered.

A patient labouring under piles should always take as much exercise as possible, riding being especially good. A gymnastic movement for the cure of hæmorrhoids is practised at the Bellevue Hospital,[*] and, it is stated, with considerable success. "It consists simply in trying to touch the toes with the fingers without bending the knees. This movement, though difficult at first, soon becomes easy. It not only strengthens and develops the muscles of the abdomen, but also those of the legs and thighs; it assists the action of other remedies, and thus aids in the cure."

In conclusion, I would sum up the most important points in the medical treatment of piles as follows : Keep the bowels scrupulously regular; adopt thorough ablution with cold water after defæcation; use moderation in quantity of food and drink taken; regular exercise; and the occasional use of some of the astringents mentioned. If this plan is followed out, tolerable comfort will be ensured, and in a few cases which are not very severe a complete cure may be looked for.

[*] *New York Medical Record*, p. 599; 1877.

CHAPTER XIX.

OPERATIVE TREATMENT OF INTERNAL PILES.

To detail the numerous operations which have from time to time been recommended for the cure of internal piles would prove a laborious and useless task; but, on the other hand, we cannot subscribe to the doctrine of Verneuil, that the treatment may be summed up in two methods: cold enemata and laxatives for mild cases, and forced dilatation for severe.

The more important operations may be usefully enumerated under the following heads: 1. The application of chemical caustics to the surface of the tumours. 2. The injection of fluids of various kinds into the interior of the growths. 3. The gradual or forced dilatation of the anal sphincters. 4. Electrolysis. 5. Ligature. 6. Crushing. 7. Excision. 8. The actual cautery.

1. The application of **chemical caustics.**— Nitric acid was originally employed by Cusack, and strongly recommended by Houston,[*] since which time it has met with very widespread support; and for a certain class of cases it undoubtedly answers admirably, viz. the small bright-red nævoid pile, the object being to destroy the spongy and highly vascular mucous membrane covering this pile, and to thus substitute for it a cicatrix. As its action is only quite superficial, it is manifestly unsuitable for use in cases where there is a very extensive new formation of tissue; and the attempt to treat such cases by means of nitric acid has occasionally brought the method

[*] *Dub. Jour. of Med. Science*, p. 95; 1843.

into undeserved disrepute. In order to use it with success, the rectum should be well cleared out by an enema, and the pile, if possible, protruded. If this cannot be done, a small speculum, preferably of silvered glass with a small aperture, should be introduced. The pile is now made to protrude into the aperture of the speculum, and fuming nitric acid spread over the surface with a glass rod or piece of stick. The little brushes made of spun glass for use with nitric acid are not suitable for this purpose, as small fragments are likely to break off and irritate the mucous membrane. In doing this care should be taken to protect the skin of the anus from contact with the acid, or it may be protected by smearing with oil or vaseline. When sufficient acid has been applied, the surplus is neutralised by the application of chalk and water or solution of carbonate soda. If confined to the mucous membrane, the application of nitric acid is almost absolutely painless; but if any is allowed to escape over the delicate skin surrounding the anus, a very considerable amount of burning pain will result. The acid application usually requires to be repeated two or three times, at intervals of a week, before the cure is complete.

Other chemical caustics have been employed for a like purpose, such as acid nitrate of mercury, Vienna paste, and butter of antimony; but none of them answer the purpose better than fuming nitric acid, which has stood the test of such a long experience. Amussat devised a special form of forceps by means of which a stick of caustic potash could be kept applied to the base of a pile for some time until sloughing was induced; and Hamilton recommends the passage of needles covered with melted nitrate of silver through the substance of the pile for a like purpose.

The **injection** of fluids of various kinds, by means

of a hypodermic syringe, into the interior of a pile, is a method of treatment which has recently been revived, more especially in America. The fluids used vary considerably. Tincture of perchloride of iron has been used by Colles ;* solution of ferrous sulphate by Van Buren ; † and liquid extract of ergot, etc. ; but the only injection which is now used to any extent is carbolic acid, either in simple solution or combined with tannic acid,‡ or combined with liquid extract of ergot.§ This method of treatment has been much more popular with American surgeons than with others ; amongst those who have especially written upon the subject, the names of Kelsey, Washburn, and Andrews must be mentioned. Kelsey, more particularly, is high in its praise, and states that he has now operated by this means two hundred times.|| He tells us that for many years it had been a common practice amongst quacks and itinerant pile-doctors, and owing to the undoubted success which they sometimes obtained, he was induced to try it scientifically, so as to determine its limits of application. It appears to act in one of two ways : either sloughing is produced if the solution used is too strong, or even sometimes with weak solutions if the patient is much debilitated ; or a certain amount of inflammatory consolidation and thrombosis of the vessels is produced, which eventuates in a subsequent shrinking and subsidence of the pile. This latter condition is to be aimed at, except in some special cases in which sloughing is especially required. Care should be taken in performing this operation that the point of the syringe is passed to as nearly as possible the centre of the tumour, and the fluid slowly injected.

* *Dublin Journal Medical Science*, p. 505 ; 1874.
† *Loc. cit.*, p. 48.
‡ Givard, *British Medical Journal*, Sept. 5, 1885.
§ Dr. Fenn, *Medical Record*, June 15, 1883.
|| Braithwaite's "Retrospect of Medicine," vol. ii. p. 45. 1885.

The point of the syringe should be kept in place for a minute or two, so as to allow the fluid to become dispersed amongst the tissues, and the needle gently withdrawn. If the injection is merely passed beneath the mucous membrane, it is very liable to cause sloughing of this membrane, leaving an ulcer slow in healing, and without curing the pile. Only one pile is to be dealt with at a time. The injections are best performed at intervals of a week, and several applications may be required for the cure of large piles. Kelsey gives the duration of treatment for severe cases as from ten to fourteen weeks, the patient being able to follow his usual occupation during that time. With regard to the solutions used, Kelsey employs carbolic acid in fifteen, thirty-three, and fifty per cent. solutions, and in some cases even pure acid. The larger and severer the piles, the stronger the solution employed. Dr. Fenn's solution consists of equal parts of liquid extract of ergot, and of a ninety-five per cent. solution of carbolic acid, and of this he injects five to ten drops into each tumour. Washburn uses solution of carbolic acid in sperm oil, of the strength one to two if he wishes to produce sloughing, and one to four if he desires to produce absorption without sloughing;* while Givard uses a solution consisting of tannic acid one part, carbolic acid two parts, alcohol four parts, and glycerine eight parts; but he states that sloughing is a usual result.†

The advantages claimed by the supporters of this operation are safety, freedom from pain, and no necessity for confinement. But even in America it has not met with by any means universal approbation. Dr. Matthews, of Louisville, declares it to be painful, insufficient, and liable to cause death by peritonitis, embolism, and pyæmia. Amongst British surgeons

* *Philadelphia Medical Reporter*, Aug. 16, 1884.
† *British Medical Journal*, Sept. 5. 1885.

this method of treatment has not met with much favour. Allingham states : * "I tried the injection plan on some few cases, but the result was much pain, more inflammation than was desirable, a lengthy treatment, and the result doubtful; certainly not a radical cure." I have no personal experience of this plan of treating piles, but I have seen very alarming collapse produced by the injection of three drops of carbolic acid into a small nævus; and as there are so many thoroughly efficient, safe, and speedy methods of treating piles, I should be slow to try one which, on the showing of those most in favour of it, necessitates a three months' treatment, while in the hands of others it has been neither devoid of danger, nor successful in producing a complete cure, and in some cases attended with very severe pain.

3. **Dilatation.** — The introduction of bougies into the rectum, using instruments of gradually increasing diameter, for the treatment of internal piles by gradual dilatation, is a very old method, being recommended by Copeland and also by Quain; but of recent years it has almost entirely given way to more definite and precise procedures.

Bodenhamer, however, still recommends it, using in addition to the ordinary bougies, rectal sounds with flexible stems.† The rationale of the treatment is that the direct pressure on the veins tends to assist the torpid circulation, and that the mechanical irritation produced by the passage of the instrument stimulates the weakened peristaltic action of the bowel.

If the reader refers to the treatment of anal fissure, he will find something of the history of forced dilatation as employed in rectal surgery (page 144),

* "Diseases of the Rectum," p. 118. Fourth edition.
† "Treatise on the Hæmorrhoidal Disease," p. 227. New York, 1884.

from which it will be seen that Maisonneuve, in his later practice, was the first to use digital dilatation, practically in the same way that it is done at the present day; and not only did he employ it for the treatment of anal fissure, but he also recommended it for the cure of internal hæmorrhoids. The method has found but little favour, however, with most modern surgeons as a means of cure in itself, although all admit its very great utility as a preliminary step to almost all rectal operations. The operations of ligature, cautery, or incision, as practised now, after a preliminary dilatation under an anæsthetic, contrast very favourably with the similar operations as performed previously: as it was then necessary to keep the patient awake, in order that he might protrude the piles for the operator. Usually, at the first touch, the piles were retracted, and the anus drawn up; and when, after repeated urging by the surgeon to bear down, he made the attempt, he was rewarded with the tightening of a ligature, or touch of the actual cautery. Is it to be wondered at that, under these circumstances, operation was much dreaded by patients, and frequently impossible for the surgeon satisfactorily to complete? Now, however, all pain is avoided by anæsthesia, and dilatation enables the surgeon to make a thorough exploration of the entire rectum, and treat whatever he may find. Hæmorrhage need not be so much feared, because, having the part completely under command, a ligature can be applied to any bleeding point requiring it: moreover, the temporary paralysis produced by the dilatation adds much to the comfort of the patient after operation. The painful spasm so much complained of formerly is not so troublesome, and retention of urine, which was so often seen, especially in the male, is now comparatively rarely met with. Still, however, some of the French surgeons recommend forced dilatation as a means of cure, on the hypothesis that continued contraction of the sphincter muscles is a cause of hæmorrhoids, which is supposed to act in the following manner : The contracted anus not becoming sufficiently relaxed during the act of defæcation, straining results, and the consequence is the protrusion of folds of congested mucous membrane, which are caught by the sphincter. This produces congestion and swelling, and, when often repeated, ends in the production of true piles ; but against this argument clinical observation shows that the sphincter is only exceptionally tightly contracted in cases of piles, while in many instances exactly the opposite condition, namely, marked atony, exists, so that the piles are continually descending through the relaxed sphincter. Consequently, although fully admitting the use of forced dilatation as a preliminary step, I fail to see how it can be expected to effect a complete cure in the average cases which come under treatment.

4. **Electrolysis** is now commonly used for the destruction of small cutaneous nævi ; and in a few cases I have found it answer admirably for the cure of small nævoid piles, such as are suitable for treatment by nitric acid, and even in some which, owing to the amount of new tissue formed, could not be expected to be cured by the very superficial slough produced by nitric acid. One or two applications will usually suffice to completely consolidate the growth. It does not necessitate the patient's confinement to the house ; and the pain is not severe if a solution of cocain is first used. The way in which I employ it is as follows : The pile being brought into view, the surface is well painted over with a solution of cocain hydrochlorate (4 per cent.), and after the lapse of five to ten minutes, four or five fine round sewing needles, mounted in a handle, are passed into the centre of the tumour, and

connected with the *negative* pole of the battery, 10 to 20 Leclanché elements being the most suitable, the other (positive) pole being applied by means of a wet sponge to the buttock. After a few minutes the surface of the pile will be seen to become white, and minute bubbles of hydrogen gas will be seen escaping round the needles. As soon as this is well marked the needles are withdrawn, and, if deemed necessary, reintroduced into another part of the same, or another pile. In a few days the piles shrivel up and disappear painlessly. If the positive pole is used the needles stick tightly in, and hæmorrhage may result from their forcible withdrawal. It has, however, in order to avoid this inconvenience, been recommended in the case of nævi to use the positive pole first attached to the needles, and then, after a few minutes, to reverse the current for a short time previous to the withdrawal of the needles. I have not, however, found this plan satisfactory, and prefer to use the negative pole all through.

5. **Ligature.**—Without doubt ligature, in some of its modifications, has been the most popular method of treating internal piles, and is by far the most frequently employed. Dating, as it does, back to the time of Hippocrates, it still retains with the majority of surgeons its popularity as one of the best methods of treatment; while with a few it appears to be employed to the exclusion of all other plans. It may be employed in various different ways; when the piles are somewhat pedunculated, the loop of a ligature may be passed round the base and tied in the ordinary way; or where the pile is sessile, a needle threaded with strong silk may be passed through the broad base, and each side strangulated separately. This plan is decidedly objectionable, because if a large vein happen to be perforated by the needle, tightening the ligature will render the venous wound

patulous, and thus permit a certain amount of hæmorrhage, while at the same time the risk of septicæmic infection must obviously be increased. Bodenhamer * recommends a method of ligaturing piles in sections, similar to what is known as Erichsen's operation for long, flat nævi. A straight or curved needle is threaded on the middle of a long piece of white silk or hemp, one-half of which has been stained black, or other easily-recognisable colour. The needle is now passed several times backwards and forwards in alternate directions through the base of the tumour, leaving loops at either side. All the white loops are now cut at one side, and all the coloured ones at the other side, and then, by tying the white ones together two and two, and the coloured in the same way at the opposite side of the tumour, the entire thickness is completely constricted. This method, however, is open to the objections above mentioned as applying to the method by single perforation, and it is but seldom employed.

Unquestionably the best way of applying the ligature is by the method introduced at St. Mark's Hospital by the late Mr. Salmon, and which, as Mr. Allingham tells us, has been practised there for nearly half a century, with a record of success quite unparalleled elsewhere. It is best performed as follows: The patient should beforehand have the bowels well relieved by taking an efficient purgative, and immediately before the operation the rectum should be well washed out with a copious soap and water injection. An anæsthetic should now be administered, and the patient placed in the semi-prone position of Sims, or in the lithotomy position. I prefer the latter, and employ Clover's crutch for retaining the patient in position. If, during the administration of the anæsthetic, the piles have become retracted within the

* *Loc. cit.*, p. 180.

anus, no uneasiness need be experienced, as they can be readily brought into view again. The anus should now be carefully stretched by the introduction and gradual separation of both index fingers. As the strain is kept up, the sphincter muscle is gradually felt to relax, and after a minute or two will be found to have lost its tendency to contract. If it is felt relaxing suddenly, great caution is indicated, as laceration is imminent. The entire rectum can by this means be explored, and the lower portion brought well into view. The surgeon should now decide on the number of piles which require removal, and in doing so it is well to remember that internal piles may be much more freely removed, without fear of stricture, than those covered by skin, and surrounding the anus. All distinct tumours, and portions of spongy and thickened mucous membrane should be removed, and the best way is to fix on each condemned piece one of the spring-catch forceps, so commonly used for hæmostatic purposes, and let them hang down in a cluster; in this way there is no danger, when afterwards the parts become obscured with blood, of any of the smaller piles being overlooked or forgotten. Taking up one of the forceps with the left hand, and gently pulling on it, the base of the enclosed pile is made tense, and an incision is made with a scissors at the junction of skin and mucous membrane; and then, by a series of snips, all the *lower* attachments of the pile are severed, and it is dissected up until it is attached only by healthy mucous membrane *above* and the vessels going down into it. During this dissection the bleeding will be but trivial, as the main vessels enter from above, and remain undivided in the pedicle. The forceps is now handed to an assistant to make light traction, and a stout ligature is placed round the pedicle and tied *very tightly;* if tied so tight that the growth is absolutely strangulated at first, much trouble and discomfort during the after treatment will be avoided. Unless complete strangulation has been assured the central portion of the ligatured mass may regain vitality as soon as the constriction by the ligature diminishes. As it cuts into the tissues, this leaves a polyp-like mass, which usually requires a subsequent operation for its removal. It is better to operate on the piles which are situated *lowest* as the patient lies, first, as by this means the further steps of the operation are less obscured by bleeding. The remainder of the forceps are now taken up one by one, and the enclosed piles similarly treated. It is seldom that more than five require ligature, and often only one or two. The ligatures may now be cut off short, and also the piles in front of the ligatures, taking care to leave enough to prevent slipping. This excision also has the advantage of showing whether the ligature has been sufficient to produce complete strangulation. The parts should now be well dusted with boric acid, a morphia suppository ($\frac{1}{4}$ grain) introduced, and a pad of absorbent cotton, with a T bandage or a sanitary towel, firmly applied to the part.

6. **Crushing.**—Without agreeing with Allingham in describing the use of Chassaignac's écraseur as "barbarous and unsurgical," I would express the opinion that the instrument is unsuitable for the removal of internal piles, for the following reasons: In the first place it is difficult to take away exactly what is wanted with it, as it has a tendency to crush more deeply than appears to be the case, while at other times it slips and does not destroy a sufficient amount of tissue; and, again, it is not an invariable safeguard against hæmorrhage, as was admitted by one of its greatest supporters, Nélaton, who stated that it was necessary to use perchloride of iron after it, in order to stop the bleeding.

Mr. George Pollock advocates* the use of a powerful crusher, by means of which he hoped so to destroy the structure of the pile that bleeding would not occur when the crushed pile was cut away; and he recommends a crusher devised by Mr. Benham (Fig. 32), by means of which a very considerable amount of force can be exercised in compressing the pile between the flat surfaces of the blades. Mr. Allingham, jun., has also introduced a screw-crusher, which is stated to answer the purpose well. And Mr. E. Downes has, I think, usefully modified Benham's clamp by making the face of the pincers curved instead of

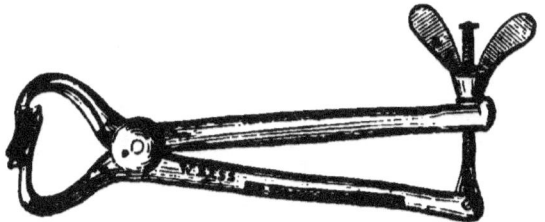

Fig. 32.—Benham's Pile Crusher.

flat.; in this way it is much more easily adjustable. The advantages claimed for this method are freedom from after pain; the absence of any extensive sloughs following the operation; and small size of wound leading to shortened convalescence. The patient should be prepared in the same way as for ligature, and after thorough dilatation each pile to be removed should be caught in the clamp forceps; they are then drawn down one by one, and the crusher applied to the base and screwed up tight. The instrument should be left in place for a minute or two, and then the portion of the pile projecting from the clamp cut away, or, as I prefer, *bruised* off by some blunt instrument, or the Paquelin cautery may be

* *Lancet*, July 3, 1880.

used for the same purpose. The clamp is now taken off, and it will generally be found that no bleeding follows. This is especially the case if the recommendation above given to bruise off the pile has been carried out, as then the vessels remain as shreds effectually occluded. If, however, a vessel of any size is seen to bleed, a fine ligature can be readily applied; in this way all the piles are removed. Care, however, should be taken not to include any skin in the clamp, and in order to avoid this, piles with broad bases may be dissected upwards first, as in Salmon's operation for ligature.

7. **Excision.**—When practised without preliminary dilatation of the sphincter, excision of piles was a very formidable procedure. Concealed hæmorrhage of severe character was a common occurrence, and even death not unfrequently followed, so that it is not to be wondered at that most surgeons soon abandoned it for the safer procedure of the ligature. Amongst the methods which were employed to obviate hæmorrhage may be mentioned the attempt of Colles (not always successful) to leave the divided stump within the grip of the sphincter; of Boyer, to retain command of the divided portion by means of a suture passed through the base; and of Druitt, who passed a long needle through the base before excision, by means of which the stump was retained outside the anus for twelve hours after the operation. None of these measures, however, proved satisfactory, and it is only within recent years that any attempt was made to place the operation on a scientific basis; but now the operation as practised by Mr. Coates, and the operation of excision as advocated by Mr. Whitehead, are recognised as amongst the best means of dealing with internal piles.

In the address on surgery delivered at the meeting of the British Medical Association at Ryde, 1881,

Mr. Coates, senior surgeon to the Salisbury Infirmary, advocates a means of removing internal piles with a specially devised clamp (Fig. 33). The patient being prepared in the way already indicated, and the catch forceps applied, each pile is drawn down, and the little clamp applied; a few sutures of the very finest catgut are now passed underneath the clamp, and.

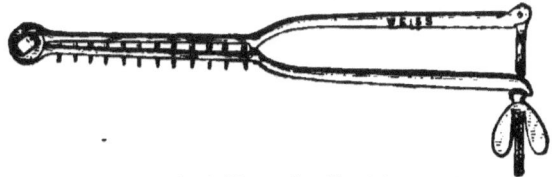

Fig. 33.—Coates' Clamp for Excision of Piles.

the pile cut off. The clamp is now opened a little, the needles which have transfixed the growth preventing retraction. Any vessel requiring it is now ligatured, and as soon as all bleeding is stopped the clamp is removed and the sutures tied. In this way a cleanly incised and evenly brought-together wound is substituted for the ligature, or for the bruised or the cauterised wounds remaining after other operations. The arrangement of the clamp is very ingeniously devised for the purpose of preventing retraction; and its usefulness for other surgical purposes will at once suggest itself to the practical surgeon.

Mr. Whitehead describes his operation * as a combination of excision with torsion; it is suitable for very severe cases. In the first case described by him there was a mass of piles protruding from the anus, which he compares in shape and size to a moderately large tomato. He divided the mass into four segments by vertical incisions through the thickened mucous membrane; each of these segments was now dissected up from the deeper coats of the rectum till it was left attached only by healthy

* *British Medical Journal*, vol. i. p. 148; 1882.

mucous membrane, and the larger vessels entering from above this were now twisted through, and the other segments dealt with in the same way; the healthy mucous membrane was brought down and sutured to the anal margin. This operation is a decided improvement on the method of Jessop, who simply twisted off the piles without preliminary dissection.*

At the meeting of the British Medical Association at Brighton, 1887,† Mr. Whitehead brought forward 300 consecutive cases of hæmorrhoids, which he had treated by excision alone, and he describes his operation as follows: The patient is to be retained in the lithotomy position; the sphincters thoroughly dilated. By means of a dissecting forceps and scissors the mucous membrane is divided at its junction with the skin round the entire circumference of the bowel, every irregularity of the skin being carefully followed. The external, and the commencement of the internal sphincters are then exposed by a rapid dissection; and the mucous membrane and attached hæmorrhoids, thus separated from the submucous bed on which they rested, are pulled bodily down, any undivided points of resistance being snipped across, the hæmorrhoids being brought below the margin of the skin. The mucous membrane above the hæmorrhoids is now divided transversely in successive stages, and the free margin of the severed membrane alone is attached, as soon as divided, to the free margin of the skin below, by a suitable number of sutures. The complete ring of pile-bearing mucous membrane is thus removed. Bleeding vessels throughout the operation are treated by torsion.

In support of this operation, Mr. Whitehead contends that internal hæmorrhoids, which are generally regarded as localised distinct tumours amenable to

* *Med. Times and Gazette*, vol. ii. p. 281; 1871.
† *British Medical Journal*, Feb. 26th, 1887.

individual treatment, are as a matter of fact component parts of a diseased condition of the entire plexus of veins, associated with the superior hæmorrhoidal vein, each radicle being similarly, if not equally, affected by an initial cause, constitutional or mechanical; and he believes that when surgical treatment becomes imperative, the extent of the mischief can only be appreciated and effectively dealt with by a free exposure of the diseased vessels; and that no operation for hæmorrhoids is complete which does not remove the entire pile-bearing area.

Although entirely dissenting from the views of Whitehead, that excision should be adopted to the exclusion of all other methods of surgical operation, I believe it to be the best form of procedure for aggravated cases.

I have now adopted excision in upwards of seventy cases, an experience which has evolved some little points of procedure which may prove of use to those surgeons who have not as yet practised this operation extensively.

In the first place it is necessary to understand clearly where the circular incision through the skin of the anus is to be situated; indeed there appears to be some considerable ambiguity about what is meant by the *margin* of the anus in describing any pile operation. If the reader will refer to Plate IV. he will see that in both figures the prolapsed mass consists of two more or less concentric rings of tissue. The inner of these is formed by the internal piles proper, while the outer is the invaginated skin of the anal canal, with possibly some external piles. Even where there are no external piles, the invaginated anal canal forms a definite thick ring round the internal piles when they become prolapsed. Occasionally as a result of forcible dilatation preliminary to operation, subcutaneous extravasation of blood makes this ring swell up to a considerable size. It is necessary clearly to understand that the anus is not merely an orifice, it is a canal about one inch long in the adult, and that when internal piles are protruded, invagination of this canal takes place. When speaking of the margin of the anus some surgeons mean the outer circumference, while others undoubtedly mean the inner circumference of this ring. It is obvious that there is a wide difference in the result of operations according to which interpretation is put on the word margin. The outer circumference corresponds with the lower outlet of the anal canal, while the inner is at the muco-cutaneous junction. I am decidedly of opinion that the best position for the incision in the operation of excision of piles lies between these two lines. In consequence of repeated prolapse of piles with invagination of the anal canal the skin lining this canal becomes somewhat redundant, so that if the incision is carried round exactly at the muco-cutaneous junction rather too much skin is left. This is not a matter of much consequence if primary union follows, but if the edges do not at once unite, the skin margin is apt to become everted and leave troublesome little tags of external piles. If on the other hand the entire skin of the anal canal is removed and primary union results, some of the mucous membrane is left exposed and causes irritation when the clothes rub against it, whereas if the primary union does not take place and the wound heals by granulation, stricture of the anus to an unpleasant extent may result.

I would recommend the circular incision to be made at the most dependent portion of the prolapsed anal canal; this leaves quite two-thirds of this structure, but removes all the little irregularities of the muco-cutaneous junction. For this purpose sharp-pointed scissors is the best instrument.

METHODS OF EXCISION.

Having made a clean cut round the entire circumference the diseased tissue is rapidly separated from the external sphincter and muscular coat of the rectum, and the separation is carried up until quite above the piles all round. Catch forceps may be applied to any large bleeding vessels. The mucous membrane is now cut through circumferentially bit by bit, one portion being accurately adjusted to the skin before the next piece is cut. I find fine catgut hardened by being preserved in absolute alcohol the best suture material; and it is a good plan to carry the needle twice, once rather deeply and then merely through the edges of skin and mucous membrane respectively, before tying the knot of the suture. If care is taken in passing the deeper portions of the suture, bleeding will be completely arrested, and ligatures to the arteries or torsion will be unnecessary. After the entire circumference has been in this way sutured, any bleeding that may remain can be completely arrested by the application of one or two additional points of interrupted suture passed deeply. The anus should now be covered thickly with boric acid, some of this powder being placed within the rectum and a firm antiseptic pad and bandage applied, the subsequent treatment being the same as that adopted after ligature or cauterisation.

There is a condition of the rectum sometimes met with which can scarcely be included under the denomination of piles, but which I can most appropriately refer to in this place. It is a condition of the rectal mucous membrane in which no distinct tumour is present, but which is attended with copious hæmorrhage and much discomfort to the patient, and which, in my estimation, is best treated by excision of the mucous membrane. It consists of a spongy and honeycombed condition of the entire surface of the mucous membrane extending up the rectum for a distance of about one inch from the anus. It is intensely vascular, and bleeds upon the very slightest provocation. I think it probable that this condition commences as an ordinary case of piles, and gradually the entire circumference of the mucous membrane becomes involved. It will generally be found upon inquiry that the disease has existed for a very long time, and sometimes we find that the patient has been operated on by cautery or ligature without permanent benefit. In one case of this kind that came under my care the patient had been operated upon four times by other surgeons; and in another the clamp and cautery, three times employed, had failed to effect a cure. In both of these cases, after thorough dilatation, I dissected away the entire ring of diseased mucosa and then sutured the healthy mucous membrane to the anal margin, with complete and, so far, permanent success.

After complete stretching of the anus, hæmorrhage can be as easily arrested by direct ligature of the bleeding vessels as in almost any other part of the body. It may be urged against Whitehead's operation, and the method I have described, that stricture of the rectum might be a likely result of the removal of the entire circumference of mucous membrane. Such, however, is not the case if the incision does not encroach too much on the skin of the anus on the one hand, or penetrate deeper than the submucosa on the other.

8. **Actual cautery.**—The use of the cautery as a means of curing piles dates back to a very remote period, and after being practically abandoned for some time, it has again assumed a definite position as one of the best means of treating this disease. This is chiefly due to the writings of Mr. Lee * and Mr. Henry Smith,† both of whom, by means of a

* See *Lancet*, vol. i. pp. 74, 541.
† *Lancet*, April 20th, 1878.

special clamp, limit the action of the cautery exactly to the part requiring removal, which method also has the advantage of retaining complete command over the stump until the surgeon is assured that all bleeding is at an end. Many varieties of clamp

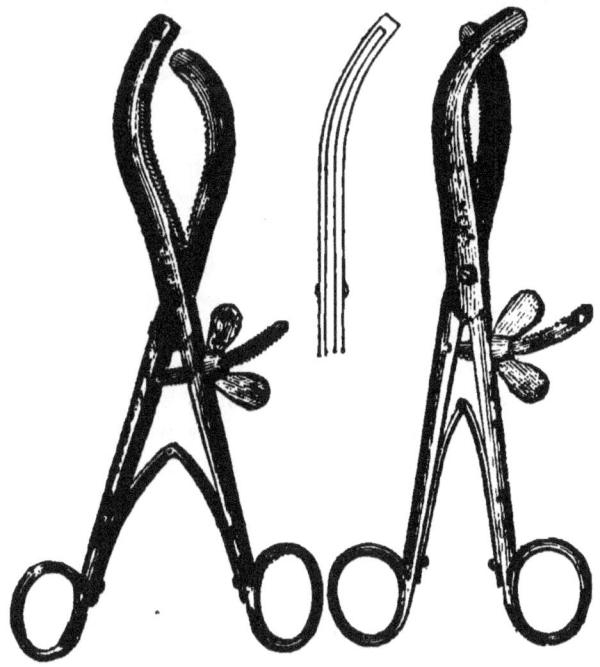

Fig. 34.—Lee's Pile Clamp.

have been introduced, the early patterns acting with a pivot at the end, so that the blades opened like a pair of compasses. This has one great objection that it compresses the pedicle unequally, so that when the blades are opened to see if the hæmorrhage is stopped, the side of the pedicle nearest the handle is apt to slip from the grasp of the instrument and continue to bleed. This difficulty is obviated in Smith's clamp by a parallel action of the blades. His instrument is, however, in my estimation, too broad for convenient application between the nates on either side; it is, moreover, provided with broad plates of ivory, in order to prevent singeing of the skin, either by radiation from the cautery or by heat conducted through the metal clamp. If, however, as is almost now invariably the case, that exceedingly ingenious instrument, the Paquelin benzoline cautery, be used, these plates of ivory are quite unnecessary, as there is no radiation likely to produce injury to the surrounding skin, and the mass of metal, even in the most lightly constructed clamps, is such that it will not become unpleasantly heated during the short time that the platinum point is in connection with it. The clamp I always use for this purpose is Lee's pattern (Fig. 34); the narrow and curved blades are readily applied to the anus; and it compresses the pedicle into a smaller space, and consequently leaves a smaller wound than any of the others.

In order to perform this operation the patient should be prepared as before described (page 276), and the same steps taken until each pile to be removed is caught in a catch forceps; they are now drawn down one by one, and the clamp applied to the base, and screwed up tight; the pile may be cut away now, leaving about one-eighth of an inch in front of the clamp to be charred; the cautery, at a dull red heat, is now passed over this several times, until it is completely burnt away down to the clamp; the blades are now gradually relaxed to see if there is any bleeding; if a vessel should spring, the clamp is screwed up, and the cautery brought into requisition again; if no bleeding is to be seen, the clamp is opened completely, and applied in the same way to the other piles until all are removed. In performing this operation, instead of cutting away the pile with a scissors or knife, it may be *dug* away with the blunt platinum point of the cautery, and when this is done there is usually less liability to subsequent bleeding.

Besides the methods of using the actual cautery above detailed, there are some others which, although not much practised, deserve a passing notice. Woillemier,* after discussing the various operations, recommends as the best the making of four vertical lines of cauterisation, involving the anus and extending about one centimetre up the rectum; they are made with iron knife-shaped cauteries heated in the fire. No attempt is made to destroy the tumours themselves, it being apparently by the contraction of the cicatrix that a cure is looked for. Woillemier points out that no fear of stricture need be entertained, because as the four lines are made at equal distances round the circumference of the anus, there is a band of normal and expansible tissue left between each cicatrix, which will ensure a sufficient amount of dilatability. In order to prevent burning by radiation, he advises that the surrounding skin be painted with collodion, and that an assistant should blow away the fumes with a bellows, lest the ether vapour become ignited by the heat of the cautery. The convalescence was somewhat long, as might be expected after such a procedure; and the author admits that it was not always successful. However admirable a similar procedure may be for the treatment of prolapse, it does not appear likely that this method as a cure for piles will ever be as popular as some of the others detailed.

Under the title "igni-puncture," Mr. Reeves details an operation which he describes as a new and immediate method of curing piles.† It consists in the thrusting a fine point of Paquelin's cautery into the centre of each pile, hoping thereby to produce complete consolidation of the whole tumour.

* *L'Union Médicale*, 1874; and *Med. Record*, p. 303; 1874.
† *Lancet*, vol. i. p. 229; 1877.

Allingham very justly criticises this method, and shows that it is neither new nor immediate. He states that it is common amongst the native doctors of India and China. It appears to be open to the same objections that apply to all forms of punctured wounds. In my own hands I have found it followed by abscess; and other observers record similar experience.

The galvanic cautery may be used for the removal of piles when it can be obtained, but the battery efficient to work it is cumbrous to carry about, and requires considerable attention to keep it in good working order. It possesses no advantages, in my mind, over the Paquelin cautery in any case, and is not at all so generally applicable.

From the large number of methods of operation which have been described, it will be obvious that the surgeon has considerable variety of choice; and if he elect ligature by Salmon's method, the use of clamp or cautery, crushing for any moderate case of piles, he cannot be far astray. For extensive piles excision is undoubtedly the best treatment; while for the small venous pile and nævoid pile, nitric acid or electrolysis will generally prove sufficient. The champion of each special operation invariably states that his particular plan is attended with greater freedom from after pain than all the others; but I think the degree of after pain met with depends rather upon anatomical grounds than upon the form of operation selected. If the anal margin is encroached upon, or *becomes inflamed* after operation, the pain will be considerable; whereas, if the operation wound lies entirely within the sphincter, and remains un-inflamed, after pain will be but trivial.

The **after-treatment** of operation must necessarily be a matter of importance. As soon as any of these operations have been performed, the surface should be well covered with boric acid; a morphia suppository of ¼ grain introduced; and a firm pad of absorbent cotton or a sanitary towel should be firmly applied to the anus. This is a decided comfort to the patient, and tends much to relieve the painful spasm of the levatores ani, which is one of the most distressing sequelæ to operations for piles. After the patient has been removed to bed, he should be watched for some time by an efficient nurse. If bleeding should occur externally its presence can easily be ascertained; and although, since the introduction of forcible stretching of the sphincter, it usually does show itself externally, even still occasionally internal bleeding will go on to a considerable extent. This is to be recognised by increasing pallor, restlessness, a sensation of heat in the rectum, and desire to defæcate; and with this there will be the typical hæmorrhagic pulse. Should these symptoms be present, the anus is at once to be examined, and the rectum, if full of clots, to be emptied. If dilatation has been properly performed in the first instance, it will usually be easy to see where the bleeding is coming from, and to apply a ligature. If, however, the bleeding is out of reach, a piece of ice may be left in the rectum; and if this fails, the rectum must be plugged in the way recommended by Allingham, or by the fiddle-shaped indiarubber bag recommended by Benton for this special purpose. After the effects of the anæsthetic have passed off, the patient may be allowed his usual diet. Stimulants, unless specially indicated, are best avoided. The bowels should be kept confined by occasional doses of opium till the fourth day, when an aperient may be given, followed by an injection of warm oil, and care must be taken to avoid constipation again occurring. About the end of the first week, ligatures, if properly applied, come away; and the sloughs following cautery and crushing about the same time; there remain, however, ulcerated surfaces, which, if the case is at all severe, will take at least another week to heal. Until the ulcers are quite well it cannot be said that the patient is perfectly recovered. There is always the danger, if he goes about too soon, that these ulcers will become chronic. This can readily be understood when the fact that the rectal veins are destitute of valves, and in cases of piles frequently dilated, is remembered. On the other hand, we see patients able to go about apparently without harm from a few days after the operation. No hard-and-fast rules can, therefore, be laid down, and the surgeon must form his own opinion on the merits of each individual case.

The immediate dangers of the operation are, indeed, slight; in fact, it can be safely asserted that no surgical procedure worthy of the name of operation is attended with a better record of success. The statistics of St. Mark's Hospital, which are much the largest available on this subject, show that the death-rate during the past forty years was only at the rate of 1 in 670 cases. A study of these statistics also shows how and with what small reason the supposed danger of tetanus following ligature came to be so universally credited and feared. According to Allingham, the total number of cases operated on up to 1881 was 4,013, of which only five suffered from tetanus; and in looking into the matter, it is seen that four out of the five cases occurred during the months of March and April, 1858; and as Curling has pointed out, tetanus was frequent in the other London hospitals during that time. Mr. Allingham's personal statistics (1,600 cases, without a single fatal result) represent certainly a very brilliant achievement, and one, as far as I know, quite unequalled elsewhere. As might be expected, the records of the general surgical hospitals, which include the period antecedent to the general introduction of antiseptic surgery, show a slightly greater rate of mortality; but even with this allowance the number of casualties has been indeed small. In my own practice I have never been even seriously alarmed about a patient operated on for piles.

Of the more directly septic poisonings resulting from the operation for internal piles, we find but very few examples; however, cases of venous thrombosis, septic periproctitis, peritonitis, and general metastatic pyæmia, are occasionally met with. Of progressive venous thrombosis, an interesting case is recorded by Ewen : * he excised a pendulous pile from a woman aged forty-eight years, and it was followed by phlegmasia dolens of the left leg, with distinct swelling and tenderness over the femoral vein.

When we consider that operations on varicose veins in the leg, on varicocele, and the ligature of nævi, especially in the pre-antiseptic period, furnished a very appreciable percentage of septic complications, it is extremely strange that such a highly vascular structure as an internal pile can be excised, cauterised, or ligatured with almost absolute impunity. This is all the more noticeable when we remember that no efficient antiseptic dressing can be kept to a part which is continually disturbed by the passage of fæces, and that the intestinal contents are in constant contact with the wound. The unquestionable authenticity of the fact, however, would appear to indicate that fæces are usually tolerably free from active pathogenic micro-organisms.

Amongst the slighter troubles following pile operation, retention of urine must be noticed; but since the introduction of forcible dilatation this is much less frequent. It is very much more common in the male sex, but it is seen sometimes in the female : it is probably due to spasm of the anterior part of the levatores ani muscles. A hot stupe over the lower portion of the abdomen is generally sufficient to overcome the difficulty, but if not, of course a catheter must be

*British Medical Journal, vol. i. p. 616; 1863.

passed. It may be necessary to repeat this for a few days, but the bladder soon recovers itself.

Convalescence may be protracted by the onset of sloughing or ulceration; the latter is especially liable to occur when the patient has been allowed up too soon; and sometimes, if a small nerve twig happens to be exposed, the wound degenerates into a true irritable ulcer of the rectum, requiring a second operation for its relief. I have, in my own practice, had two instances of this. If sloughing should become at all extensive, it is a serious complication, as severe secondary hæmorrhage may result if the slough happen to open a large artery; and if the destruction of the deeper rectal coats ensue to any great extent, very formidable stricture of the rectum may result.

Abscess occurring in the ischio-rectal fossa, and possibly terminating in fistula, is also an occasional complication.

If the skin surrounding the anus has been too much encroached upon, stricture of the anus will be an almost inevitable result, but if due caution is exercised this should never occur, except as a result of severe phagedænic ulceration or extensive sloughing.

Although it has been necessary to enumerate all these possible complications, I again repeat that no operation in the whole range of surgery is attended with less risk, if due caution is observed, than the removal of internal piles.

It remains now to consider what prospect there is of permanent recovery, and whether a rectum once cleared of piles is likely again to become diseased. Much, of course, will depend upon the form of operation selected, and the way in which it is carried out; but, perhaps, more will depend upon whether the piles have been but a symptom of some permanent obstruction of the rectal circulation, as disease of liver, uterine complications, stricture, etc. Frequently there is a slight amount of bleeding the first few times the bowels are moved after operation, and the patient is thereby much distressed. The surgeon, therefore, should always caution a patient about this probability, at the same time assuring him that it is a matter of no consequence, and that it will almost certainly soon subside. From my own experience of cases operated on some few years ago, which I have been able to keep under observation, I can safely say that when the operation has been properly performed in uncomplicated cases, absolute recovery is the rule, and recurrence quite exceptional. One interesting point I have frequently observed, and I have little doubt but that the observation is familiar to other surgeons : when a few small piles have been removed from an intensely congested rectum, with spongy mucous membrane, and when there is an opportunity afforded of seeing this case some months afterwards, it will usually be found that the mucous membrane has quite regained its normal condition; that its surface is smooth; and that the great congestion has disappeared. Now, it has always appeared to me that the mere removal of the mechanical irritation of a few small piles was not in itself sufficient to account for this change. Can it be that the formation of a cicatrix, necessarily following the removal of a pile, has any effect upon the rectal circulation? As is well known, the formation of a cicatrix in iridectomy and other similar operations has a very marked effect in lowering intra-ocular tension in cases of glaucoma. Many have been the exclamations put forward to account for this result, and perhaps one of the most plausible is that of Exner,* who has been able to show, by a series of carefully conducted injections, that direct communications between somewhat large arteries and veins are formed in the stump of

* Schmidt und Graefe, "Handbuch," Band v. p. 121.

iridectomy wounds. Hitherto I have not had an opportunity of examining pathologically cases of piles cured by operation, but when the undoubted fact is remembered that the presence of an operation cicatrix in the rectum does relieve congestion, it appears only reasonable to assume that a similar explanation to that given by Exner may be accepted in this case.

CHAPTER XX.

BENIGN NEOPLASMS OF THE RECTUM AND ANUS.

THE word "polypus," which has descended from a remote period of surgical history, is used with much vagueness in reference to all parts of the body; but in none is this more noticeable than in the rectum : by what process I know not, it has come to be used as a term for any neoplasm attached by a narrow pedicle. And as the greater number of benign neoplasms of the lower bowel answer this definition, the term "rectal polypus" is held by many writers to include the various pathological formations included under the heading of this chapter. As, however, the term is generally used, it would be inconvenient to discontinue it, and when used it must simply be held to imply that the growth alluded to is more or less pedunculated. Pedunculated growths are not confined to the rectum, but are found all through the intestinal tract. According to Leichtenstern,* their relative frequency in the various situations of the intestine is as follows :

Rectum	...	...	...	75	Ileum	...	...	...	30
Colon	...	...	...	10	Jejunum	...	...	5
Cæcum	...	...	...	4	Duodenum	...	...	2
Ileo-cæcal valve	...	2				

* Ziemssen's Cyclopædia, vol. vii. p. 634.

The principal forms of benign neoplasms met with in the rectum belong to one or other of the following varieties of outgrowths (or so-called polypi), viz.: (1) adenoma; (2) fibroma; (3) papilloma; (4) teratoma;

Fig. 35.—Adenomatous Rectal Polypus; natural size (Pozzi).

(5) lipoma; (6) cystoma; (7) enchondroma; (8) angioma; (9) myoma; (10) lymphoma.

1. **Benign adenoma** is one of the most frequently met with forms, and is especially found in young children, but at the same time it cannot be said to be a common affection: unless, indeed, as is possibly

the case, many instances undergo spontaneous cure without coming under the notice of the surgeon. In size these polypi vary from that of a pea to that of a walnut, although sometimes they may attain a much larger size, as in Fig. 35, in which the growth measured eight centimetres in height by seven centimetres in width, and was removed by Trélat from an adult man. The case is reported in full by Dr. Pozzi, in an interesting paper on rectal adenoma, published Oct. 25, 1884, in the *Gazette Médicale de Paris*, and I am

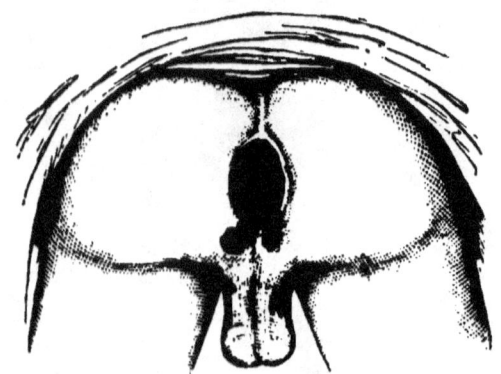

Fig. 36.—Double Rectal Polypi producing Prolapsus Recti.*

indebted much to him for the use of the illustrations. The shape is usually more or less globular or pyriform, although occasionally these outgrowths may be quite cylindrical. I removed one of the latter variety recently, which was about an inch long and a quarter inch in diameter; the end was blunt-pointed, the whole growth resembling the extremity of an earth-worm. The surface is usually irregularly lobulated, and when looked at closely, appears roughly granular. It is attached to the wall of the gut by a narrow pedicle of mucous membrane, which contains the vessels for the supply

* Richmond Hospital Collection.

Chap. XX.] *POLYADENOMATA.* 299

of the growth. These are sometimes of considerable size, so that arterial pulsation can be felt with the finger; the colour is usually bright red; or if the growth has been caught by the sphincter, it may be livid from congestion, or even gangrenous. Although usually single, adenomata may sometimes be more numerous, being sometimes double (Fig. 36), while in comparatively rare instances the entire surface of the colon may be covered with them. Such cases are described by Luschka,* Billroth,† and Richet.‡ Cripps mentions two cases as having come under his personal observation, in which a brother and sister suffered from this condition.§ And he states that a careful search through the pathological museums of London has only resulted in finding three specimens, those at the Middlesex and King's College Hospitals being peculiarly remarkable (Figs. 37 and 38).

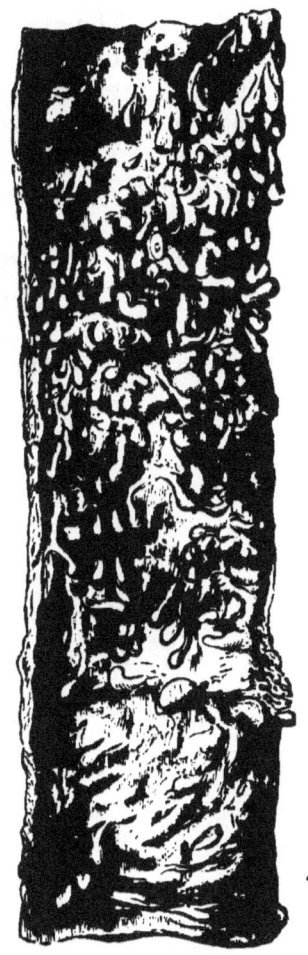

Fig. 37.—Multiple Polypi of Rectum. ‖

I have had experience of one case only, a male aged

* Virchow's *Archiv*, vol. xx. p. 133.
† Langenbeck's *Archiv*, 1869.
‡ "Traité Pratique d'Anat. Méd. Chirug." Fourth edition, Paris, 1873.
§ *Loc. cit.*, p. 276.
‖ Museum of Middlesex Hospital.

fifty years. The symptoms were very trivial; a digital examination was instituted on account of a very slight bleeding, which revealed the fact that the entire mucous membrane within reach of the finger was studded with numerous polypi that had probably been there for a very long time; no treatment was adopted.

Fig. 38.—Multiple Polypi of Rectum.*

The most usual position for the origin and consequent attachment of the pedicle is about two inches above the anus. As the pedicle gradually elongates, the tumour comes to touch the anus, and at last to be protruded during defæcation. When this stage is reached, the pedicle may be caught by the sphincter, and the growth strangulated; or by the forcible retraction of the rectum, or the passage of a hard motion, the pedicle may be ruptured, and the

* Museum of King's College Hospital.

neoplasm voided. It is probable that this happens tolerably frequently, as, the tumour being devoid of sensation, the patient may be alike unconscious of its existence and its loss. A woman brought her child, aged four years, to me at Sir Patrick Dun's Hospital, saying that she had just passed a lump of flesh. Upon examination this proved to be a well-marked adenoma of about the size of a large walnut; the pedicle had become torn away at its attachment to the growth, and had evidently been very narrow. The amount of bleeding was slight, and a digital examination failed to find any trace of the pedicle. In another instance a medical student felt something protrude from his rectum after defæcation, and as it was absolutely devoid of sensation, he caught it between his finger and thumb and pulled it away, thinking it was a foreign body: then followed a little bleeding, but not to any dangerous extent. I afterwards examined this growth, and found it a typical adenoid tumour, about the size of a filbert-nut.

This is the form of polypus which is undoubtedly the most common in children. Why so, it is not easy to say; but we have a parallel example of their greater liability to benign neoplastic formations in the frequency with which simple warts are met with on the hands of young persons. The greater observed frequency of these growths in the rectum, as compared with the rest of the intestinal tract, very probably is to some extent more apparent than real, as it is in the former position alone that diagnosis during life is possible. As, however, the rectum is more exposed to direct irritation, it would be more reasonable to suppose that the mucous membrane there would suffer most. That adenomata may result from direct irritation is, I think, indicated by the frequency with which small polypoid adenomata are met with in connection with ulcerating rectal cancer. (*See* page 306.) Recently their connection with parasitic irritation has been noticed by Dr. Belleli,* of Alexandria. The bilharzia hæmatobia, which lives in the portal, the mesenteric, and splenic veins, and the veins of the rectum and bladder, deposits ova that block up the small vessels of the bladder and rectum, in the former case giving rise to what is known

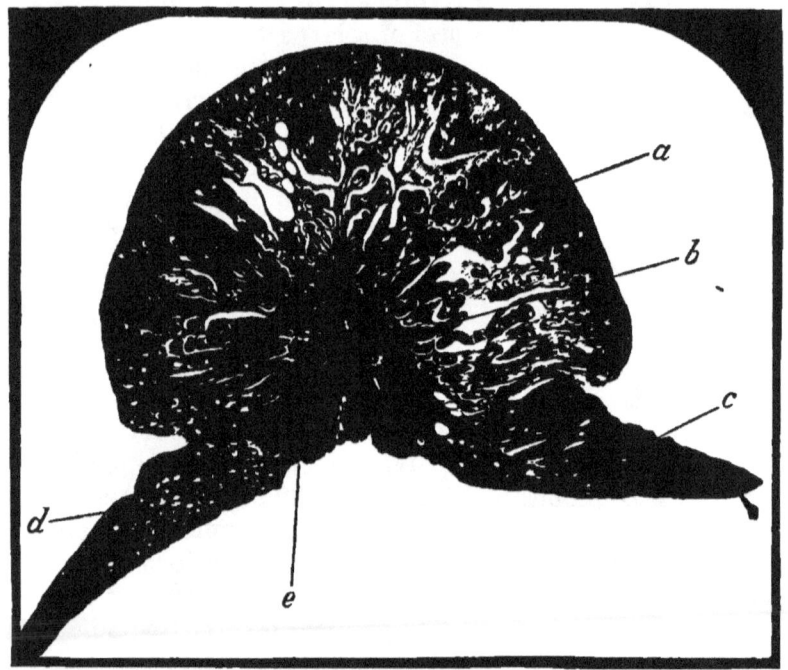

Fig. 39.—Adenoma of Rectum. (× 10 diameters.)
a, Glandular structure; *b*, hilum of connective tissue and vessels; *c*, mucous membrane cut longitudinally; *d*, mucous membrane cut transversely; *e*, solitary gland.

as "endemic hæmaturia," common in Egypt and other parts of Africa, and in the latter causing hæmorrhage with the stools, and other dysenteric symptoms. Dr. Belleli states that ova with polypi occur throughout the entire intestinal tract, but are especially frequent in the rectum; and in one case of adenoid polypus,

* *Progrès Médicale*, No. 30, 1885.

Chap. XX.]   STRUCTURE OF ADENOMA.   303

the size of an apple, numerous groups of ova were to be seen between the little masses of epithelial cells.

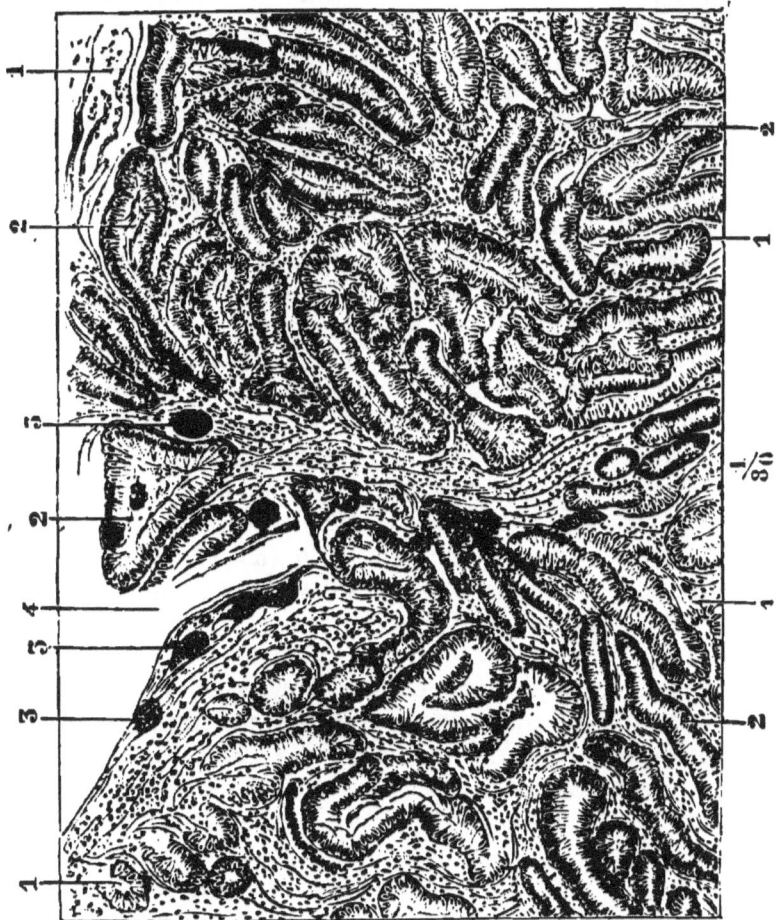

Fig. 40.—Section of Adenomatous Polypus (× 80).   Pozzi.

1. Connective tissue between glandular *culs-de-sac;* 2, glandular *culs-de-sac* cut in various directions; 3, blood clots; 4, irregularities of the surface of the tumour.

Further observation, however, appears to be necessary in order to determine whether there is any causal relation between this trematode parasite and rectal polypi.

The pedunculated adenoma, when occurring in the young, may be taken as the type of a purely benign growth, and when removed is almost certain not to recur again. In the adult, however, Van Buren states that he has seen* polypoid growths of this nature which returned after removal as undoubted epitheliomata; but this experience does not appear to have been borne out by other observers.

Formation of adenomata.—The illustration (Fig. 39), which was taken from a child who came into hospital with a prolapse at the apex of which two small polypi were forming but as yet no pedicles had appeared, shows well the method of formation of an adenoma. A transverse section has been made through the centre of the hilum and through the mucous membrane on either side of the polyp; on the right side (at c) the section has passed vertically through mucous membrane, while on the left side (d) the mucous membrane has been cut horizontally; the section also traverses a solitary gland (e). The figure is from a microphotograph kindly made for me by Professor J. A. Scott. The first step in the production of polypus is an increase in length and hypertrophy of the follicles of Lieberkühn (c, Fig. 39), and at the same time the cylindrical cells lining these tubes become elongated and enlarged. The result of this is that the mucous membrane at this point buckles up and projects as a tumour. As the follicles become increased in number and size they become more and more convoluted (Figs. 40, 41). The tumour now offers some resistance to the passing mass of fæces, and becomes dragged down, the circular fibres of the bowel also assisting to expel the tumour. The pedicle of mucous membrane is thus formed.

* *Loc. cit.*, p. 329.

HISTOLOGY OF ADENOMA.

If a section be made passing through the hilum of one of these little growths, it will be found that,

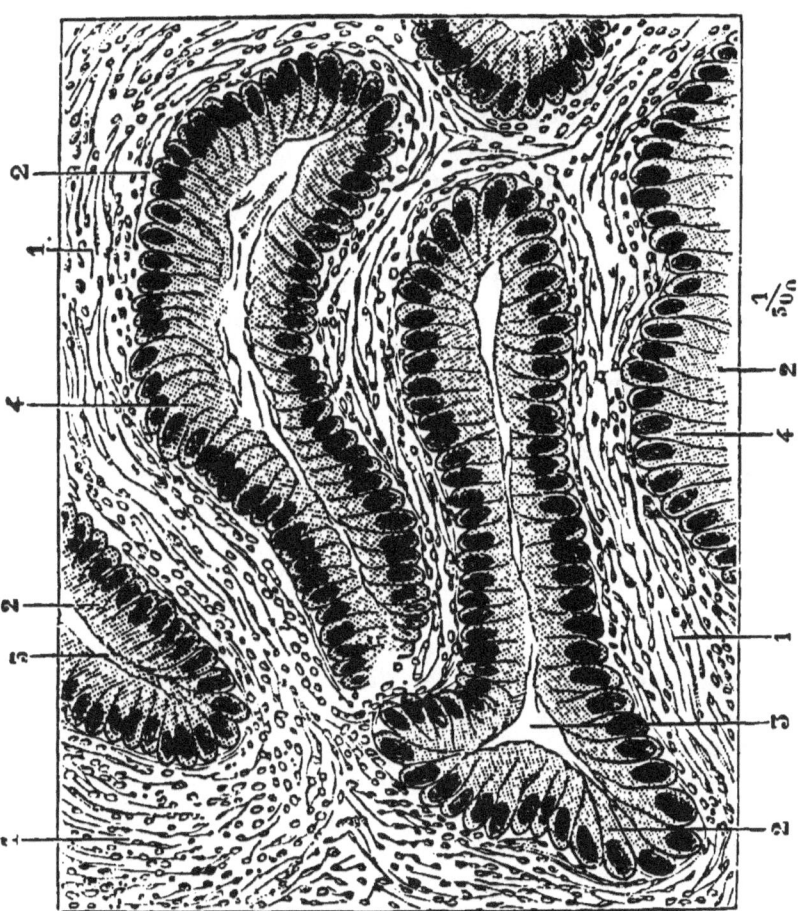

Fig. 41.—Section of Adenomatous Polypus (× 300). Pozzi.
1, Connective tissue between glandular acini with numerous embryonic cells: 2, hypertrophied follicles; 3, cavity of the follicle; 4, enlarged and irregular shaped epithelial cells.

starting from that point, bands of connective tissue radiate in all directions to the circumference of the polypus, and between these are numerous secondary offshoots ; the little spaces thus left are packed with Lieberkühn's follicles considerably larger than normal; the section appears as if these were closed cavities, but in all probability this appearance is produced by the line of section traversing convoluted tubes ; and this probability is rendered greater by the fact that numbers of these tubes open on the surface of the polyp, the roughly granular aspect of which is due to the numerous orifices.

The idea of the progress and growth of a rectal adenoma, conveyed in the above description (*i.e.* the idea of tendency to outgrowth into the lumen of the bowel in the direction of least pressure ; and the production and elongation of the pedicle as encouraged by the cumulative action of the constant intestinal peristalsis) would appear to apply more or less in general to the pathological history of all benign neoplasms occurring in the rectum (or perhaps I might more correctly say, in all parts of the intestinal canal), and it would seem to supply us with the rational explanation of the reason why these growths almost universally assume the polypoid or pedunculated character ; whereas, those neoplasms which are of a malignant type have no such proclivity, on account of the obvious contrast they present in their more marked invasion of, and adhesion to, surrounding tissues, and their consequent much smaller liability to the influences of mobility.

It is for this reason, too, that the names " benign neoplasm " and " polypus " have come to be almost, if not altogether, synonymous and interchangeable terms, as applied to tumours of the rectum. It is in this sense that the relation of the two terms to one another, as employed in this work, should be understood.

Complications.—In the upper portions of the intestine, rectal adenoma is not an uncommon cause of intussusception, and similarly in the rectum prolapse is sometimes met with as the result of reflex spasm of the muscular coat of the bowel in an effort to extrude the growth. Obstruction is rarely caused by simple adenoma. Esmarch, however, gives a case in which an enormous mass, 4 lbs. in weight, blocked the whole lower bowel.* Ulceration of the surface of the growth, though occasionally noticed, is uncommon.

Symptoms.—Unless when it is attended by some of the above complications, polypus, when situated high up, is marked by no definite symptoms; when, however, it is attached low down, or comes to touch the anus by gradual elongation of the pedicle, its presence gives rise to some sensation of fulness and distress in the lower bowel; and, as a result, constant efforts at expulsion, with tenesmus, are present; there is more or less discharge of glairy mucus; and if the growth has come low enough to be nipped by the sphincter, bleeding will result: in fact, bleeding is frequently the first indication of the presence of a polypus; and, in children especially, rectal hæmorrhage is strong presumptive evidence of adenoma.

The **diagnosis** of this disease is usually easy, piles being the only other rectal disease liable to be mistaken for it. In the latter case the patient is usually older, the tumours multiple, attached lower down, and by a broader base; the surface smooth, and not so distinctly lobulated; and the mucous membrane of the termination of the lower bowel more congested and involved. In examining a patient digitally for suspected polypus it is well to follow the advice of Mollière, and pass the finger first up to its fullest extent, and then gradually to withdraw it, sweeping it round the entire rectal surface. In this way the pedicle will be hooked by the finger, and the polypus found; if, however, the examination is conducted from

* *Loc. cit.*, p. 176.

below upwards, the tumour may be pushed up out of reach.

Treatment.—Small polypi, situated near the anal margin, may be with safety twisted off with a catch forceps; but those attached higher up should always be ligatured and snipped off. This is especially necessary where the tumour is large and the pedicle thick, as, in addition to containing a tolerably large vessel, the entire thickness of the rectal wall *may* be invaginated in it, carrying with it, possibly, a process of peritoneum.

2. **Fibroma.**—Tumours of the connective tissue type are sometimes found in the rectum, attached by pedicles, and constituting some of the varieties of the so-called fibrous polypi. The small growths so frequently found in connection with fissure, and situated at its upper extremity, are of this nature; but in many cases I have no doubt that the origin of fibrous polypus is directly from internal piles. If a series of cases of internal piles are examined it will be found that fibroid change gradually takes place and becomes more marked in proportion to the duration of the disease, until finally the dilated veins entirely disappear, nothing but soft fibrous tissue remaining. When this has occurred the tendency of the growth is to become pedunculated. Fig. 42 represents a case of this kind: in the upper portion is a typical fibrous polypus attached by a broad pedicle to ordinary pile structure; it was removed with four other similar growths from the rectum of a woman who had suffered from piles for upwards of twenty years. I have a series of pile sections taken from a large number of different cases showing all gradations of this fibroid change by which an internal pile becomes a fibrous polypus. The clinical evidence of this change is in the colour and texture of the pile, both of which gradually become modified. Where very old piles exist the bright red colour is lost, and when the fibroid change is

complete they assume a yellowish-white appearance, the mucous membrane becoming more like skin, at the same time they become firmer and more solid to the touch. The removal of the growths is to be conducted on the same lines as operations for internal piles.

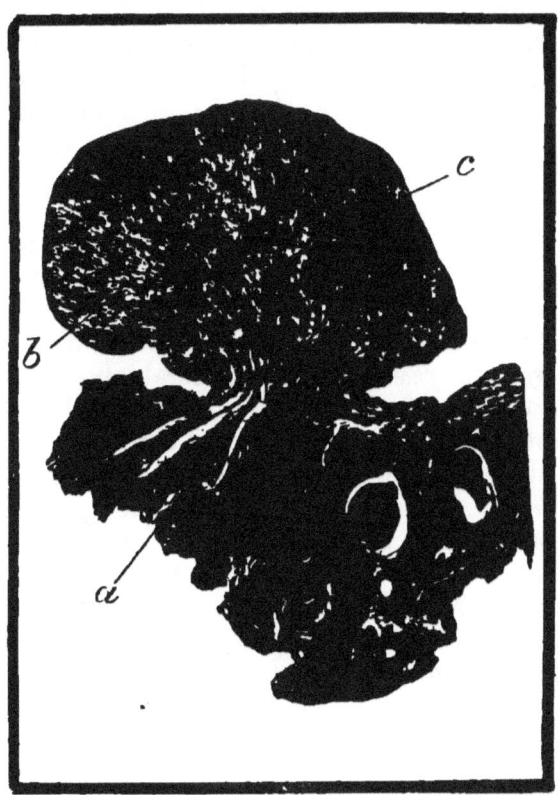

Fig. 42.—Fibrous Polypus of Rectum. (× 2½ diameters.)
a, Hæmorrhoidal structure; b, fibrous tissue; c, altered mucous membrane.

Fibrous polypus is sometimes found to assume large size, as in the case recorded by Mr. A. A. Bowlby.* A woman, while straining at stool, felt something come down which she was unable to return.

* Transactions of the London Pathological Society, 1883.

An examination showed a tumour the size of a fœtal head, attached by a pedicle 1½ inch in diameter to the anterior wall of the rectum, about 4 inches up. The base was transfixed and ligatured, and the mass cut away. It weighed 2 lbs. all but one ounce; and section showed it to be composed of loose connective tissue, the meshes of which contained a viscid fluid. A case is also recorded by Dr. Barnes,* in which a tumour in the rectum, the size of an orange, obstructed labour, and was removed by the galvanic cautery. It proved to be a connective tissue tumour, in part cavernous.

3. **Papilloma.** ["Villous tumour of rectum" (Curling); "Villous polypus" (Esmarch); and "Granular papilloma" (Gosselin).]—Under these various terms is described a remarkable but rare form of rectal growth, resembling in general appearance the villous tumour of the bladder, with this slight difference, however, that the lobes in the bladder tumour are more filiform, while in the rectum they are flattened or club-shaped (Fig. 43). They are composed of the papillæ of the mucous membrane, which have proliferated freely, and are covered with cylindrical-celled epithelium; they are attached to the wall of the bowel by a more or less broad pedicle, but occasionally are sessile. They give rise to a great deal of bleeding and mucous discharge; and when low down may be troublesome, owing to their protrusion at stool. They are to be removed in the same way as other polypi, *i.e.* by ligature and subsequent excision. Possibly, in some cases, the villous polypus breaks away, and is voided in the same way that the adenomatous variety undoubtedly occasionally is. These tumours are considered to be usually quite benign, but Allingham † states that he has met with two cases in which

* *British Medical Journal*, p. 551; April 12, 1879.
† *Loc. cit.*, p. 317

PAPILLOMA OF RECTUM.

recurrence, as a well-marked epithelioma, took place after the removal of villous tumour.

Papillomata of the anus differ from the variety just mentioned by having a covering of squamous instead of cylindrical epithelium. They sometimes

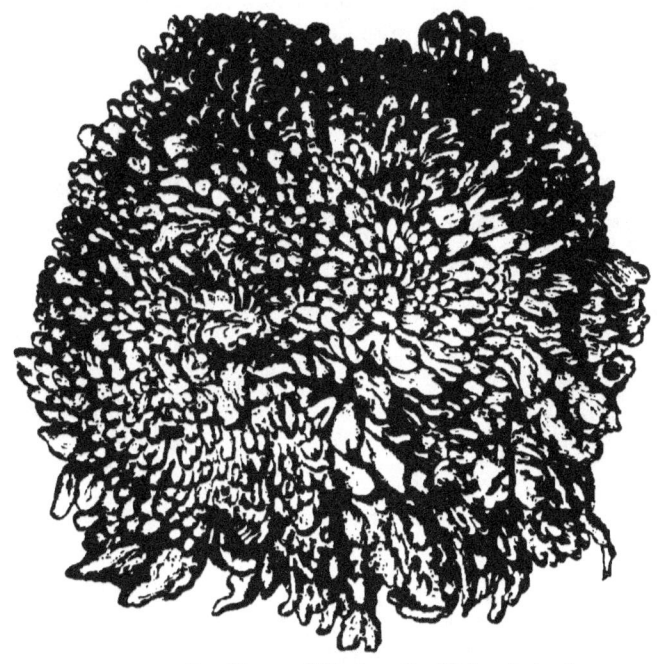

Fig. 43.—Papilloma of Rectum.* (Natural size.)

probably originate, like warts on the penis, from the irritation of gonorrhœa or other acrid discharge, while at other times they certainly appear to form without any such apparent cause. I have removed a mass of this kind (Fig. 44), which weighed nearly half a pound, from the anus of a young woman, in whom there was not the slightest suspicion of venereal contamination. They are easily removed

* Museum of University College Hospital.

with scissors, and as the attachment is external, bleeding points can be taken up without difficulty, and pressure applied to stop general oozing. Care, however, must be taken not to remove too much skin from the anal canal, as otherwise stricture may result.

Upon microscopical examination anal papillomata present similar characters to those met with in other regions of the body, where the covering is of scaly

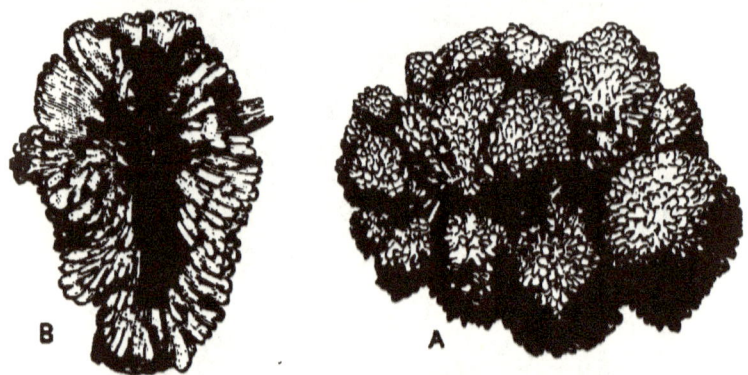

Fig. 44.—Anal Papilloma.
A, Surface of anal papilloma; B, section of anal papilloma.

epithelium (Fig. 45). The "cauliflower-like" appearance is due to the numerous sulci between the papillæ; the epithelial covering extends to the deepest ramifications of these sulci, so that in cross-section it sometimes appears as seemingly isolated spots.

4. **Teratoma, or dermoid cyst.**—Although tolerably commonly met with in the sacro-coccygeal region, and in the pelvis in connection with the genital organs, growths of this kind, originating in the rectum proper, are of extreme rarity. The most remarkable case recorded is that by Dr. Port.* A girl, aged sixteen, had suffered for three months from painful straining and difficulty in obtaining an evacuation.

* Transactions of the Pathological Society, vol. xxxi. p. 307 ; 1880.

Chap. XX.]   HISTOLOGY OF PAPILLOMA.   313

A polypoid tumour of large size came partly out at the anus when the patient wanted to pass a motion, and a lock of long hair repeatedly made its

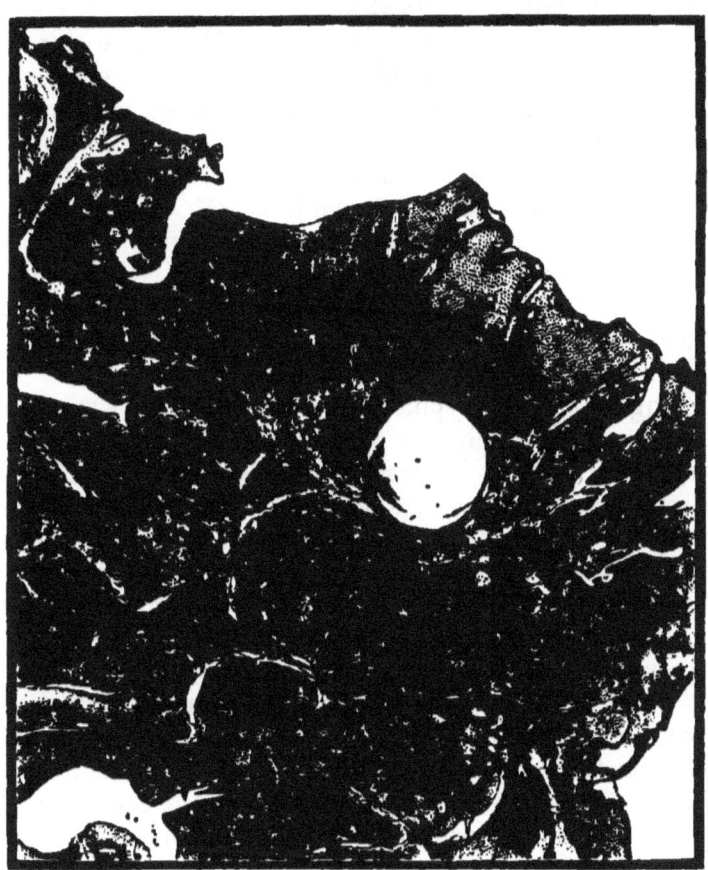

Fig. 45.—Papilloma of Anus. (× 15 diameters.)

appearance, and could only with difficulty be replaced. It was found that the tumour was attached to the rectal wall by two pedicles about three inches above the anal orifice. The tumour afterwards became completely extruded, when the pedicles were ligatured and the growth removed. It measured 2½ inches by 2 inches by 1½ inch, and the bulk of it was found to be made up of fibrous tissue, with numerous fat cells, and embedded in it were two masses of bone, one hard and the other spongy. The coverings of the tumour showed all the characteristics of ordinary skin, *i.e.* epidermis, papillæ, hair follicles, and sebaceous glands. The microscopical examination proved also the existence of numerous bundles of muscular fibres below the cutis. A canine tooth was observed growing from the tumour, near the pedicles.

A somewhat similar case is described by Dr. Danzel.* A woman, aged twenty-five, complained of hairs protruding from the anus, which she pulled out when they became too long. They were found to spring from a polypoid growth about the size of a small apple, which was situated in the front of the rectum, about 2½ inches from the anal margin. The tumour was removed ; and, besides the lock of hair, a tooth was found on the outside, and microscopical examination demonstrated a bony capsule containing brain substance in the interior ! Mollière records a case of small tumour, of about the size of an almond, the surface of which was covered with normal skin.

The tumours of the sacro-coccygeal region, although occasionally obstructing the rectum by pressure, are beyond the scope of this work to consider in detail.

5. **Lipoma** occurs in the interior of the rectum, as a more or less pedunculated growth, and a considerable number of cases have been put on record. Unless of a size sufficient to give rise to obstruction or prolapse, lipomata do not appear to be characterised by any definite symptoms. Claude Bernard records a case† of a woman, aged eighty-three years, who suffered from constipation. Upon passing her finger

* Langenbeck, *Archiv f. klin. Chirurg.*, p. 442 ; 1874.
† Quoted by Mollière, p. 525.

up she felt the growth, and detached it, with complete relief of her symptoms. The growth was of about the size of a pigeon's egg, and composed of pure fat. Esmarch * mentions a case as occurring in the Clinic at Prague, in which an extensive invagination and prolapse were produced by a small lipoma. The growth having been removed, the prolapse was cured. Virchow† quotes a similar case from the practice of Sangalle, in which two submucous polypi, which were pedunculated, and about the size of hen's eggs, produced an invagination of the colon, and finally a prolapse. A third case of like nature is reported from Langenbeck's Clinic by Bose ; ‡ and cases are on record in which the tumours have been expelled by unaided efforts of the patient to defæcate.§ This, in all probability, is due to the fact that rotation of the pedicle has taken place, which causes rupture or strangulation. Virchow has shown‖ that a similar rotation, and final separation, sometimes takes place on the outside of the gut in the pedicles of fatty tumours occurring in the appendices epiploicæ, and he thus attempts to account for the occurrence of free lipomata in the peritonæal cavity. In some of the recorded cases the pedicles of rectal lipomata have been noticed to contain a tolerably large funnel-shaped process of peritonæum. This, taken with the fact that all the recorded cases appear to have descended from the sigmoid flexure, or upper part of the rectum, would possibly tend to show that these growths had originated in one of the appendices epiploicæ that had become inverted : against this theory, however, is the fact that similar tumours are sometimes found in the small intestine. In colour they are

* *Loc. cit.*, p. 154.
† "Pathologie des Tumeurs," vol. i. p. 379. Paris.
‡ Esmarch, *loc. cit.*, p. 154.
§ Castilian, Mollière, *loc. cit.*
‖ *Loc. cit.*, p. 380.

described as being redder than the ordinary lipomata, but otherwise they present no characters different from the same growth occurring in other parts of the body.

The removal of these growths can be carried out in the same way as that of other pedunculated tumours : namely, by ligature and snipping off of the tumour. It is essential to remember how frequently there is a prolongation of peritonæum into the pedicle of these growths. This would become a real source of extreme danger if simple excision were practised, or if the écraseur was used. Broca records* a very instructive and warning case, in which he removed a polypus from the rectum by means of an écraseur. The patient died in forty-eight hours, and an autopsy revealed the fact that a circular opening had been made communicating with the peritonæal cavity, which had permitted the escape of fæces, and so induced a fatal peritonitis. Voss reports a case † in which he successfully removed a lipoma from the rectum of a woman, aged forty-seven years. It was as large as a goose's egg, and attached by a pedicle, the origin of which could not be felt. He made an incision over the equator of the tumour, and without difficulty shelled it out of its capsule.

Ligature, therefore, may be relied on as the most satisfactory and safest way of dealing with this pedunculated growth, as it obviates all danger of hæmorrhage, and should the pedicle contain a peritonæal pouch, it is not at all likely that any serious consequences would follow, as adhesion of the opposed serous surfaces would soon follow the application of a ligature.

In the neighbourhood of the anus fatty tumours have been noticed, some of which have gradually encroached upon the lumen of the bowel. They are sometimes pedunculated, while at others

* " Traité des Tumeurs," vol. ii. p. 536. Paris, 1869.
† *London Medical Record*, p. 200 ; 1881.

they are more diffused in the subcutaneous tissue. One of the most interesting cases of this kind is recorded by Robert,* in which the tumour sprang from the ischio-rectal fossa and was removed, and was at first mistaken for a perinæal hernia. It occurred in a riding-master, aged forty-five years, and measured ten centimetres by seven. The operation at first consisted in cutting down upon the tumour layer by layer, as in the case of a hernia, but as soon as its true nature was evident it was followed into the ischio-rectal fossa and extirpated.

6. **Cystoma.**—Serous cysts have occasionally been spoken of as occurring in the rectal wall, but I have been unable to find any such case in which the pathology was made out with exactness. Sometimes, however, small cysts filled with viscid fluid are met with in infiltrating adenomata, and possibly larger ones, having a glandular origin, may occur.

Atheromatous cysts are sometimes met with in the region of the anus. Those recorded have been of small size, and surgically unimportant. They could of course, if necessary, be easily removed by the same means had recourse to elsewhere.

The fact that hydatid tumours have occasionally been met with in the pelvis should be borne in mind when investigating the causes of obstruction due to pressure on the rectum from within the pelvis.

7. **Enchondroma.**—A case of cartilaginous tumour of the rectum has been put on record by M. Dolbeau. It was removed from a young man, aged twenty-seven years. It was of the size of a small nut, freely movable, and situated at the entrance of the anus, where it gave rise to little trouble; the mucous membrane in the neighbourhood was eroded. Microscopic examination showed that it was in part composed of fibro-cartilaginous structure, but the greater

* "Annales de Thérapeutique," October, 1884, quoted by Kelsey.

part was adenomatous.* Van Buren gives a case which he considered was one of chondromatous tumour, but the case was not verified by minute examination.

8. **Angioma.**—The most remarkable example of this extremely rare condition is that described by Mr.

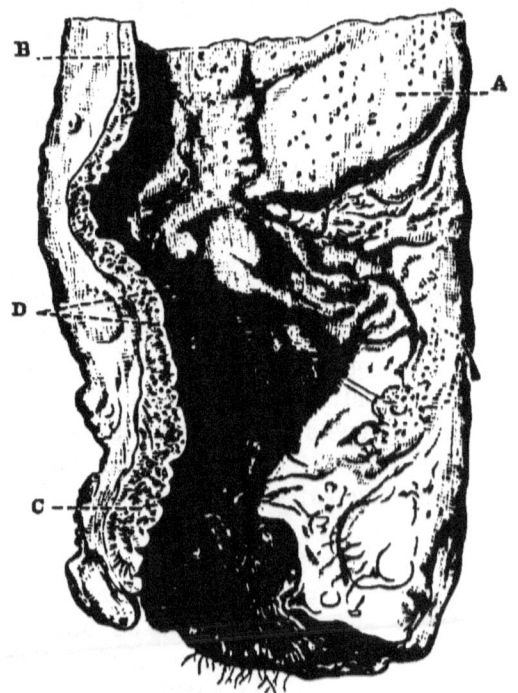

Fig. 46.—Angioma of Rectum (Barker).
A, Mucous membrane; B, section of mucous membrane; C, nævoid structure; D, ulcers from which the fatal hæmorrhage took place.

A. E. Barker † (Fig. 46). A healthy man, aged forty-five years, stated that since boyhood he had difficulty in obtaining a motion when he was at all constipated, and that at these times there was bleeding from the bowel. Sometimes he remained free from these symptoms for

* "Bull. de la Soc. Anatomique," vol. v. p. 6, second series.
† "Medico-Chirurgical Transactions," vol. lxvi.

several years at a time, his bowels as a rule being regular. A careful examination of the wall of the bowel showed three shallow ulcers on the rectal mucous membrane; they were seated on some smooth longitudinal folds of the wall of the gut, of a yellowish colour, and suggesting a quantity of fat in the submucous tissue. The ulcers, though shallow, exuded continuously a considerable quantity of blood. Their base, however, presented a peculiar mottling of a purplish colour, as also did the surface of the irregular folds alluded to, the whole picture giving rise to the suspicion of a nævoid mass in the wall of the bowel. The statement of the patient that similar bleeding had occurred on and off since boyhood seemed to lend support to this view. The patient was afterwards admitted into University College Hospital; the bleeding became of daily occurrence, and very copious; and, in spite of all treatment, he died of hæmorrhage. At the post-mortem examination the wall of the rectum in its lower four and a half inches was found much thickened by a nævoid growth in its walls, which gave a purple colour to the mucous membrane. There were three or four prominent longitudinal folds, each three-quarters of an inch or more in width; the two largest were on the left side of the bowel.

Fig. 47.—Section of Angioma of Rectum, showing cavernous structure (Barker).

These were the folds felt during life, one of them just to the left of the middle line in front. Two ulcers, one of them about the size of a threepenny-piece, the other larger and somewhat irregular, were situated about two inches from the anus. The tumour everywhere presented the character of cavernous nævoid tissue (Fig. 47). At the same meeting of the Medico-Chirurgical Society Mr. H. Marsh related the case of a girl, aged ten years, who had suffered repeatedly from rectal hæmorrhage. With the aid of a speculum, the patient being under chloroform, a nævoid growth was seen in the lower part of the rectum, completely surrounding the bowel. Several applications of Paquelin's cautery relieved the symptoms considerably, but did not cure the growth.

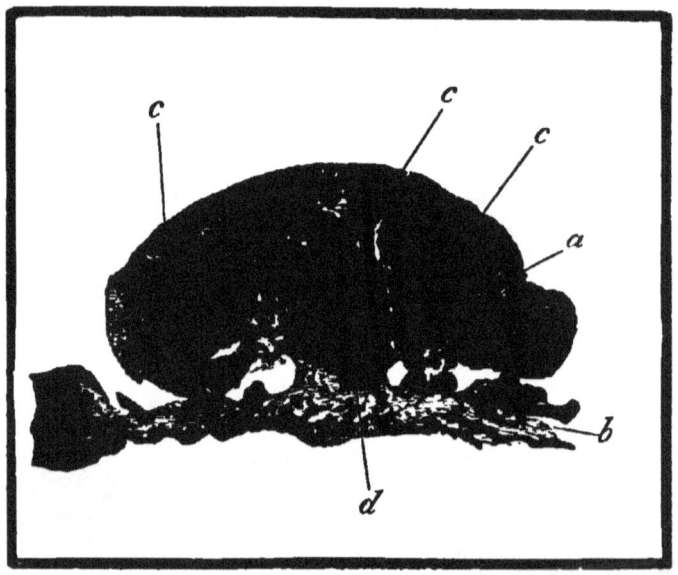

Fig. 48.—Lymphoma of Rectum. (× 5 diameters.)
a, Remains of mucous membrane; *b*, muscular coat of bowel; *c*, solitary glands much enlarged; *d*, pedicle.

In the case already referred to as recorded by Dr. Barnes also,* the large polypoid tumour removed is stated to have been in part cavernous.

9. **Myoma** is rarely met with in the rectum; it forms a firm more or less pedunculated growth, originating in the muscular tunic, and having the mucous membrane freely movable over it. I have some sections kindly placed at my disposal by Dr. Macan from a case in which he enucleated a tumour of this

* *See* page 310.

nature from the wall of the rectum; it resembles closely the similar tumours so commonly met with in the uterine wall. Probably many of the cases described as hard fibromas are really of this nature. According to Allingham, there is a specimen of this tumour in the Hunterian Museum.

10. **Lymphoma.**—Fig. 48 represents a section of a polypus, from a boy aged six, the subject of a prolapse of the rectum, at the apex of which two small growths, attached by short pedicles, were seen and removed. The general character resembled that of ordinary adenomata, but, upon minute examination, these were found to consist almost entirely of lymphoid tissue. Presumably they originated in the solitary glands of the mucous membrane. In Fig. 48 apparently three of the solitary glands (c) have grown together and greatly increased in size; they appear to have pushed their way through the mucous membrane, small remnants of which may still be seen at several points (a) in the section.

Infiltrating lympho-sarcoma has been observed in several portions of the intestinal tract, but pedunculated growths—like the one illustrated—unassociated with obvious lymphoid disease in other parts of the body, is, as far as I know, previously unobserved.

CHAPTER XXI.

MALIGNANT NEOPLASMS OF THE RECTUM AND ANUS.

Of the various new growths which are found in the rectum, and which are clinically malignant, cylinder-celled epithelioma, or, as it is sometimes called, "malignant infiltrating adenoma," is unquestionably the most common.

In the following chapter we will discuss, as far as is at present possible, the pathological differences between the various forms of rectal tumour exhibiting malignancy; but as it is sometimes quite impossible to differentiate these varieties *clinically*, it will be convenient to retain the term "cancer," using it in its broadest sense, as synonymous with all the forms of malignant tumour, whether histologically of epithelial or connective tissue origin. It is not necessary here to discuss the vague and speculative theories which have from time to time been put forward to explain the ætiology of cancer. Much has been written and said upon this subject, but nothing definite has been arrived at; and it still remains an inscrutable mystery why it is that tissue in all respects apparently identical with normal epithelial structure should overstep its natural limits of growth and development; extend widely into neighbouring regions; appear as metastatic growths in other situations; break down and suppurate as a result of excessive and exuberant growth; recur after wide removal; and, lastly, produce that constitutional disturbance and rapidly-progressing marasmus known as the cancerous cachexia.

In order to arrive at some idea as to the frequency of rectal cancer, both relatively to the examples of the same disease in other parts of the body, and more particularly in other parts of the intestinal tract, it becomes necessary to refer to large statistics. It is, however, quite useless to collect for this purpose a simple record of cases published in periodical literature, the returns of large hospitals alone affording reliable information. Leichtenstern* has carefully collated the following figures from the returns of the K.K. Allgem. Krankenhaus at Vienna. Out of 34,523 deaths at the hospital between the years 1858 and 1870 were 1,874 cancers of different kinds, equal to 5·4 per cent.; and of 4,567 cancers at the same hospital, 143 were of the

* Ziemssen's "Cyclopædia of Medicine," vol. vii. p. 635.

rectum, and 35 of other parts of the intestines; the former were therefore 3 per cent., the latter 0·76 per cent. of the whole, and the former 80 per cent., the latter 20 per cent. of all cancers of the intestines. Mr. W. R. Williams* has collected a large series of equally reliable statistics from the Middlesex Hospital, St. Bartholomew's Hospital, St. Thomas's Hospital, and University College Hospital. Out of 5,556 cases, he gives the following table, showing the frequency in some of the more important organs :

	Male.	Female.	Total.
Breast	13	1,310	1,323
Uterus	—	1,160	1,160
Tongue	384	64	448
Rectum	130	127	257
Skin of face and neck, including rodent ulcer	161	89	250
External genitals	126	102	228
Lip	221	2	223
Intestines, etc	23	26	49

It will be seen from this table that the results obtained at the London General Hospitals are practically the same as those observed at Vienna. Referring to the records of the Brompton Cancer Hospital, as given by Jessett,† we find that out of a total of 1,908 cases of cancer admitted, 58 were suffering from cancer of the rectum, or slightly more than 3 per cent. One would expect that a slightly smaller proportion of rectal cases would present themselves at this special hospital; because persons suffering from cancer of the rectum would be more likely to apply to a general hospital, under the impression that it was some other form of disease that they were suffering from; whereas, persons suffering from some of the other and more easily recognisable forms of cutaneous cancer would

* *Lancet*, May 24, 1884.
† "Cancer of the Alimentary Tract," p. 238. London, 1886.

gravitate to the Brompton Hospital. It may, therefore, be taken as sufficiently accurate that in 3·5 per cent. of all cases of cancer the disease is situated in the rectum; and in 80 per cent. of cases of intestinal cancer the disease is located in the lower bowel. In the records of St. Mark's Hospital, as given by Allingham,* out of 4,000 cases of rectal disease 105 were examples of cancer.

The degree in which apparently similar forms of carcinoma exhibit the clinical features of malignancy varies notoriously with the situation in which the disease develops. Thus, for instance, epithelioma of the tongue is extremely malignant; whereas, the same disease situated upon the lip is, at any rate in the early stages, one of the most benign of the unequivocal epitheliomata; similarly, epithelioma on the scrotum is very much more satisfactory to deal with than the same disease when occurring on the penis. Compared with other regions of the body, it would appear that the rectum is one in which the average intensity of the malignancy is not very great, the disease for a long time not passing the limits of the intestinal wall. Allingham, whose experience on this subject is so extensive, puts the average duration down at two years, the most rapid terminating fatally within four months of the earliest symptom of its invasion; while the longest duration noticed by him was four years and a half. It is, however, quite impossible to estimate accurately the duration of this disease, as the symptoms during the early stages are so slight that they may be scarcely sufficient to attract the attention of the patient. This will be a matter of familiar observation to all surgeons. It not unfrequently happens that a patient comes to us complaining of some slight diarrhœa or other mild rectal trouble, and an examination unexpectedly reveals the fact that he is the victim of cancer so very

* *Loc. cit.*, p. 3.

extensive that it must have obviously existed for a very considerable period. And, again, the life of the patient is not unfrequently sacrificed by the accidental complications of the disease, such as intestinal obstruction, or involvement of the bladder, rather than by the progressive marasmus, which is the usual mode of termination of cancer of other regions.

Some authors state that as the result of their experience a greater number of males suffer from rectal carcinoma, while others assert that the opposite is the case. The large statistics of Williams, however, show that there is extremely little difference in the relative frequency. Out of 257 cases, there were 130 males and 127 females.

Although essentially a disease of middle life and old age, rectal cancer has been met with several times under the age of 20 years. The earliest age that I have seen recorded is that noticed by Allingham* as having occurred in the practice of Mr. Gowlland at St. Mark's Hospital, in which a boy not 13 years of age suffered from cancer of the rectum; while Allingham gives a case of his own in which a boy of 17 years died of what is described as encephaloid of the rectum. Considering the vagueness of the term encephaloid and the frequency with which it is applied to rapidly growing tumours of the sarcoma type, it appears possible in the absence of detailed microscopic examination that this tumour was sarcomatous, and it is well known that tumours of that type are not very uncommon in early life. A case of cancer is recorded by Godin[†] at 15 years, Quain one at 16 years, and Cripps one at 17 years. Schœning[‡] describes two cases as occurring at the Rostock clinic. In the first, a girl, aged 17, presented typical symptoms of rectal cancer. She was stated to have suffered from rectal prolapse at the

* *Loc. cit.*, p. 270. † Quoted by Mollière.
‡ Deutsch. Zeitschrift f. Chirurg., Bd. 22, Hft. 1 and 2; 1885.

age of seven, and the more severe symptoms began to manifest themselves at the age of 16 years. The tumour was excised, and presented the microscopic characters of undoubted carcinoma. The disease recurred, and proved fatal in two months. The writer concludes that she suffered from adenoma at the age of seven years, which subsequently began to infiltrate and become malignant. In the second case, a girl, of 17, presented herself with a tumour the size of a fist, very hard, and encroaching on the pelvic organs, and affecting the inguinal glands. As the tumour could not be removed, the constricting tissues were divided. A portion removed proved the tumour to be an alveolar, cylinder-celled carcinoma partly undergoing cystic degeneration.

CHAPTER XXII.

THE PATHOLOGY OF MALIGNANT NEOPLASMS OF THE RECTUM AND ANUS.

THE older method of classification of tumours into benign and malignant, although of great practical utility, was soon found to be insufficient; for although the difference between typical varieties was sufficiently obvious, cases were met with on the border land between the two which it was impossible to refer to either with certainty, and for these the class of semi-malignant tumours was introduced. Since the clinical classification has given place to the histological, it does not appear that the exact limitation of the groups is thereby rendered, in some instances, more definite, and this is notably the case in cancer of the rectum. The clinical differences between the simple adenoma, or mucous polypus, of the rectum, and cancer of that organ, are sufficiently obvious: the simple adenoma generally occurring in young persons; being attached by a long pedicle; not tending to recur after removal; or to affect the constitution: the cancer, on the other hand, is sessile; tends to infiltrate deeper parts; to break down and ulcerate; to profoundly affect the constitution; to recur after removal, and produce metastatic growths of similar character at a distance from the original site. Now when these growths are examined under the microscope, they both consist essentially of the same tissue, namely the glandular structure of the mucous membrane, such as is normally found lining the Lieberkühn follicles of the intestine; the only difference being that in the benign form there is a tendency to project into the lumen of the bowel, and to draw down a pedicle of *normal* mucous membrane, while in the cancer the wall of the intestine is from the very first infiltrated with the new formation. First the muscularis mucosæ becomes perforated, then the submucosa invaded, and subsequently the muscular coat itself is infiltrated (*see* Fig. 49), so that the only histological difference between these growths is really one of situation, and of relation to surrounding tissues, not of structure.

As might be expected, cases are occasionally seen in which it is impossible to say to which class, whether adenoma or adeno-carcinoma, the growth should be referred; so that histologically as well as clinically the limits of classification are not very distinct. As the carcinomata originating in any structure are principally composed of the epithelial elements similar to those normally present in the immediate vicinity, it follows that those commencing in the intestinal mucous membrane should consist principally of adenoid tissue, and such has been found to be the case. Mr. Harrison Cripps,

328   THE RECTUM AND ANUS.   [Chap. XXII.

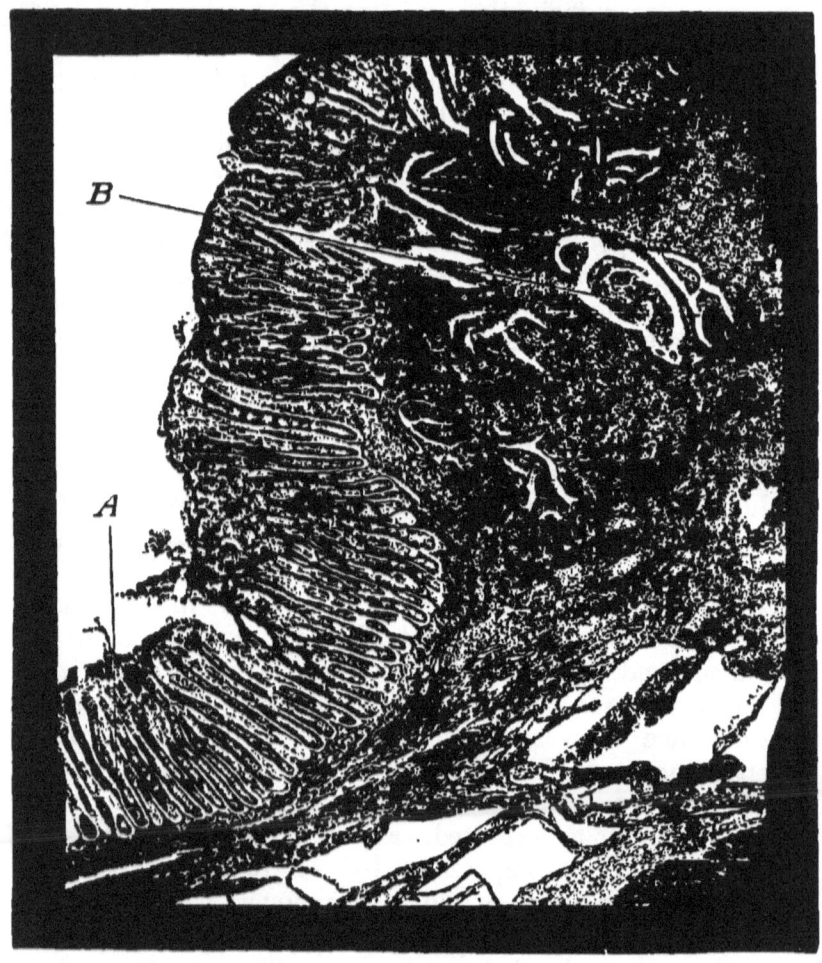

Fig. 49.—From Micro-photograph of Margin of Cancerous Nodule.
(× 10 diameters.)
A, Normal mucous membrane; B, infiltration of muscular structure by glandular tissue.

whose investigations of the histology of rectal cancer have been extremely extensive, embracing a careful investigation of sixty separate examples of the disease, says: "In the rectum I have failed to discover any growths or tumours which pathologists designate as scirrhous or medullary cancers, or as belonging to the different varieties of sarcoma. Considering the eminence of many careful observers who have applied such names to these growths, it would be quite unjustifiable to assume that such distinctive structures never form the entire bulk of the tumour; but I feel bound to state that, with perhaps a more than average opportunity of examining such growths from the rectum, I have been myself unable to discover tumours composed entirely of the distinctive features appertaining to these diseases." *

Mr. Treves† expresses a somewhat similar opinion, that the form of cancer found throughout the entire intestinal tract is cylinder-celled epithelium, and he quotes from a monograph by M. Haussmann,‡ who says: "We will give, then, cancer of the intestine the following definition: cancer of the intestine is cylindrical epithelioma of that organ." Putting aside for the present the question of sarcoma, the occurrence of which in the rectum is undoubted (and of which I have had an instance in my own practice), let us consider what is meant by the terms scirrhous and medullary cancers at the present day. If the former is taken only to indicate that in which, with the epithelial development, there is a very considerable hyperplasia of the connective tissue, forming firm and hard masses of tissue, the so-called stroma, which frequently manifest a tendency to contract and pucker the invaded tissue, then, without question, scirrhous cancer does occur in the rectum. Similarly, if medullary cancer or encephaloid is taken to mean that the tumour is of rapid growth,

* "Diseases of the Rectum and Anus," p. 317. 1884.
† "Intestinal Obstruction," p. 268.
‡ "Thèse de Paris," No. 228. 1882.

soft structure, and that the epithelial cells are more or less embryonic in character, and the connective tissue ill developed, then this form of tumour is also present in the rectum. But if, on the contrary, these terms are taken to represent distinct types of carcinoma, the epithelial elements of which essentially differ from those found in the organ or tissue in which the carcinoma originates, then not only are these forms not to be found in the intestine, but they cease to have an existence in any other part of the body. As, however, these terms have been used with much vagueness, it is better to dispense with them altogether. All observers, however, appear to be agreed that at any rate by far the most common form of intestinal cancer is the columnar-celled epithelioma, or adeno-carcinoma, or infiltrating adenoma, as it is variously termed. The term cylindroma, which has been frequently used as a synonym for this disease, is misleading in the extreme, having been introduced by Billroth for a special form of tumour quite unconnected with this form of cancer.

When cancer primarily attacks the anus, as might be expected, the bulk of the tumour is composed of scaly epithelium, and the growth resembles that met with in the lip. (*See* Plate I.) Cripps has, however, stated that when a cancer, originating in the interior of the rectum and of the adeno-carcinomatous variety, invades the anus, the character of the epithelial cells varies, and comes to resemble the ordinary scaly type. This is remarkable, as the metastatic reproductions of intestinal cancer in other organs correspond very accurately to the original histological character of the growth; as, for instance, the multiple tubercles in the liver, which are such a common sequela to rectal cancer, when examined under the microscope present the same follicular structure so characteristic of adeno-carcinoma. My colleague, Dr. Purser, has given me a section of a tumour in the lung, secondary to a carcinoma of the sigmoid flexure, and in it the reproduction of the original character of the tumour was most marked, little masses of epithelial cells closely resembling Lieberkühn crypts being surrounded by normal lung tissue.

There is some considerable variation in the macroscopic characters presented by adeno-carcinoma when present in the rectum; these differences being chiefly influenced by the rate of growth and the direction in which the tumour principally extends. These varieties have been distinguished by Cripps as the "tuberous," "laminar," and "annular."

The **tuberous adeno-carcinoma** presents itself as a considerable-sized mass projecting into the lumen of the bowel, obviously implicating the mucosa, which can be traced into it, but not moved freely over it. Associated with it may be other smaller masses. It is not necessarily very hard to the touch, and, in the early stages, does not extend beyond the limit of the rectal wall, as is demonstrated clinically by the free mobility of that organ in the pelvis. This form tends to ulcerate very rapidly; first the mucous membrane on the surface necroses; then the centre of the mass breaks down, exposing the muscular layer of the intestine, and leaving a crater-like cavity surrounded by the infiltrated mucosa and submucosa; at last the intestinal wall is perforated, and the pelvis becomes invaded, the bladder or urethra may be opened, the vagina ulcerated into, or the nerves of the sacral plexus involved in the neoplasm, or even the bony wall of the pelvis may become implicated. This variety may be taken as the type of the more rapidly growing adeno-carcinoma. It is the form most frequently met with in younger subjects. It may produce obstruction by the bulk of the growth, but does not usually do so by producing contraction of the intestinal wall, as the other and more chronic forms do. This no doubt is the variety alluded to by most of the older writers under the head of medullary or encephaloid cancer; though in all probability the same term was applied to some of the sarcomata, which from rapid growth and large tumour formation resemble closely the tuberous adenocarcinomata in their clinical aspects.

The laminar form.—This, according to the investigations of Mr. Harrison Cripps, is the commonest variety. It occurs as a layer of adenoid growth spreading laterally in the submucosa, of a thickness of about a quarter of an inch or less, while the area over which it extends may be considerable. It has a tendency rather to extend laterally than vertically, so that in time the entire circumference of the gut may be involved. Although principally situated in the submucosa, it is obvious that the mucous membrane is attached to, and incorporated with, the growth; and in the same way, the muscular tunic of the intestine is adherent to the tumour deeply. As the tumour advances in growth there is a considerable development of connective tissue in the outer walls of the intestinal tube, which subsequently undergo contraction, producing the puckering and cicatricial constrictions which have given origin to the use of the word scirrhous in connection with this disease. As in the former variety, ulceration of the mucosa soon occurs; which may be followed by perforation of the rectal wall into any portion of the genito-urinary system; or at other times the new formation will be more rapid than the ulcerative action, and the result will be the spreading of a fungating mass into the rectum.

The annular form is that in which the neoplasm surrounds the rectal tube without extending vertically to any great degree. It would appear to be one of the most chronic forms, and naturally attended with much contraction, forming the true "malignant stricture."

Besides the infiltration of surrounding structures, rectal adeno-carcinoma tends to reproduce itself in other parts of the body; and like all the group of the carcinomata, the lymphatic glands become implicated with extreme frequency. When, as is usually the case, the disease is situated entirely within the rectum, leaving the anus free, the first to be involved will be the pelvic and lumbar glands; and sometimes these are seen to be of very large size, the glands along the iliac vessels being sometimes quite as large as hen's eggs, and capable of recognition during life by abdominal palpation. Next in order, the lumbar glands are enlarged; but the lymphatics of the groin only become implicated as a consequence of involvement of the external skin of the anus, or when in an advanced stage of the disease a very widespread lymphatic implication follows the primary enlargement of the pelvic glands in cases of adeno-carcinoma. Next in frequency to the lymphatic system, the new growths are liable to be found in the liver, probably the most frequent cause of disseminated hepatic cancer being the form of disease under consideration. As is usual with metastatic tumours, the secondary growths reproduce with singular exactness the histological characters of the original tumour. Involvement of the peritonæum also is not unfrequent, the metastatic growths appearing like grains of boiled sago over the surface, and matting together, when extensive, the coils of small intestine. Secondary deposits have also been found in the pancreas, lungs, etc.

The essential histological characteristic of adeno-carcinoma is the fact that in this disease the adenoid tissue perforates the muscularis mucosæ, and develops in the submucosa and muscular coat (Fig. 50). It is this characteristic alone which serves to establish the accurate diagnosis between the malignant and

Fig. 50.—Cylinder-celled Epithelioma of Rectum (× 10 diameters).
A, External muscular coat of bowel; B, internal muscular coat of bowel; C, masses of adenoid tissue separating the bundles of muscular fibre of the internal muscular coat.

non-malignant forms of adenoma, and in this respect the case is exactly analogous to difference between a wart and an epithelioma on the skin proper. In fact, the essential element in the production of a carcinoma is the development of epithelium beyond its natural superficial limits. For fuller detail of the histology of rectal cancer the reader must be referred to the work of Mr. Cripps.* It is, of course, seldom that the very earliest stage of rectal carcinoma can be investigated, as no important symptoms are usually produced until the disease has made considerable progress, so that it is impossible to state what the initial change is. Cohnheim has propounded a very ingenious theory, by which he attributes an embryonic origin to all tumours, and considers that an embryonic rudiment is left during development, and that at some later period this may undergo proliferation. He bases one of his arguments in support of this theory, on the frequency with which cancer occurs at the places where, during development, diverse epithelial formations pass one into another, as the lips, rectum, stomach, cervix uteri. Certainly the fact that, in a large majority of cases, rectal cancer commences at a place corresponding closely to the site of junction of the proctodæum and mesenteron would appear to favour this view.

In order to investigate the method of growth, it is necessary to examine the spreading margin of the tumour; that which projects into the rectal lumen being the most suitable for demonstrating the mode of growth, the deeper parts being altered by the way in which the neoplasm is disseminated between the normal structures, and mixed with the débris of atrophic tissues. The central parts are also unsuited for minute examination, as fatty degeneration and breaking down of the tissues is usually taking place there. If the spreading margin be examined, it will be found that it is raised above the level of the adjacent membrane, and sometimes overhangs it to some extent, but it will always be found to be attached by a

* *Loc. cit.*, p. 308.

broad base, and incorporated with the structures forming the rectal wall; it is, however, never distinctly pedunculated as in the case of the simple adenoid growth. It is quite true that we sometimes find small pedunculated adenomata in the rectum in conjunction with adeno-carcinoma, but they usually appear as if

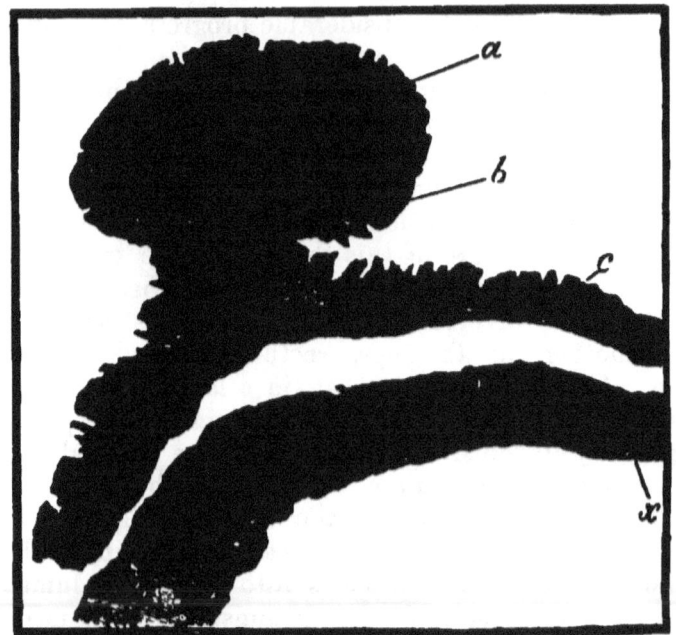

Fig. 51.—Adenoma of Rectum from Case of Rectal Cancer.
(× 12 diameters.)
a, Glandular structure; *b*, connective tissue and muscularis mucosæ; *c*, healthy mucous membrane; *x*, muscular coat of bowel.

they were due simply to the irritation of the discharge from the cancer. Fig. 51 represents a section of a simple adenoma which existed in conjunction with undoubted cancer from a specimen kindly given me by Dr. Patteson. And although a few cases are recorded in which a malignant form of disease has followed the removal of a simple adenoma, yet they are so rare that the rule may be adopted, that pedunculation is a very strong argument against malignancy. According to Cripps, if the surface of a growing margin be examined with a low microscopic power, it will appear like "an ant-hill thickly studded with fungi. Upon closer inspection these bodies are seen to be projections from the surface of the tumour." Upon making sections, the Lieberkühn follicles are found much increased in size, being three to four times longer than normal, while the individual cells are also much increased in length, sometimes being ten times longer; *i.e.* one-hundredth of an inch. The follicles may be lined by a single layer of columnar epithelium, only leaving a central cavity. In other instances the central cavity is absent either by approximation of its walls, or by a growth of offshoots from the epithelial walls. These offshoots consist of a central stroma of retiform tissue, upon which a bipinniform arrangement of cylinder cells is seen to fill up completely the cavity. The question arises whether these cavities are shut sacs, or only cross-sections of convoluted tubes of dilated Lieberkühn crypts. Cripps appears now to take the latter view. Where the sections are taken from very rapidly growing and soft tumours it will be found that the typical cylinder cells will not be formed, the whole aspect being more embryonic in character; the cells being rounder and less defined, the way in which they are disposed, and the tendency to follicular formation, however, leave no doubt of their connection with the adeno-carcinomata.

Colloid or gelatinous cancer.—The writings of Cruveilhier * have been frequently quoted as showing that this form is the most frequently met with in the intestinal tract. As Cripps, however, justly remarks, an examination of museum specimens does not tend to show that this disease was more common

* "Traité d'Anatomie Pathologique Générale," p. 64, *et seq.*
w—23

formerly than at present, and certainly an examination of recent specimens tends to indicate that colloid must be considered one of the rarer forms of intestinal cancer. In the reports of cases read before the various pathological societies formerly, the terms were used with much vagueness, and probably applied to very different forms of growth. In the rectum it may occur as a definite tumour, or as a diffuse infiltration, and is characterised by the translucency of its substance. The stroma contains, instead of closely packed masses of epithelial cells, a more or less clear jelly. According to Ziegler,* "the colloid or gelatinous texture of the tumour is due to mucoid or colloid change affecting the cancer cells. It begins with the formation of clear globules in their interior; the cells then perish, and the globules coalesce with each other and with the larger gelatinous lumps already formed. In this way a large homogeneous mass is ultimately built up. It is not uncommon for all the cells over a wide area to perish in this manner, so that the stroma is the only formed constituent remaining; in other spots cell groups may still be found encircled by colloid masses; in others there is no colloid substance at all."

Some, no doubt, of the gelatinous rectal cancers might be with greater precision designated as *carcinoma myxomatodes*. Professor Purser has kindly given me a very beautiful microscopic section of a colloid cancer of the upper part of the rectum, which shows the new growth perforating the muscularis mucosæ, its development in the submucosa, and infiltration and separation of the bundles of the muscular coat. In the greater part of the section nothing but the stroma is left, while in a few places cells containing globules of colloid matter still remain (Fig. 52).

* "General Pathological Anatomy," p. 224. London, 1883.

Of colloid cancer I have had two well-marked cases in my practice. The first formed a tubular stricture about four inches long, commencing immediately inside the anus (Fig. 53). The lower portion

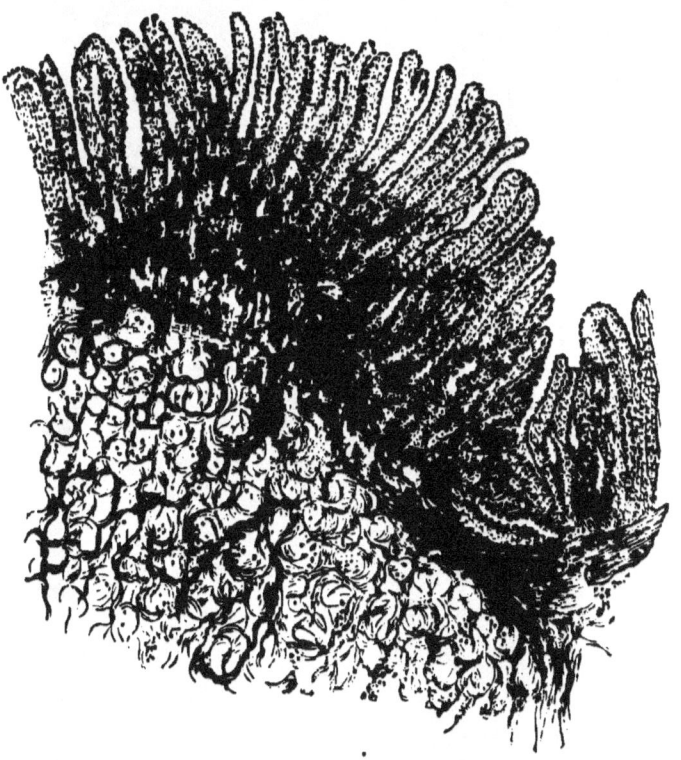

Fig. 52.—Colloid of Rectum (× 50), showing perforation of Muscularis Mucosæ by new growth.

was much ulcerated and the intestine above considerably dilated; it was removed from a woman aged thirty by trans-sacral incision. She made an excellent recovery and so far remains free from recurrence.

The second case was that of a man, aged sixty, in which the disease appeared as a nodule at the upper extremity of a cicatrix, following a very extensive

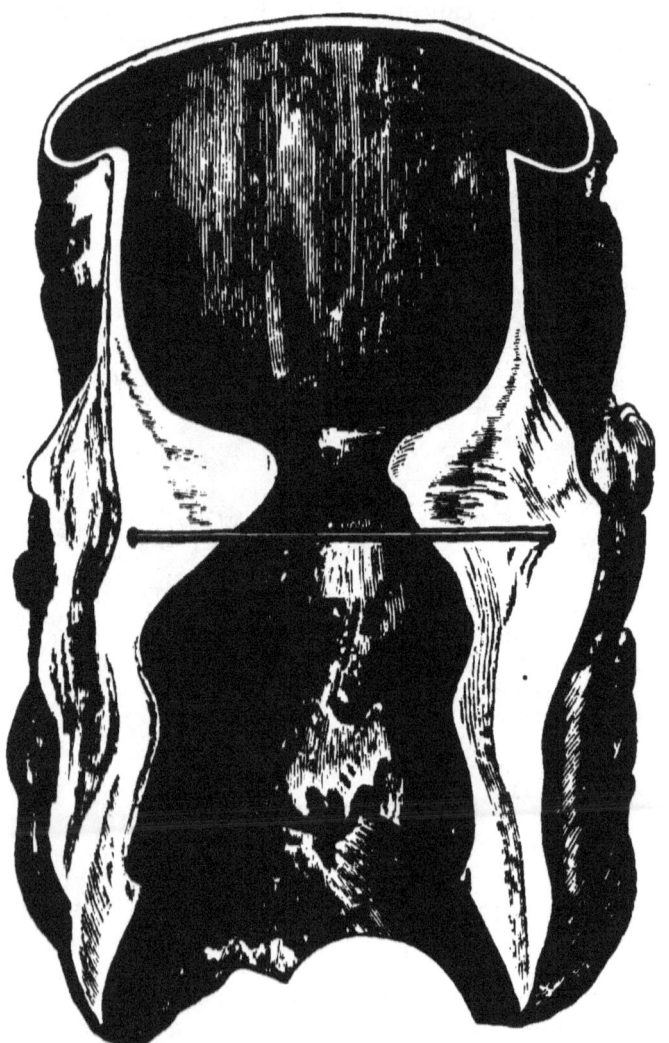

Fig. 53.—Case of Colloid Cancer of Rectum, natural size, removed by trans-sacral Incision.

operation for rectal fistula; this was also successfully excised.

SARCOMA OF RECTUM.

The second great class of malignant neoplasms, coming, in order of frequency, after the carcinomata, are those tumours the bulk of which is composed entirely of embryonic connective tissue, but **sarcomata** are rare in the intestinal tract. Mr. Cripps states* that he has been unable to find in his extended examinations of rectal growths any of the characteristic structure belonging to the different varieties of sarcoma.

In the Museum of the Royal College of Surgeons

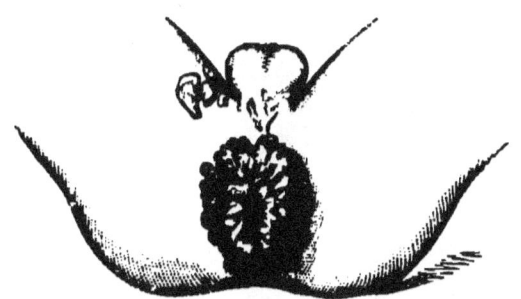

Fig. 54.—Large Sarcomatous Tumour of Anus and lower part of Rectum, with Secondary Tumours on the inside of the Thigh.†

of Ireland are two very remarkable examples of sarcomatous growths: In the first (Fig. 54) there is projecting from the anus an enormous mass which measures five inches by four; it is much lobulated on the surface, presenting somewhat the appearance of an ordinary papilloma of this region. It differs however, in this, that the individual lobules are much larger, and the intervening depressions much shallower; a small group of secondary growths appears near the scrotum, in the skin of the thigh, and the disease extends up into the rectum for a distance of about two inches. There does not appear, however, to have been any obstruction, as the tube was quite pervious behind the growth. There is, unfortunately,

* *Loc. cit.*, p. 318.
† Museum, Royal College of Surgeons in Ireland.

no very reliable history with this specimen. Dr. P. S. Abraham, the late curator of the museum, kindly undertook a detailed examination, and he made microscopic sections from the mass inside the rectum, from the external growth, and from the secondary formations. In all of them the appearances were practically identical: there was no trace of proliferating mucous membrane; almost the entire of the sections consisting of small spindle cells, with but little fully developed connective tissue.

Fig. 55.—Sarcomatous Infiltration of Rectum, producing long tubular stricture.*

The second specimen (Fig. 55) is one in which a long tubular rectal stricture exists, commencing about one inch inside the anus, and extending upwards for a distance of five inches. All the coats of the bowel appear to be lost in the growth which surrounds the intestine evenly, and which measures one inch in thickness at the middle portion. Above the neoplasm, the intestine is widely dilated, showing very clearly that during life the degree of obstruction must have been considerable.

* Museum of Royal College of Surgeons in Ireland.

Dr. Abraham has made a careful examination of this specimen also, and the structure appears to be almost identical with the former, and undoubtedly is an example of spindle-celled sarcoma.

These two specimens illustrate remarkably the very different macroscopic appearances which may be produced by tumours of the connective tissue type.

A well marked case of alveolar sarcoma of the rectum is reported by Billroth.* The patient was aged fifty-six, and there was a three years' history of difficult defæcation, the growth prolapsing when the bowels moved: it increased so much in size that it could not be replaced. It extended into the rectum for a distance of 6 c.m., and there was a well-marked constriction where it was embraced by the sphincter. There were no enlarged glands or evidence of secondary tumours. The growth was excised, and microscopic examination showed it to be a well-marked example of small-celled alveolar sarcoma.

In connection with Hodgkin's disease, several examples of **lympho-sarcoma** of the intestines have been recorded. In these instances it appears that the tumour originated in the adenoid tissue of the mucosa and in Peyer's patches, and they do not hitherto appear to have produced obstruction or other important symptoms. Dr. Carrington† has detailed a case in which a tumour of this kind weighing one-and-a-half pounds, occupied the cæcum without producing symptoms. I have had personal experience of one case in which a tumour, apparently of this nature, was situated lower down in the intestinal canal, and gave rise to rectal obstruction. I have already alluded to this case in another connection (page 59). The patient, an old man of sixty, came under my care in

* Quoted by Esmarch, "Handbuch der allgemeinen und speciellen Chirurgie," Pitha u. Billroth, Band iii. p. 183. 1882.
† *British Medical Journal*, vol. xi. p. 773; 1883.

June, 1884, at Sir Patrick Dun's Hospital. He complained of piles and difficulty in getting the bowels to move. Upon making an examination a tumour was felt in the hollow of the sacrum, obstructing the rectum. The mucous membrane was freely movable over it, and the tumour itself was movable in the pelvis. As I thought, from the movability of the mucous membrane over it, that it was outside the intestine, I attempted its removal by linear proctotomy. When reached it was found to be very soft, and it broke down under the finger. As much as possible of it was removed. The patient, however, died of septic periproctitis. At the postmortem, it was found that the neoplasm infiltrated and thickened considerably the posterior portion of the muscular wall of the rectum, the new growth in parts above where it was removed being one and a half inches thick. The mucous coat was entirely unaffected, and freely movable over the growth, which appeared to have originated in the muscular coat of the bowel. The pelvic and lumbar glands were all very much enlarged, but with this exception there did not appear to be any general disease of the lymphatic system. Microscopic examination was kindly made by Dr. Abraham, who states that it was in all respects similar to the descriptions given of lympho-sarcoma.

Melanotic sarcoma.—Primary melanotic cancer of the rectum is extremely rare; and according to Virchow this is the only portion of the intestinal tract in which it has been found. Mr. Treves, however, states that there is a specimen of apparently primary melanotic growth arising from the ileum at the London Hospital Museum; and there is an example of melanotic growth in the colon in the Museum of Trinity College, Dublin, but as there were apparently (from the history) similar growths in other parts of the body, it is improbable that the intestinal growth was the primary one. In November, 1884, I brought before the surgical section of the Academy of Medicine in Ireland, the following typical case of melanotic sarcoma of the rectum.

The patient, who was sent to me by Dr. J. K. Barton, was admitted into Sir Patrick Dun's Hospital Sept. 16, 1884. She was a tolerably healthy-looking woman, aged sixty years. Eleven months before admission into hospital she first felt a lump coming down when she was at stool, and difficulty in obtaining an evacuation, with occasional hæmorrhage. A month later she was in a Dublin hospital, where, she stated, a pile which appeared externally was removed. From that time she remained free from bleeding and pain for six months. Towards the end of May, 1884, she suffered from flatulence and indigestion, and the bowels, which had for a long time been costive, became more so, relief being only obtained after the use of strong purgative medicines, and then with considerable straining and pain. There was some slight discharge of bloody mucus occasionally, but not to any great extent. A month later she became conscious of a growth in the rectum, which partly protruded when the bowels moved; lately this had increased much in size. The pain during defæcation was considerable, and was referred to a point immediately above the symphysis pubis, and she was much troubled with pruritus ani.

Upon making an examination the anus appeared normal, and the sphincter was not unduly relaxed. About an inch from the anal verge, on the anterior aspect of the rectum, two distinct and tolerably hard tumours could be felt. By passing the finger well up, the superior limits of both could easily be made out, and below them a smaller mass was to be felt. With one finger in the rectum and the other in the vagina, it was easily determined that there was no abnormal adhesion between the two canals. The rest of the rectum, as far as it could be examined with the finger, appeared normal; and no enlarged glands could be felt in the hollow of the sacrum; nor was any evidence to be found of engorgement of the liver or other abdominal viscera.

I removed the growth by the usual method, Clover's crutch being employed to keep the patient in the lithotomy position. The anus was first stretched, and an incision carried from its margin back to the coccyx, dividing the posterior wall of the bowel to the extent of about 1½ inches. The angles of the incision were held asunder, and a good view obtained of the interior of the rectum and the origin of the tumours. An incision was next carried round the anterior two-thirds of the wall of the gut, about half an inch below the attachments of the growths, and well above the external sphincter. The wall of the intestine was now carefully dissected from the vagina, until it was evident that the healthy bowel could be felt between the finger and thumb above the highest limits of the disease. A curved incision was made with scissors well free of the mass, and the whole removed. Hæmorrhage was not as severe as might have been anticipated, only two ligatures being required. There was a little oozing from a point deep in the incision between the vagina and bowel, to which the benzoline cautery was applied, and, finally, a deep suture was passed to arrest it. No attempt was made to suture the divided portions completely. The wound was thoroughly well washed with a solution of corrosive sublimate, 1 in 2,000, and a sanitary towel wet in the same solution was applied.

The progress of the case was quite satisfactory. The temperature never reached 100°; indeed, for a few days it was subnormal, during which time she

Chap. XXII.] *MELANOTIC SARCOMA.* 347

was much depressed. The bowels moved on the fourth day, and again on the eleventh; each time she had complete control of the motion, and has not since suffered from incontinence, except when she has diarrhœa. When I saw her, nine years after the operation, there was not the slightest evidence of

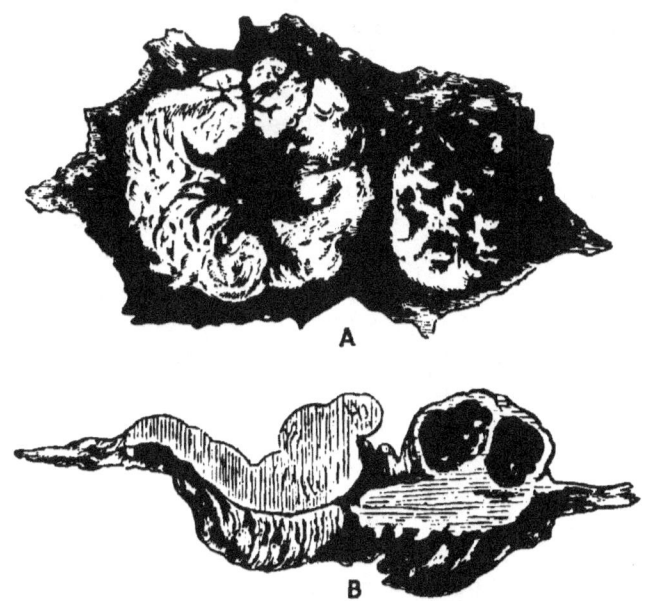

Fig. 56.—Melanotic Sarcoma of Rectum.
A, Surface view; B, section.

recurrence; the bowels moved naturally every day; and she was earning her living as a cook.

Upon examination of the structures removed it was found that a good margin of healthy tissue surrounded the disease. The piece measured about 3 inches in breadth and $2\frac{1}{2}$ inches in length, and consisted of the anterior two-thirds of the rectal tube (Fig. 56). A section carried through both of the principal growths shows that the greater portion of the smaller one is of a sooty black colour, while the larger one is quite white. The third and smallest one is also melanotic. Microscopic examination, kindly made by Dr. Abraham, shows that the growth is a typical sarcoma, much pigmented. In no part of it was there to be found any evidence of proliferation of the gland tissue of the mucosa, and, as far as could be made out, the disease originated in the submucosa.

At a meeting of the Société de Chirurgie, January 28,* 1880, M. Nepveu delivered a lecture on the subject of rectal melanosis, and gave statistics of the cases which have previously been recorded, from which it appears that but ten instances had been noted. In only five of these was there any microscopic examination detailed, and all of these were instances of sarcoma. In two the position was immediately within the anus, once at the sigmoid flexure, and the rest were situate at the anus. In all the cases recorded the disease ran a rapidly malignant course; and in four which were submitted to operation, recurrence was not long delayed.

Virchow has pointed out the remarkable fact† that intestinal melanosis, which is such an extremely rare disease in the human subject, is met with frequently in the horse.

Ossifying cancer of rectum.—As far as I am aware, the case put on record by Mr. Wagstaffe of this form of neoplasm is unique.‡ The following is a condensed summary of the case as recorded: The history pointed to disease of the rectum for about twenty years, when symptoms of obstruction came on. This was followed by pelvic suppuration and death.

* "Mémoires de Chirurgie" (quoted by Kelsey). Paris, 1880.
† "Pathologie des Tumeurs," vol. ii. p. 281. Paris, 1867.
‡ Transactions of the Pathological Society of London, vol. xx. p. 176.

OSSIFYING CANCER. 349

At no time could any tumour be distinguished by examination with the finger in the rectum, nor by manipulation above the pelvis, but the history pointed distinctly to obstruction of the bowel in this region. Upon examination of the pelvic viscera, post-mortem, a tumour was found in the back of the rectum, of about the size of a walnut. It occupied nearly the whole calibre of the rectum, but the disease involved more or less the entire circumference of the intestine, upon a level rather above the larger mass. A small opening large enough to admit a goose-quill was found in the sigmoid flexure, about twelve inches above the cancerous growth, and communicating with a circumscribed abscess cavity within the peritonæum, and this again communicated with the rectum below the obstruction. When first laid open the surface of the cancer presented a nodulated red appearance, but the larger or posterior mass was roughened in its lower half by numerous sharp spicules of bone which projected from its surface. Section showed the growth involving the thickened muscular coat, as a hard contracting mass; and from its base firm fibrous bands ramified into the neighbouring fat, just as from the base of an ordinary scirrhous tumour. That portion which projected into the cavity of the rectum was softer, and its lower part was occupied throughout by numerous spicules of true bone. On the surface, the softer structures having sloughed away, the bony constituents were exposed. The growth did not extend to involvement of the sacrum, which was perfectly healthy, and the other bones of the pelvis were also free from disease. The other viscera were examined and found healthy. The ulceration in the sigmoid flexure appeared simple in character. The solid portion of the growth was composed of cellular and nuclear structures embedded in granular matrix. Bands and fibres, composed almost altogether of nuclei, ramified in the growth, and could be traced as continuous with the osseous portions. It appeared that the nuclei became darker, more granular, and harder in outline as the examination was carried towards the ossified parts; the intervening matrix became more fibrous, and the processes of bone branched out into this. The bony spicules contained numerous lacunæ, whose size was about that of the ordinary nuclei of the growth. They were of various forms, generally branching, and were arranged with no regularity, but in the manner usually found in adventitious bony deposits in tumours.

Cancer of the anus is not very commonly met with; if it originates at that aperture, it is of the usual squamous type of carcinoma, and does not present any characteristics to distinguish it from the same disease in other parts of the body (Plate I.).

Secondary cancer of the rectum, most commonly following cancer of the uterus, does not require here any separate consideration.

CHAPTER XXIII.

SYMPTOMS OF RECTAL CANCER.

As in cancer of other parts of the body, pain is a prominent symptom at a certain period of the disease; but in the early stages it is in many instances exceedingly slight; this is so as long as the disease is confined to the interior of the rectum, and before the anus or the pelvic contents have been encroached upon. So slight is the pain, that in some instances patients consult a surgeon on account of some slight discharge from the anus or sense of uneasiness in the rectum, and an examination reveals the fact that a very extensive neoplasm is present which must have existed for months previously. I recently was consulted by a gentleman who complained of slight œdema of left arm, pain down right leg; with some loss of sensation over the area supplied by the anterior crural nerve. He never had diarrhœa, the bowels moved every day without pain, but on one or two occasions he had passed a little blood; he had a hard mass in the right iliac fossa, enlarged inguinal glands on the same side, and enlarged glands in the left axilla. He ridiculed the idea that he had anything wrong with his rectum, nevertheless I found upon making a digital examination that the whole pelvis was filled with a mass of rectal cancer. This is no doubt due, as Hilton has pointed out, to the characteristics of the upper part of the normal rectum, *i.e.* its great distensibility and little sensibility, conditions the physiological reason of which is obvious. In the immediate neighbourhood of the anus these conditions are reversed, and, as might be expected when this region is involved in the disease, the pain experienced is extreme. It will be within the experience of most surgeons to have met with cases of malignant disease of the rectum in which for months or even years trivial pain alone is complained of. Sooner or later, however, pain becomes a prominent symptom, and is frequently very intense. In no locality, not even excepting the tongue, is the suffering sometimes more severe. The pain may be due to four distinct causes; and the character of the suffering in each case is quite distinct: 1. The disease may involve the anus, where, owing to the abundance of cutaneous nerves and continued motion of the part, the pain will be severe. 2. As the cancer extends beyond the limits of the intestinal tube, the nerves of the sacral plexus may be encroached upon, which may result in violent neuralgia, or in painful cramps of the muscles of the lower extremity. It is well to bear this always in mind, as not unfrequently an attack of (so-called) "sciatica" has been the first indication of a cancerous rectum. 3. Obstruction, when situated in the rectum or lower part of the sigmoid flexure, is followed by a considerable amount of pain; which is always that of a paroxysmal character, and associated with frequent efforts to defæcate. 4. Implication of the bladder will be, of course, associated with considerable pain, especially if the disease has progressed so as to form a fistula, and permit the flow of fæces into the bladder, or of urine into the rectum.

Bleeding is a symptom which is seldom altogether absent; and on the other hand is not often severe. It commonly follows the passage of hardened fæces, and may be taken as an indication that ulceration has commenced. A certain amount of discharge also is a common result, frequently blood-stained and abominably fœtid. At a later stage this discharge, mixed with thin fæces, comes away through the patulous anus, the relaxed sphincters having lost all power of control. The skin about the neighbourhood becomes excoriated, constituting by no means the least of the miseries to be endured by the sufferer.

Diarrhœa may alternate with constipation, or be continuously present, and is often the earliest symptom which attracts attention. Every case of diarrhœa, or so-called dysentery, which has become at all chronic, should be examined by the rectum, and in not a few the cause will be found to be a malignant growth. I have several times seen cases which had been treated for diarrhœa for considerable periods which owed their origin to this cause, and the importance of making an early examination in these cases cannot be over-estimated. Early diagnosis is of greater importance here probably than elsewhere, the great majority of cases not coming under the notice of the surgeon until the disease is so far advanced that the hope of successful operative interference is past.

As has been before pointed out, narrowing of the intestinal tube, sufficient to retard the passage of fæces, may be due to two distinct causes in cancer: either the neoplasm may by its exuberant growth obstruct the calibre of the bowel; or in the more chronic form the cicatricial contraction may form a true stricture of the gut. In either case the symptoms will be similar. Stricture of the rectum produces symptoms in some respects differing from those met with in obstruction of the intestine higher up. The continuous straining and tenesmus which is so marked in the former is absent in the latter; while vomiting of fæcal matter, which comes on tolerably soon when the small intestine is completely stenosed, may not appear for a very long time when the rectum is occluded. In some of the recorded cases complete obstruction was continuous for many weeks or even months before continuous and fæcal vomiting supervened.

Cancerous obstruction, which may have existed for some time, may eventually give way, and an exit be established for fæces through the rectum again, or by an alternative route. In the first instance, the neoplasm may slough to such an extent that the bowel will become pervious again, or, as in the case recorded by Wagstaffe,* ulceration of the bowel above the obstruction may lead to perforation and the formation of stercoral abscess, which may again open into the bowel below the cancer, thus affording a new, though not very efficient, route for the fæces. In the case of the celebrated Talma, as recorded by Quain, this also appears to have been the case.

Where an opening of sufficient size forms into the vagina, the more urgent symptoms of obstruction may be relieved, but the patient is left in

* See page 348.

a truly miserable state; but where the opening takes place into the bladder, no sufficient exit for fæces will be by this means provided, and the urgency of the obstruction will continue; while at the same time the other symptoms will be much aggravated. Opening into some of the pelvic viscera by ulceration in this way may be due to breaking down of the neoplasm itself, or it may be due to the distension and irritation of fæces above the obstruction; the ulceration then being of a simple character. This form of stercoral ulceration may take place at a long distance above the seat of obstruction, several cases being recorded where the cæcum has given way and produced a fatal peritonitis, in consequence of the dilatation due to rectal cancer. At other times nature has attempted to overcome the obstruction by the formation of an artificial anus at some part of the cutaneous surface, but such cases are of extreme rarity, and likely only to give a very inefficient relief to the obstructed gut. Dieffenbach records a case* in which it was necessary to evacuate, by means of free incision in the buttock, an enormous quantity of fæces extravasated from a cancerous rectum; and Smith† gives a case in which an extravasation in this way found its way into the hip joint.

As has been elsewhere stated, the glands first affected, if the disease does not implicate the anus, will be the pelvic and lumbar systems. The former may be felt through the walls of the rectum, and the latter occasionally by deep abdominal palpation.

When secondary tumours have formed in the liver, there may be indication of its increase in size; and possibly, if the abdominal wall be thin, the surface may feel irregular and knobby.

Œdema of either leg is a symptom not uncommonly

* Quoted by Leube, Ziemssen's "Cyclopædia," vol. vii. p. 437.
† "Surgery of the Rectum." 1872.

present in the later stages, and is usually of grave import as indicating an involvement of the iliac vein in the disease. In common with all forms of cancer, the peculiar cachexia soon becomes obvious, and if hæmorrhage has been at all abundant it comes on more rapidly. I think the sallow skin which is so characteristic is more marked in this form of cancer than in others. The onset of bladder implication is indicated by frequent and painful micturition; and fistula is of course soon rendered obvious after it has occurred.

In a case to which I have already alluded (page 104), which I saw under the care of the late Mr. B. W. Richardson, the first symptom which aroused suspicion was turbid urine, from which a sediment settled. Upon examination by the microscope particles of striped muscular fibre and other fæcal débris were to be seen, and a digital examination demonstrated a rectal stricture high up. Leube * notes a case in which the secondary involvement of the ureter in rectal cancer produced a large hydronephrosis.

The duration of symptoms may, in difficult cases, materially assist the diagnosis. If there is a history of rectal trouble slowly increasing for years, it is highly probable that the disease is not malignant.

Digital examination.—Whenever the symptoms of rectal cancer exist at all, a complete digital examination should be made. In the majority of cases, within a short distance of the anus the surgeon will feel a hard nodular and irregular surface, which may surround the entire circumference of the bowel, or be more particularly confined to one side of it. When stricture exists, the tumour frequently is felt projecting into the lumen of the bowel, and conveying to the finger a sensation almost exactly resembling that of the os uteri. Should the finger not encounter anything abnormal, the patient should be

* Ziemssen's "Cyclopædia," vol. vii.

made to stand up, and the digital examination should then be repeated, the patient at the same time being told to bear down. In this way a tumour which was not within reach by the ordinary method may occasionally be explored. Should nothing still be felt, and the symptoms clearly point to rectal disease, the patient should be etherised, and a careful bi-manual examination instituted, with the patient in the lithotomy position. This method is also of use in determining the height to which neoplasms, that are easily recognisable below, extend upwards. The existence of malignant disease having been determined, it is essential, with a view to treatment, to determine the following points: First, the distance to which the disease extends upwards; this may be done with the finger alone, by the bi-manual method, or by a ball-ended probang. Secondly, the movability of the rectum upon the other pelvic structures is of use in estimating whether or not the disease has spread past the limits of the intestinal tube. And thirdly, a careful examination should be made to feel, if possible, any enlarged glands, which may sometimes be felt in the hollow of the sacrum through the rectal wall. In examining a case of this kind the greatest care should be employed, as in several recorded cases the attempt to pass a probang, or even a roughly made digital examination, has been followed by rupture into the peritonæal cavity.

In the female additional information may be gained by vaginal examination, the extent of the growth being sometimes easily determined through the recto-vaginal septum; while the fixity or freedom of the uterus is a point of great importance to make out.

Diagnosis.—There are but two conditions with which rectal cancer is likely to be confounded: viz. tumours external to the intestinal tube; and non-malignant stricture.

In the case of the former the diagnosis is easy if the disease is within reach of the finger. The fact that the mucous membrane is freely movable, and that the neoplasm is unquestionably outside the bowel, will render the matter clear. Hilton records a case * in which the presence of enlarged glands, which could be felt through the rectum, in a case of chronic ulceration, had given rise to the opinion that the case was one of cancer; when the ulcers healed the swelled glands disappeared, showing that they were simply due to irritation. In the same way uterine tumours, or even the fundus of a retroflexed uterus, by pressing on the rectum and causing obstruction, have given rise to an erroneous diagnosis of rectal cancer.

To distinguish between the malignant and non-malignant strictures is a matter of great difficulty. In this the duration of symptoms will prove of much service, the onset and progress of the non-malignant being extremely slow. The sensation conveyed to the finger will also be different. The ordinary stricture is smoother and more regular, and there is generally an absence of the nodular and protruding masses so characteristic of cancer. Cripps has also drawn attention to the fact that in the malignant form there is usually a portion of tolerably healthy mucous membrane between the cancer and the anus, whereas in the non-malignant stricture this portion is generally more or less infiltrated.

The diagnosis between squamous epithelioma of the anus and papillomata is sufficiently easy; as in the latter the skin surrounding the tumour is not involved, the neoplasm being in some instances even pedunculated, whereas in the epithelioma there will be considerable infiltration of the true skin.

* "Lectures on Rest and Pain," p. 294. 3rd edition, edited by Jacobson.

CHAPTER XXIV.

TREATMENT OF RECTAL CANCER.

THE medical treatment of cancer of the rectum presents two chief points which must be borne in mind by the surgeon : first, to ensure that the bowels are kept sufficiently free to obviate the occurrence of fæcal accumulation above the disease ; and, secondly, to supervise the use of morphia and other narcotics. In order to relieve pain, morphia, either hypodermically in the form of suppositories, or internally, is frequently used somewhat recklessly, with the result that there is superadded to the miseries of the rectal cancer the mental suffering and total inability to bear physical pain of the morphia habit, so that unless used with a very sparing hand, opium, instead of rendering the remainder of life more comfortable, will add to its suffering.

The use of bougies, or any dilating instrument, is attended with extreme danger, several cases of fatal rupture having been induced by this means.

The operative treatment of cancer of the rectum may, with advantage, be classed under two heads: the one necessarily palliative, as directed only to the relief of the prominent symptoms of intestinal obstruction and pain; the other having for its object the complete removal of the disease.

Of the former, three operations are at present practised where extirpation is inadmissible, and of these colotomy must still be ranked in the first place, although there seems to be a tendency amongst operating surgeons to make use, as far as possible, of other plans of treatment, even where the symptoms of severe obstruction are manifest. Of these procedures the most important is linear proctotomy, or external rectisection, which has, chiefly owing to the writings of Verneuil,* Panas,† and Kelsey, obtained a recognised place in surgery as a treatment for malignant stricture, and at the Copenhagen Congress ‡ Verneuil spoke strongly in favour of this procedure as replacing both colotomy and excision. In many cases he considers it preferable to the former as being less dangerous, equally efficient, and more convenient; and he considers complete removal by excision impossible. Those surgeons who practise excision confine colotomy to the cases in which it is impossible to extirpate the whole mass, and so the cases in which opening of the colon is now practised would be incapable, in consequence of their extension, of relief by the linear proctotomy of Verneuil. The operation, therefore, must be compared with extirpation alone, and I think that the results now gained by the latter procedure will decide most surgeons in selecting it. For the treatment of non-malignant stricture linear proctotomy is an admirable method, and for a description of the operation the reader is referred to the chapter on that subject. The suggestion of Kelsey to make two vertical incisions posteriorly, and remove the mass of neoplasm from between them, gives more room certainly, but it is open to the same criticism as the more simple operation.

The third form of palliative operation is the removal, with a scoop or the fingers, of as much as possible of the cancerous mass. Such cases are described by Allingham, Cripps, and Volkmann;

* *Gaz. des Hôp.*, October and November, 1872; and *Gaz. Hebdom.*, March 27, 1874.
† *Gaz. des Hôp.*, December, 1872.
‡ "Compte Rendu," par C. Lange, Secrétaire Géneral, tome ii., section de Chirurgie, p. 21.

and the result was a removal of the obstruction. From the recorded cases it appears that when the mass was thoroughly broken down and removed hæmorrhage was not excessive, and sometimes even partial cicatrisation has been known to follow. Sir Joseph Lister states[*] that he has seen in the practice of Simon of Heidelberg great advantage follow the scraping of epithelioma of the rectum with the sharp spoon.

The radical cure of cancer of the rectum may be attempted by a free excision from the perinæum, or through the sacrum, and where the disease is situated high up, the operation, to which Marshall has applied the term colectomy, may be performed. This subject is, however, rather beyond the scope of this work; but it will be found fully dealt with by Mr. Treves in his book on intestinal obstruction.

Excision of the rectum is now a thoroughly established operation, and although at first it met with a great deal of opposition in England, it is now pretty generally adopted as the best treatment in selected cases. Originally performed by Faget in 1763, it does not appear to have attracted much attention till 1833, when Lisfranc again brought it into notice; but its establishment, as at present practised, is due to the German surgeons.

In order to arrive at a just conclusion as to the advantages of extirpation of the rectum, it is necessary to review the course which rectal cancer runs when not subjected to operation. It would appear, from a consideration of a large number of statistics, that the average duration of life is about two years from the appearance of the first symptoms, and during that time the condition of the patient is truly miserable. Where obstruction is present, the constant straining is a source of perpetual pain and annoyance to the

[*] *Lancet*, May 20, 1882.

patient, and even when this symptom is not present the continued mucous and bloody discharge, the extreme pain suffered when the disease encroaches on the bladder, the anus, or the nerves of the sacral plexus, combine to render this disease one of the most distressing that can possibly come under the observation of the surgeon; and it is little to be wondered at that any operation which can hold out a chance of remedying this condition should readily be grasped at by both surgeon and patient. We must, however, consider the question from more than one point of view: first, as to the immediate risk to life; second, as to the probability of complete cure, and, if so, the condition in which the patient will be left; and, lastly, supposing recurrence to take place, how long will it be delayed, and what will be the course of the secondary disease. I am convinced that a careful and unbiassed consideration of the facts bearing on these questions will serve to convince the impartial observer that they are not only sufficient to justify the operation in suitable cases, but that it is the duty of the surgeon strongly to recommend it.

In the interval that has elapsed since the first edition of this work was published many additional cases have been recorded, more particularly by the German surgeons, showing a somewhat diminished death-rate, mainly due to improved methods of operation; but the mortality must always remain somewhat high, as it is impossible completely to obviate stercoral fouling of the wound by any means at present at our disposal.

Let us proceed to consider these questions in detail. First, as to the immediate risks of the operation. In trying to estimate the mortality of any operation, more particularly one which has only of recent years been extensively practised (as is the case with the operation under consideration, in Great

Britain at any rate), it is manifestly useless to collect all the cases published in the journals and from these deduce statistics, as there is a strong tendency amongst surgeons to publish isolated successful cases, while their fatalities are not so accurately recorded. Consequently we must only place reliance upon the experience of those surgeons who give the results of the total number of operations which they have performed.

I have collected 175 cases in which, I think, we may be satisfied that the conditions necessary for faithful statistics have been carried out, so that we may take the result as fairly reliable. These give a death-rate of 16·5 per cent.; and when we take into consideration the nature of the operation and the disease for which it is performed, we may consider this a fairly good result.

Upon looking to the cause of death in these cases, we find that in upwards of 80 per cent. periproctitis, or peritonitis, is stated to have been the chief factor in producing the mortality.

Although the full details of Listerian dressings are inapplicable to these operations, a great deal can be done in the way of antiseptic treatment to obviate the above preventible complications; and the more fully we appreciate the advantages of closing the deep parts of the wound completely by sutures, thorough drainage, frequent washings with antiseptic solutions, and dusting with boric acid, the more likely are we to reduce the death-rate still further. Volkmann states that amongst his early cases he lost a great number from septic inflammation, but since he has adopted better methods of wound treatment his results have greatly improved. He advocates continuous irrigation of the wound with an antiseptic fluid, such as solution of salicylic acid, or carbolic acid, until granulation is established. Billroth, between the years 1860 and 1876, lost 13 out of 33 cases, and all the cases died of septic periproctitis and peritonitis.* Cripps† gives twenty-three cases within his own experience, of which four died. The statistics given by Heuck‡ of the practice of Professor Czerny for a period of six years appear to be the best hitherto recorded. Of twenty-five patients operated on, only one died as a direct result of operation.

In many respects the history of rectal extirpation resembles the early history of ovariotomy; and it is highly probable that with increased care in wound treatment and operative detail the rate of mortality will be materially lessened. It is, therefore, at present premature to be guided too much by statistics.

Let us now consider what are the probabilities of complete cure; or, if recurrence takes place, how long will it be delayed? Billroth, in 1881, had only two cases in which the patients lived two years after the operation; and Allingham speaks with great caution, apparently not considering that life is even prolonged by the operation; on the other hand, Cripps found, that out of twenty-three cases, in nine the disease recurred after periods varying from four months to two years, and he was able to trace six that remained well at periods varying from two to four years. Curling§ had one case in which there was no return after six years; Velpeau records two cases which were well after ten years; and Chassaignac has had similar experience; but, probably, the best results obtained by any one are those of Volkmann.∥ He states that three times he has had complete cures, and several cases of very late recurrence:

* "Clinical Surgery," New Sydenham Society, 1881.
† Loc. cit., p. 397.
‡ Archiv für klinische Chirurgie, Band xxix. Heft 3.
§ "Diseases of Rectum," p. 164. 1876.
∥ Sammlung klinischer Vorträge, May 13, 1878.

once after six years, once after five years, and once after three. One died of carcinoma of the liver eight years after operation without local recurrence, and one case remained well eleven years after the removal of a very voluminous and high-reaching mass; in this case local recurrences in the shape of hard nodules in the cicatrix occurred twice, and were removed. In Czerny's experience, according to Heuck, nine were alive at the time of publication of the paper, and free from relapse. Of these, two had survived the operation longer than four years, one had been operated on three years and nine months before, three were well after intervals of at least two years, one at the end of twelve months, and two at the end of six months; while in fifteen cases (60 per cent.) there was a local recurrence within one year.

Dieffenbach records thirty cases in which the patients lived many years after operation, but this statement is usually looked upon with suspicion.

In my own practice one patient remains perfectly well and free from return nine years after operation, while another has continued six years without recurrence.

Although the total number of cases is as yet small, and the opportunity of judging whether many of the apparent cures will be permanent is insufficient, the results hitherto recorded will compare most favourably with the records of operation for cancer in other parts of the body, notably the tongue and breast, both as regards the prolongation of life, and the possibility of complete cure.

As to the condition of the patient after recovery from operation, we must remember the horrible disease for which that operation was performed, and compare the condition before and after its removal. When the sphincter has not been removed, the amount of incontinence is usually trivial, and it is only when there is diarrhœa that any trouble arises. This is generally easily met by the use of an antiseptic pad. When the entire lower end of the rectum has been removed a considerable amount of control often is maintained, but even in the worst cases of incontinence met with after ablation of the rectum the result compares favourably with the usual artificial anus following colotomy, and is vastly preferable to the state of a patient suffering from advanced rectal cancer.

Recently attempts have been made to lessen the amount of incontinence following these operations. Willems* separates a band of fibres of the gluteus maximus from the rest of the muscle by blunt dissection, and brings the divided end of rectum out through this opening before suturing to the skin: in this way he attempts to confer a voluntary control over the new anus. Gersunt† proposes to give a sphincter-like action to the divided end of the bowel by torsion. The divided end is grasped with two pairs of catch forceps, and twisted in its long axis until a considerable elastic resistance is experienced in attempting to introduce the finger into the bowel; the free end of the gut is then sutured to the skin with the twist maintained. In his first case the torsion was carried to 180°, and in a second to 270°. In both cases continence was good after operation. I have adopted this highly ingenious method in one case (No. 13), and believe it to be a distinct addition to the operation.

A more troublesome sequela of operation than incontinence is stricture, which in many of the recorded cases appears to have given a very great deal of trouble. In those cases where it has been found impossible to draw down the gut and suture it to the skin, the extensive surface heals by granulation, and the orifice gradually becomes constricted; even in the

* Centralbl. f. Chirurg., May 13, 1893.
† Centralbl. f. Chirurg., July 1, 1893.

hands of some of the most skilful surgeons, treatment by means of tubes, incision, or even colotomy has been subsequently required. If, however, a small strip of mucous membrane can be retained down to the anus, or the mucous membrane brought down and sutured to the skin, as in the procto-plastic operation of Amussat for imperforate rectum, this trouble is not likely to arise. The freedom from incontinence which some of these patients enjoy is very remarkable. In a case of my own there is a slight prolapse of mucous membrane which occludes the anus, and prevents escape of fæces, except during defæcation. As O'Beirne pointed out long ago, the rectum in health is empty, except immediately before the act of defæcation.

Recurrence of the disease usually takes place as nodular masses in the cicatrix; or in the deep lumbar glands, liver, or other internal organs. When occurring in the cicatrix, a secondary operation is often attended with good results. And even where not suitable for removal, these secondary growths are usually much less painful than the primary disease, owing to the destruction of the sensory nerves of the region at the time of operation. Death from internal cancer is also considerably less painful than that from unchecked cancer of the rectum.

The most complete and accurate directions as to the selection of cases and the details of operations for excision of rectal cancer when situated low down are those given by Volkmann.* He classifies the cases met with under three heads: 1. Where there is a localised nodule of disease which can be removed by dilatation of the anus, and the wound closed by suture; this is not attended with difficulty unless situated high up. 2. Where the greater proportion of the rectal circumference, including the

* *Sammlung klinischer Vorträge*, May 13, 1878.

anus, is diseased; in this case the anus must be surrounded by an incision extending into the ischio-rectal fossa, the rectum dissected up, and amputated above the seat of disease. Volkmann, in the paper alluded to, recommends the bringing down and suturing of the divided rectal tube to the skin, drainage tubes being inserted between the stitches. 3. Where the disease is altogether above the anus, involving the entire circumference of the bowel. A deep posterior incision to the coccyx is the first essential procedure in this instance; the rectum is then incised round its circumference above the external sphincter, the bowel dissected up and amputated. This operation is open to an objection not applicable to the other two, namely, that as the blood-vessels supplying the lower portion of the rectum are of necessity divided, gangrene of this portion is apt to occur.

The field of operation has recently been much extended, and cancers extending farther up the bowel can now be successfully dealt with. (*See* page 374.)

The following description of the **operation of perinæal excision** includes the principal points to be borne in mind. In order to prepare a patient for operation, a dose of purgative medicine should be given for a couple of nights before, and the bowel well emptied by a copious enema on the morning of the operation. The patient should be retained in the lithotomy position by means of Clover's crutch, and an incision carried deeply from the back of the anus to the coccyx. This is an exceedingly important part of the operation, as it gives full room for further manipulations; and has been called, not inaptly, by Allingham, the "key" of the operation. If the entire circumference of the bowel, including the anus, is diseased, incisions should be now carried well clear of the disease round the anus, and deeply into the ischio-rectal fossa, the attachments of the levatores ani divided, and the dissection carried upwards posteriorly and at the sides. This can be readily accomplished, but in front there is always considerable difficulty owing to the close attachments of the rectum to the bladder and urethra in the male, and the vagina and uterus in the female. In the former the presence of a full-sized sound in the urethra will prove of much assistance, and in the latter the occasional introduction of the finger into the vagina will serve a like purpose. For dissecting the intestine free, a pair of blunt-pointed scissors will be found the most convenient instrument; and assistance may be gained by the use of a blunt hook, using it in the same way that a strabismus hook is used to hook up the ocular muscles in an enucleation of the eyeball. If the disease has not implicated the anus, or if a vertical strip of mucous membrane be unaffected, the preceding operation should be so far modified as to leave as much normal tissue as possible, care being always taken that at least one quarter of an inch of healthy tissues surrounds the disease upon all sides. The dissection having been carried up to healthy tissue above the disease, the rectum is to be amputated. For fear of hæmorrhage this has frequently been done with the écraseur, the Paquelin cautery, or even the ligature; but as the part is so well under control bleeding need not be feared, and the section can be made much more cleanly with a pair of curved scissors. A number of catch forceps should be at hand to secure vessels as they are divided, but there is not likely to be any free bleeding until the last section is made, and then the arteries can be picked up and tied, generally without difficulty. An important question now to decide is whether any attempt should be made to bring down the gut and suture it to the skin wound. Cripps strongly advocates leaving the wound to granulate without the application of any sutures, his objection to the stitches being that they cut out before union takes place, and that while in place they produce little pouches outside the gut in which fluids will collect, and become septic; while leaving the

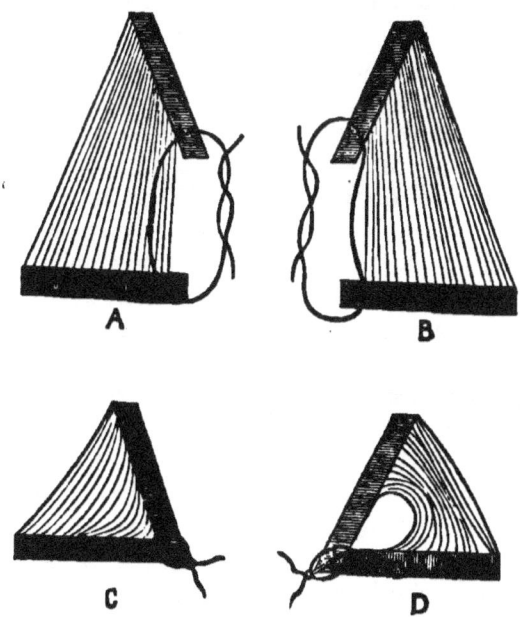

Fig. 57.—Diagram showing the method of passing Sutures.
A, Suture passed deeply; c, the same suture closed; B, suture passed through bowel and skin only; D, the same suture closed.

wound entirely open, with the patient in the recumbent position, ensures absolutely free drainage. Other operators give similar advice; while Volkmann and Czerny recommend stitching, so as to diminish wound surface as much as possible, and by joining mucous membrane to skin to obviate the tendency to stricture. It appears to me that a great deal depends upon the way the sutures are put in. If they are simply put through skin, and then through the gut,

Y—23

they will, when closed, make a cavity outside the rectum; but if they are passed deeply through the surrounding pelvic structures as well, these cavities cannot be formed, and as the strain will be then divided over a larger surface, the tension will be taken off the gut so much that they will be much more likely to hold (Fig. 57). If two such sutures are passed on each side they will bring the gut well down if it has not been divided very high up, and a number of superficial sutures should then be put in to complete the adjustment of the skin and mucous membrane. I consider the deep closing of the wound a most important element in the operative detail. After the operation an attempt may be made to keep back the fæces by plugging the gut with iodoform gauze or other similar antiseptic material, a catheter passed through the plug being left in the bowel to permit of the escape of flatus. If the gut will tolerate this plug, and it can be left *in situ* for a week or ten days, it will prove of enormous advantage, by permitting a complete adhesion of the gut to the perinæal wound to take place. The diet for the first fortnight after the operation should be carefully regulated, so as to leave as little solid residue as possible.

Complications of the operation.—Wound of the peritonæum is of frequent occurrence, and careful anatomical measurements have been made to determine the distance of the peritonæal pouch from the anus, so as to define the limits within which the rectum can with safety be removed. Such measurements are, however, comparatively useless. In the first place the measurements must vary with the different positions of the anus and the amount of fluid in the bladder, and the relation of the serous covering to the bowel in health is no criterion whatever as to its state when the rectum is diseased, as the constant straining, if there is obstruction, and the dragging due to contraction of the cancer, will materially alter the normal relations. The only important anatomical point to remember in this connection is the fact that, in the female, the peritonæal pouch descends about an inch farther than it does in the male. Allingham states* that he has met with peritonæum within two inches of the anus in a female, and removed five inches of gut in a man without ever having seen it. In order to render wound of the peritonæum less likely to take place, I have tried fully distending the bladder before operation, which I find has a much greater effect in lifting the peritonæum away from the bowel than the converse proceeding of distending the rectum has of clearing it away from the suprapubic region for lithotomy. This is manifestly of greater use in the male than in the female, and the distended bladder is more easily recognised and protected during operation than the empty one. Wound of the peritonæum does not, however, appear to be such a serious complication as some surgeons have thought. According to Heuck † this accident occurred in eleven of Czerny's cases ; in six of these the wound was sutured, while in the remainder the rent was left open, care being taken to join accurately the margin of the divided bowel to the skin, thus preventing extravasation. As Cripps, however, has pointed out, direct involvement of the peritonæum in the cancerous growth is a very serious complication, as it indicates such an implication of the lymph paths that recurrence of the disease cannot long be delayed.

Implication of the other pelvic structures is a very serious complication ; and, when extensive, must be held to contra-indicate operation. A slight involvement of the recto-vaginal septum, however, can easily be dealt with, the vaginal opening being either closed

* *Loc. cit.*, p. 281.
† *Loc. cit.*

at the time of operation, or by a subsequent plastic procedure. Where, however, the bladder, prostate gland, and urethra are much involved, the prospect of useful interference is small indeed; although a case is recorded by Nussbaum in which a man was reported well three years after the removal of rectum, prostate, and neck of bladder;* but such extensive operations have not since been frequently imitated.

Amongst the modifications of excision, the combination of colotomy with it is one of the most important. In a paper read at the Société de Médicine of Lyons, in May, 1884,† M. Maurice Pollosson advocates the combination of laparo-colotomy with extirpation of rectal cancer. He selects the left iliac region as the site for the operation, because there more readily than in the lumbar region can he close up the lower segment of the bowel, which closure he regards as a point of essential importance in the operation. This he does by invaginating some millimetres of the lower free end after dividing the bowel clean across, and closing up the opening completely by means of five or six catgut sutures, which thus bring into close apposition the serous surfaces. The artificial anus is completed by suturing the upper extremity of the bowel carefully into the wound. After the patient has recovered from this operation, he proposes to extirpate the cancerous mass, which, by virtue of the preliminary operation, is practically removed from its relations as a part of the digestive tract, and converted into a pelvic tumour. Operating under the conditions so brought about, it is possible to apply the principles of antiseptic surgery much more thoroughly and efficiently than in the conditions existing without such a preliminary operation. In most cases he believes that it would be advisable to allow the patient to

* Bayr., *Ärtz. Intelligenzblatt*, November, 1868; quoted by Van Buren. † *St. Louis Courier of Medicine*, July, 1884.

recover from the effects of the first operation before performing the second; though in certain cases he thinks that circumstances might be such as to make it better to go on and extirpate the cancerous mass at once after establishing the artificial anus.

Mr. James E. Adams has recommended* the performance of lumbar colotomy as a preliminary measure in all but the very slightest cases, and as soon as the patient had recovered from this to excise the rectal cancer from the perinæum. The advantages of diverting the fæces from the wound during healing, and from the recurrent growth, should it take place, are sufficient in his opinion quite to justify the additional operation. In a case in which he performed the double operation, and in which the patient was under observation for two years subsequently, he states that the advantages were very obvious. Although a recurrence took place six months afterwards, the patient was quite unaware of its existence. The annoyance was so trivial, he contends, that by adopting the course indicated any patient might pass through all the phases of this horrible and fatal malady with scarcely any pain at all.

I should not be disposed to adopt either of the above modifications except in very special circumstances, as the advantages of retaining the fæcal outlet in the perinæum are very great.

The removal of cancers situated high up.
—There is a class of cases which, as Volkmann has well described, are too high for removal from the perinæum, and too low for removal by laparotomy. Dr. P. Kraske of Freiburg† communicated to the German Surgical Congress a method which he had worked out on the cadaver. According to him access to the upper part of the rectum is made far easier by splitting the

* *British Medical Journal*, Aug. 15, 1884.
† "Annals of Surgery," vol. ii. p. 415; 1885.

soft parts in the middle line from the second sacral vertebra to the anus, dividing the muscular attachments to the sacrum as far as the ends of the bone on the left side; excising the coccyx, and then dividing from the sacrum the attachments of the two sacrosciatic ligaments, and drawing away the left edge of the wound. Still further access to the upper portion of the rectum is gained by chiselling away a bit of the lower left side of the sacrum. If the bone be divided in a line beginning on the left edge at the level of the third posterior sacral foramen, and running in a curve concave to the left through the lower border of the third posterior sacral foramen, and through the fourth to the left lower corner of the sacrum, the more important parts, especially nerves, are not injured; and the sacral canal is not opened. The upper portions of the rectum thus become so accessible that the rectum can be brought into full view and amputated without difficulty up to where it passes into the sigmoid flexure. Further, this procedure admits resection of the upper rectum with preservation of its lower end. Kraske tried this method on the dead body in a case of high rectal cancer; and then twice on the living subject. Once in a debilitated woman, aged forty-seven years, the cancer commenced a short distance above the anus, while its upper end could not be felt. The rectum was amputated (with avoidance of the sphincter) where it was wholly surrounded by peritonæum. The patient made a good recovery. His second case was in a man, aged thirty-seven years; the lower extremity of the disease could just be reached with the finger. A portion of the lower bowel was spared, though divided posteriorly. The gut was pulled down, and the anterior two-thirds united by suture. The lower (posteriorly open) portion was closed later by a plastic operation.

Kraske's paper has opened up a field of operation

in cases which before were considered quite inoperable, and his method has now been frequently adopted and modified in several important details, particularly by the German surgeons. The method adopted by Bardenheuer* appears at once the simplest and most satisfactory. An incision is carried from the back of the anus to the middle of the sacrum; the muscles divided from the sacrum and the sacro-sciatic ligaments cut through; the sacrum is now cut through transversely at the level of the third sacral foramen, and the posterior surface of the rectum cleared in the superior pelvirectal - space (above the levatores ani with their fascial coverings). As the whole hand can now be introduced into the pelvis the bowel can be explored up to the sigmoid flexure if necessary. If the disease is situated entirely above the attachment of the levatores ani the resection of the diseased bowel with circular suture of the intestine can be effected, or where the anus is involved this portion can be extirpated and the upper lumen of the bowel brought down and sutured to the more convenient portion of the incision.

I have adopted the trans-sacral extirpation of rectal cancer in four cases, in all of which a successful result was obtained. Of two of these (Nos. 10 and 12 in table, page 379) I will give the details, as they may be taken as typical cases of the operation for high-situated cancer. A woman, aged thirty, had suffered from symptoms of rectal cancer for eight months with gradually increasing obstruction, which at the time of her admission to the hospital had almost become complete. Digital examination revealed a ragged, ulcerated surface, entirely surrounding the rectum, commencing immediately above the anus, and extending higher than the finger could reach. The tumour was freely movable in the pelvis, and its upper limits were made out by bimanual examination.

* Volkmann, *Klin. Vorträge*, 296.

An incision was made from the back of the anus to the middle of the sacrum, and the muscles and ligaments divided from the bone. The sacrum was now divided through the fourth segment, an American pruning shears being used for this purpose, which did the work much more rapidly than a saw, and without the crushing of an ordinary bone forceps. The piece of bone was removed and the rectum cleared with fingers and scissors. This portion of the operation is much more easily done from above than from below the levator ani muscle. As the anus was involved it was surrounded by an elliptical incision and the entire rectum dissected free. In doing this the peritonæum was opened anteriorly. The rectum was divided close to its junction with the sigmoid flexure about one inch above the highest limit of disease. As the strain would have been too great to bring down the gut to the normal position of the anus it was brought out at the upper angle of the wound, where it was secured without undue tension in the position formerly occupied by the lower end of the sacrum and upper bone of coccyx. The peritonæal wound was adjusted by fine catgut suture, and the entire perinæal wound, including the former position of the anus, was carefully closed by a series of deep sutures passing under the bottom of the wound, and a few superficial stitches where required to adjust the skin accurately. Primary union ensued through the entire wound without a single drop of suppuration. The patient suffers little inconvenience from her sacral anus as she has a very considerable power of continence. The tumour removed is illustrated. (*See* page 340.) It proved to be a colloid carcinoma.

The second case (No. 12 in table) was that of a woman, aged forty, who had suffered from rectal cancer for one year with considerable obstruction. Examination showed that the disease was situated well above the anus, its upper limits could not be determined, but it moved freely in the pelvis, and was unattached to the uterus.

The lower end of the sacrum and the coccyx were removed as in the last case, but the anus was not incised, the tumour was then isolated from the surrounding pelvic structures with the greatest facility. The bowel was divided at the junction of the sigmoid flexure with the rectum, three-quarters of an inch above the disease; and below, it was again divided a similar distance from the disease, leaving about one inch of healthy gut above the anal canal. The sphincters and lateral portions of the levatores ani muscles were unmutilated. In removing the tumour the peritonæum was freely opened. The divided ends of the bowel were readily brought together and sutured round the entire circumference, a plug of iodoform gauze was introduced into the rectum through the anus, and the middle of the incision was kept open, through which the space surrounding the bowel was also lightly plugged with iodoform gauze. The bowels moved for the first time on the eighth day, when the posterior portion of the freshly united rectum gave way and some of the fæces escaped by the wound; the union, however, remained perfect in front and laterally. The fistula thus formed gradually closed by granulation, and two months after operation finally healed. Digital examination shows that the line of union is smooth and devoid of narrowing, and defæcation is absolutely normal.

This appears to be a result as near perfection as it is possible to obtain; but hitherto the success attending circular rectorraphy has not been large, apparently due to the fact that in the first attempts at defæcation after operation, portion of the freshly united intestine is apt to give way in the effort to dilate the anal canal; in this respect circular rectorraphy contrasts unfavourably with the similar operations on the small intestine and colon, where no such strain has to be encountered. This difficulty may be met in one of two ways; the incision may be carried through the sphincters into the anus, and the wound thus formed kept open until union of the intestine above is firm, when it can be closed by a plastic procedure; or, what I should prefer, and intend for the future adopting, is to suture carefully the front and sides of the rectum, but to leave in the first instance a sufficient opening for defæcation posteriorly; if then the fistula thus formed did not close by granulation, as in the case above recorded, it could readily be occluded by a plastic operation, when the rest of the circumference was firmly united.

Subjoined is a table of cases of excision of cancer from my own practice. The cases subjected to excision constitute only a small proportion of the entire number of cases of cancer of the rectum coming under notice; the vast majority when presenting themselves having such extensive pelvic adhesion that they were deemed inoperable, except where obstruction was marked, when colotomy was undertaken.

TABLE OF CASES.

CASES OF RESECTION OF RECTUM FOR MALIGNANT DISEASE.

Name.	Date.	Sex.	Age.	Hospital.	Private.	Method.	Pathology.	Result.	Observations.
1. B. McL.	Nov. 1, 1884.	F	60	H	—	Posterior linear incision.	Melanotic sarcoma.	R	Perfectly well and able to earn living as a cook 8 years after operation.
2. Dr. N. ...	Oct. 5, 1886.	M	65	—	P	do.	Cylinder-celled epithelioma.	R	Perfectly free from recurrence 6 years after operation.
3. W. F. ...	Nov. 21, 1887.	M	62	H	—	do.	do.	D	Septic peritonitis.
4. M. D. ...	July 16, 1889.	F	48	H	—	do.	do.	R	Vaginal septum implicated and removed. Recurred in suture points 6 months afterwards.
5. Mr. McK.	Oct. 18, 1889.	M	70	—	P	do.	do.	R	Died suddenly a year subsequently. No evidence of recurrence.
6. Mrs. F. ...	Jan. 5, 1890.	F	60	—	P	do.	do.	D	Died of shock same day.
7. Miss H. ...	Feb. 22, 1890.	F	45	—	P	do.	do.	D	Septic peritonitis.
8. Mr. A. ...	Feb. 18, 1891.	M	55	—	P	Transverse division of sacrum.	do.	R	Recurred as abdominal carcinoma a year afterwards.
9. O. R. ...	Mar. 19, 1892.	M	70	H	—	do.	do.	R	Continence good.
10. B. C. ...	Feb. 2, 1893.	F	30	H	—	Perineal incision.	Colloid cancer.	R	Occurred in cicatrix of fistula operation.
11. Mr. A. ...	May 15, 1893.	M	60	H	—	Transverse division of sacrum.	do.	R	Circular rectorrraphy. Union perfect. Defæcation normal.
12. A. L. ...	Sep. 14, 1893.	F	40	—	P	Perineal incision.	Cylinder-celled epithelioma.	R	Torsion of bowel. Continence good.
13. Mrs. D. ...	Oct. 11, 1893.	F	50	—	P	do.	do.	R	

CHAPTER XXV.

COLOTOMY.

THE history of this operation is in many respects one of interest. Although it is upwards of a century and a half since it was first proposed, it is only within the last thirty years that it has been practised upon at all a large scale, as a means of obviating death in one of its most painful forms, viz. by intestinal obstruction; and its present position as a recognised operation is mainly due to the efforts of English surgeons, notably Curling, Bryant, and Allingham. Apparently the first suggestion of the operation was made by Littré in the year 1710, but it does not appear that he performed colotomy; and it was not till sixty years afterwards that the operation was actually performed on the living subject by Pillore of Rouen, who opened the cæcum in the right inguinal region. The dread of wounding the peritonæum, which, with our ancestors, was so great, suggested to Callisen the possibility of opening the descending colon, where it was uncovered by peritonæum in the left loin; but failing in this intention on the dead body, he does not seem to have attempted it on the living. In 1797 Fine of Geneva opened the transverse colon by an incision in the umbilical region. Subsequently Amussat published six cases, in which he was able to open the colon without wounding the peritonæum, five of these cases terminating successfully; and since then the operation of lumbar colotomy has borne the name of this distinguished surgeon, and is universally recognised as "Amussat's operation."

As the object of this procedure is to provide an

alternative outlet for the intestinal contents, through which more or less incontinence of fæces is a necessary result, it follows that the condition of the patient afterwards is by no means pleasant to himself or those about him : it is, therefore, only to be undertaken in cases in which the indication is very clear, and after the patient has been fully told of the inevitable result of the operation. In the case of imperforate infants, where perinæal incision has failed, it is the duty of the surgeon to lay the case fully before the parents, telling them that life may be possibly saved by means of colotomy, whereas without it death is certain. The onus of deciding for or against colotomy should be thrown on the parents in cases of imperforate rectum; but where the patient is an adult, suffering from obstruction, he alone must decide. In the latter case the pain and distress are usually so extreme that the sufferers generally gladly accept the conditions, and when the case is successful in relieving urgent symptoms, are loud in their thanks for the relief obtained. The surgeon has no right in these cases to act of his own motion as the arbiter between life and death ; and, if he fail to recommend colotomy in urgent cases, is, in my mind, as guilty of dereliction of duty as if he refused to sanction tracheotomy for the relief of a patient suffering from obstructed glottis.

It is necessary to indicate clearly the conditions calling for this operation, and they may be conveniently grouped under the following heads : 1. Congenital malformations that cannot be relieved by perinæal incision. 2. For the relief of distress attending recto-vesical fistula. 3. For obstruction, the result of (*a*) pressure of tumours; (*b*) cancer of the bowel; (*c*) non-malignant strictures, which are of such an extent as to preclude perinæal operation. 4. As a means of treating extensive ulceration by providing physiological rest to the part.

Operation of lumbar colotomy, Amussat's operation.—This operation, which used to be the one most frequently recommended, and has until recently been generally adopted in all cases of obstructive disease of the rectum in the adult, is one of some little difficulty to the inexperienced operator; and it is therefore essential to bear in mind the anatomical landmarks which indicate the position of the descending colon, in order to avoid the accidents that have not unfrequently happened during its performance. Allingham has directed special attention to this subject.* He says: "The anatomical guide to the position of the ascending or descending colon is the free edge of the quadratus lumborum muscle, but this is by no means always easily found, and consequently it is better to substitute a more certain and unmistakable guide; and this may be obtained by marking a spot on the crest of the ilium fully half an inch posterior to a point midway between the two superior spinous processes. From more than fifty dissections and the experience of over eighty operations of my own and others, I can confidently assert that the colon is always normally situated opposite this point. Before operating I mark this spot with ink or iodine paint, and I have always found it, when the superficial structures are divided, a most useful landmark and guide to the exact position of the intestine."

The vertical incision of Callisen, and the transverse incision of Amussat, gave place to the oblique incision as recommended by Bryant, and for it the following advantages are claimed: More room is afforded for manipulation; the incision taking the course of the vessels and nerves lessens the liability to their injury; that following the ordinary integumentary fold when the patient is recumbent, it

* "Diseases of the Rectum," p. 302. Fourth edition.

LUMBAR COLOTOMY.

facilitates repair and tends to prevent prolapse of the bowel.

Before operating the bowel should be as completely emptied as possible by means of laxatives, and an enema, if the obstruction is not complete. The patient should lie in the semi-prone position, with a small, hard pillow under the opposite loin. An incision should now be made parallel to the last rib on the left side, midway between this bone and the crest of the ilium; the centre of this incision, which should be about four inches in length, should correspond with Allingham's point. The incisions are now carried deeply, some fibres of the latissimus dorsi and posterior edge of external oblique muscles being divided; and the edge of the quadratus lumborum muscle and lumbar fascia next looked for. The fascia when found should be freely divided, and probably also a little of the outer edge of the quadratus lumborum. The fascia transversalis is now met with and divided, and the subperitonæal fat exposed. If the gut does not now present in the incision, two pairs of dissecting forceps should be taken, and with them the little masses of fat pulled asunder, and the search prosecuted, the most usual mistake being that of looking for the bowel too far forward. If any difficulty still be experienced, the body should be rolled a little forward; and at the same time air may be injected into the rectum by means of Lund's insufflator. This will usually have the effect of rolling the bowel into the wound. The colon thus reached and having been identified, deep sutures may now be passed through the entire thickness of the abdominal wall, and through the posterior wall of the colon. A longitudinal incision is now made in the gut, the loops of the sutures hooked out, cut, and tied, and as many more sutures as may be necessary completely to adjust the skin and mucous membrane put in. Immediately after opening there is often a free discharge of fæces; while at other times, especially if the bowel has been first well cleared, no fæces may pass for several days. In one of my cases nothing passed till the sixth day; and my friend, Mr. Thomson, tells me that in a case upon which he operated nothing passed for seventeen days. It is a considerable advantage when this is the case, as it permits of healing of the wound to take place quietly without disturbance. A pad of tenax, or other absorbent antiseptic material, should be kept applied, and changed whenever the bowels move; and it will much conduce to the comfort of the patient and those around him if large doses of charcoal are administered internally, which, in addition to making the motion harder, tend to remove the odour. I have recently tried for this latter purpose naphthaline in two-grain doses, given wrapped up in wafer paper, and found it answer admirably as a deodoriser.

At first there is no control whatever over the motions, and the patient is very miserable if there is any tendency to diarrhœa; but, later, if the bowels are kept moderately costive, there is usually but one motion in the twenty-four hours, and, although the patient has no power to restrain it, he knows when it is coming sufficiently long beforehand to make the necessary preparation, and with an absorbent pad comfortably adjusted he is then tolerably comfortable. At any rate, the freedom from pain and frequent straining contrast now most favourably with the antecedent miseries of rectal obstruction.

Accidents during and consequent on operation.—Wound of the peritonæum is of frequent occurrence; and when there is a meso-colon present, its injury is inevitable. If the opening is at all free, prolapse of small intestine is likely to take place, and considerably complicate the operation. In such a case the proper course would be, having reduced the prolapse, to plug an aseptic sponge into the wound, and prosecute the search for the colon, and as soon as it is found the peritonæal wound must be carefully closed before attempting to open the bowel.

In some cases the operator has failed entirely in finding the colon, some portion of the small intestine being opened in its place. It is hard to imagine how this accident has occurred on the left side, because the small intestine could only be reached after the peritonæum has been opened, and in these circumstances the appearance of the longitudinal bands and appendices epiploicæ is so characteristic of the large intestine, that with ordinary caution the distinction should easily be made. On the right side the mistake of opening the duodenum instead of the colon seems to me to be a much more real danger. One of the most experienced colotomists living has candidly admitted that this accident has occurred in his practice.

During the after-treatment one of the greatest dangers is the occurrence of diffuse inflammation and suppuration along the areolar spaces of the abdominal wall. I have seen very extensive sloughing of the skin of the loin follow an operation of this kind.

Although it is impossible, from the nature of the wound, to follow strictly the rules of antiseptic surgery, much may be done in this direction with corrosive sublimate solution and iodoform; but the most important of all measures, I believe, is the accurate suturing of the mucous membrane to the skin, thus preventing extravasation of fæces.

During the after-treatment also, a collection of fæces in the lower segment of the gut is a troublesome complication. When occurring to any extent it may produce, as Bryant has pointed out, symptoms of intestinal obstruction, notwithstanding the fact that an outlet for fæces exists higher up; and even where this is not the case the irritation in cases of malignant ulceration defeats to a great extent the object of the operation, while in cases of vesico-intestinal fistula the trouble is even more exaggerated. In order to remedy this, several suggestions have been made: (1) to pass the sutures deeply, so as to include the entire thickness of the bowel instead of its posterior aspect only; as this, however, necessitates the passage of the sutures across the peritonæal cavity it does away with the sole advantage claimed for lumbar colotomy. (2) It has also been attempted to bring out a knuckle of intestine at a very acute angle, in the hope that in this way a spur might be formed similar to that which is found in artificial anus following hernia. (3) The only proceeding, however, by which the advantages of lumbar colotomy can be combined with absolute closure of the lower segment is by means of the operation recommended by Mr. Thomas Jones.* He detached the mucous membrane from a prolapsed portion of gut, and from the lower margin of the colotomy opening, turning it on itself, and attaching the raw surfaces by means of the catgut, and afterwards brought together the surfaces denuded of mucous membrane. No fæcal matter passed beyond the opening after this procedure had been carried out. In a case of recto-vesical fistula in which lumbar colotomy had been performed I found this method answer admirably.

Of course it is obvious that no attempt to close the lower opening should be contemplated when there is a possibility of establishing at some future date the normal exit for fæces.

Another troublesome after-complication is prolapse of the bowel through the artificial anus. This is frequently due to the continued expulsive effort trying

* *British Medical Journal*, April 24, 1886; p. 782.

to get rid of the accumulation of fæces in the lower portion of the bowel. It is to be treated by the adjustment of a well-fitting pad, and, if possible, by the closure of the lower orifice.

In common with all other extensive wounds of the abdominal parietes, hernia may occasionally occur. Of this accident Mr. Simpson records an instructive example.* Upwards of four years after the operation of colotomy, the patient felt something suddenly give way while coughing, and a tumour appeared immediately below the artificial anus, and he died in two days. At the post-mortem the tumour was found to contain a large loop of ileum in part gangrenous.

Operation of inguinal colotomy (Littré's operation).—Under this head is usually described the operation of opening the cæcum, or sigmoid flexure, by an incision in the right or left groin. It is with the latter alone that we are at present concerned in considering the treatment of rectal disease. It is, of course, obvious that in this procedure the peritonæum is necessarily injured. It is performed by making an incision through the muscular coats of the abdomen parallel to Poupart's ligament, and then drawing forward a loop of large intestine, securing it to the wound, and opening the bowel between the points of suture; the subsequent treatment being similar to that of lumbar colotomy. The operation has been pretty generally selected in preference to the retro-peritonæal procedure in cases of imperforate rectum, because in these cases the colon is very hard to find from behind, and is frequently attached by a meso-colon, which would necessitate peritonæal wound.

Statistics of the older methods of colotomy.—The most comprehensive record of cases of colotomy hitherto published is that by Dr.

* *British Medical Journal*, May 23, 1885; p. 1039.

W. R. Batt;* and the following is his analysis of cases, slightly condensed: Of a total of 351 operations, 154 were performed for malignant disease, 20 for fistula, 52 for imperforate anus, 40 for obstruction, 72 for stricture, 4 for ulceration, and 9 for miscellaneous causes. The recoveries were 215, deaths 132, equal to a mortality of 38 per cent., the result of four cases being unrecorded. Of these the number of operations performed by Amussat's method was 244: 165, or 68·2 per cent., recovered; 77, or 31·8 per cent., were fatal; and the result in 2 cases is unrecorded. After Littré's method 82 operations were performed: of these, 38, or 46·9 per cent., recovered, and 43, or 53·1, proved fatal, the result in 1 case being unrecorded. After Callisen's method 10 were operated upon, 2 of which recovered, 7 were fatal, and in 1 the result is not stated. Four cases were performed by Fine's method, all of which are recorded as having been successful. In one fatal case a T-shaped incision was adopted, while in 10 the method of operating was not stated. Of these, 6 recovered and 4 proved fatal. Of the total number, 160 were males, 147 females, and in 44 the sex was not given. Of the 154 cases operated on for malignant disease, 105, or 68·4 per cent., recovered; 48, or 31·6 per cent., were fatal; and in one case the result is not stated. The patients in 72 instances were males, in 74 females, and in 8 the sex was not mentioned. Following Amussat's method were 124 cases, of which 91, or 73·5 per cent., recovered, and 33, or 26·5 per cent., were fatal. According to Littré's method there were 23 cases, with 12, or 52·2 per cent., recovering, and 11, or 47·8 per cent., proving fatal; of the 4 cases following Callisen's method all proved fatal, and Fine's case recovered. Of the 2 in which the method was not stated, 1 recovered and 1 died. The ages of the

* *American Journal of Medical Sciences*, Oct., 1884; p. 423.

patients were as follows: 20 to 30 years, 22 ; 30 to 40 years, 22 ; 40 to 50 years, 30 ; 50 to 60 years, 29 ; 60 to 70 years, 18; over 70 years, 2; while in 31 cases the age was not given. With regard to the duration of life after operation in malignant cases, Dr. Batt has published the following details of cases in which the patients recovered from the immediate effects of operation : 13 died within six months, 15 between six months and one year, 10 died between one and two years, 8 died between two and three years, and one died four and a half years after operation. Of 20 cases operated on for fistula, 18, or 90 per cent., recovered, 2 alone proving fatal. Following Amussat's method were 17 cases, with 15 recoveries, and 2 deaths ; and by Littré's method one case, which terminated favourably. The method of operating is not mentioned in two cases which recovered. Of the 52 cases operated on for imperforate anus, 24, or 47·1 per cent., recovered ; and 27, or 52·9 per cent., were fatal ; and the result in 1 case is not stated. Following Amussat's method were 12 cases, 6 recovering and 6 ending fatally ; and following Littré's method were 34 cases, 17, or 51·5 per cent., of which recovered ; 16, or 48·5 per cent., ended fatally ; and there is one case in which the result was not given. Five cases were operated on after Callisen's method, 1 of which recovered and 4 died. In 1 fatal case the method is not mentioned. Of 40 operations for obstruction, 19, or 50 per cent., recovered ; 19 died ; and in 2 the result was not mentioned ; 24 were performed by Amussat's method, of which 13, or 59 per cent., recovered, 9 terminated fatally, and in 2 the result is not mentioned. Eleven cases were performed after Littré's method, of which 3 recovered and 8 proved fatal ; and 3 are recorded by Fine's method, all of which recovered; in one case in which the method is not stated, and in one case in which a T-shaped incision was made, a fatal result followed. Of the 72 cases operated on for stricture, 41, or 57 per cent., recovered, and 31, or 43 per cent., ended fatally. Following Amussat's method were 59 operations with 35 recoveries, or 59 per cent., and 24 deaths. After Littré's method were 10 cases, with 4 recoveries and 6 deaths. Callisen's method was performed in one case which recovered, and two cases are given in which the method is not mentioned, one terminating in recovery, the other fatally. Of the 4 operations performed by ulceration, 3 terminated in recovery and 1 in death. All were performed after Amussat's method. And of 9 patients operated on for miscellaneous causes, 5 recovered and 4 died. Amussat's operation was performed in 4 cases, Littré's in 2, and in 2 the method is not mentioned.

Arranged in a tabular form showing the various forms of operating, we find the following convenient summary condensed from Batt:

Form of Operation.	Cases.	Result not ascertained.	Recovered.	Died.	Mortality per cent. of terminated cases.
Amussat . .	244	2	165	77	31·8
Littré . . .	82	1	38	43	53·1
Callisen . .	10	1	2	7	77·7
Fine . . .	4	...	4	...	0·0
Not stated .	11	...	6	5	45·4
Total . . .	351	4	215	132	38·0

According to these statistics, the mortality of inguinal colotomy is 20 per cent. greater than that of the retro-peritonæal operation. According to Erckelen's statistics,* the mortality shows a difference of 10 per cent. in favour of Amussat's operation. I

* *Archiv f. klin. Chir. Langenbeck,* p. 41; 1879.

think, however, that it will be admitted that statistics of this kind, which are collected from published cases, are at all times misleading; but especially is this the case in the instance at present under consideration, for in the first place the inguinal operation has been selected in a large proportion of the total number as a treatment for imperforate rectum, and frequently not adopted until after an extensive exploration from the perinæum, when the patient was nearly exhausted. And, again, as these statistics contain the records for many years back, they embrace a period when peritonæal surgery of all kinds was in a very different condition from that in which it is at the present day, so that I think the time has come when the relative merits of both operations may be fully discussed without our being too much influenced by the results obtained under the older methods of wound-treatment, and at the present day the verdict in favour of anterior colotomy is amongst modern surgeons almost universal.

The unquestionable advantages of laparo-colotomy are these: 1. It permits a thorough exploration of the abdominal cavity, which may enable the surgeon in some instances to perform a more radical operation for the complete removal of the disease, and if removal is impracticable it ensures that the opening is made above the seat of obstruction instead of below, as has happened with the lumbar operation. 2. The large intestine is found with ease and certainty. 3. A complete operation for closure of the lower lumen when considered necessary can be much more readily and completely carried out, thus making the artificial anus a *terminus*, and not a lateral outlet to the intestine. 4. A shorter distance of intestine intervenes between the opening and seat of disease. 5. The abdominal wall being thinner in front, the extent of wound surface is less, and the finer skin in the front abdominal wall permits a much more accurate coaptation of skin and mucous membrane. 6. The position of the wound is much more convenient for the patient, and it is interfered with less by the clothing. 7. A largely decreased death-rate. So, then, the sole disadvantage of laparo-colotomy is that now exploded surgical bugbear—wounding of the peritonæum; and it must be remembered that even in the hands of skilled colotomists wound of the peritonæum in the lumbar operation has not unfrequently taken place. It is manifestly easier to deal with a peritonæal wound, advisedly and carefully made, than with an accidental opening at the bottom of a rather deep incision.

Delayed opening of the intestine. "Operation à deux temps."—The unequivocal advantage which has been found to follow the division of the operation of gastrotomy into two stages has naturally suggested a similar manner of proceeding in colotomy; and cases have recently appeared in which both laparo-colotomy and lumbar colotomy have thus been performed. In a communication made to the Clinical Society of London by Mr. Davies-Colley,[*] three cases are recorded in which the lumbar operation was performed in two stages, the intervals being one, four, and six days respectively. It would appear from these cases that the procedure necessary to retain the loop of bowel in the wound was attended more or less with symptoms of intestinal strangulation, and in order to minimise this result as much as possible Mr. Davies-Colley has devised a form of clamp, in which two pairs of ivory studs placed on steel bars are made to grasp the bowel at two places; and by this means the loop of intestine is held without being strangulated, until sufficient healing of the wound has taken place to obviate any risk of extravasation. In two out of

[*] *Lancet*, March 21, 1885.

the three cases this instrument was used, and the resulting constitutional symptoms are described as being trivial.

The analogy between gastrotomy and colotomy, however, scarcely holds, because the mere fact of retaining a small portion of the stomach wall in the abdominal wound can have no direct influence one way or the other on the œsophageal disease for which the operation has been undertaken; while in the case of the colotomy a certain amount of obstruction will probably have existed before the operation, which will be increased by the drawing out the loop of intestine, and to the symptoms of obstruction may be added more or less those of strangulation.

I operated on a case of rectal cancer in November, 1884, by laparo-colotomy, having chosen this operation deliberately as in view of modern surgery preferable to lumbar colotomy. A loop of the sigmoid flexure was drawn out, emptied by pressure upwards from the seat of obstruction, and caught between two clamps: it was now divided between the clamps, the wall of the lower segment was inverted so as to bring the peritonæal surfaces into apposition, and carefully sutured up, and the upper orifice was stitched to the wound. This patient lived for two years and a fortnight after the operation. Until a short time before her death she was able to go about and attend to the artificial anus without help; the bowels moved but once a day, and she was conscious that the motion was coming sufficiently long beforehand to make all the necessary preparations. I have now performed laparo-colotomy fifteen times altogether, and all of them recovered completely from the operation, with one exception, which was complicated by general peritonæal adhesions, complete obstruction and great meteorism, extreme difficulty being experienced in clearing the colon. I might have increased this number considerably had I advised the operation as a routine practice in cases of rectal cancer too advanced for excision. I have only recommended colotomy in cases where obstruction was a prominent symptom, or where some complication, as vesical fistula or extreme pain from involvement of anus, was present. I am satisfied that the majority of cancer cases which have passed the period for useful excision will go on to their fatal termination quite as easily and slowly without colotomy. If, however, obstruction or other complication arises, it is the surgeon's duty to recommend the operation.

The writings of Mr. Cripps and Mr. H. Allingham have done much to popularise this operation, and I cannot conceive anyone who has had experience of anterior colotomy again resorting to lumbar colotomy, except possibly in cases of extreme meteorism, and even in these circumstances if the incision is kept small, and a sponge at once introduced into the peritonæum, prolapse of small intestine can be readily controlled.

Much has been done to simplify this operation. It can now be done with the greatest ease and rapidity, while the mortality inherent to the operation has been reduced to the most insignificant proportions. Instead of making the inguinal incision recommended by most authors, I am convinced that the left linea semilunaris is the best position. It is the thinnest portion of the abdominal wall. As frozen anatomical sections show, the sigmoid flexure usually lies against this line, and presents in the wound the moment the peritonæum is opened. No muscle is wounded, so that the wound can be brought accurately together, lessening the risks of abdominal hernia and extensive prolapse afterwards. If the incision does not extend lower than a line joining the umbilicus with the middle of Poupart's ligament, the deep epigastric artery will not be injured.

Method of performing laparo-colotomy. —I have now quite discarded the use of any clamp, and conduct the operation as follows : Make a short

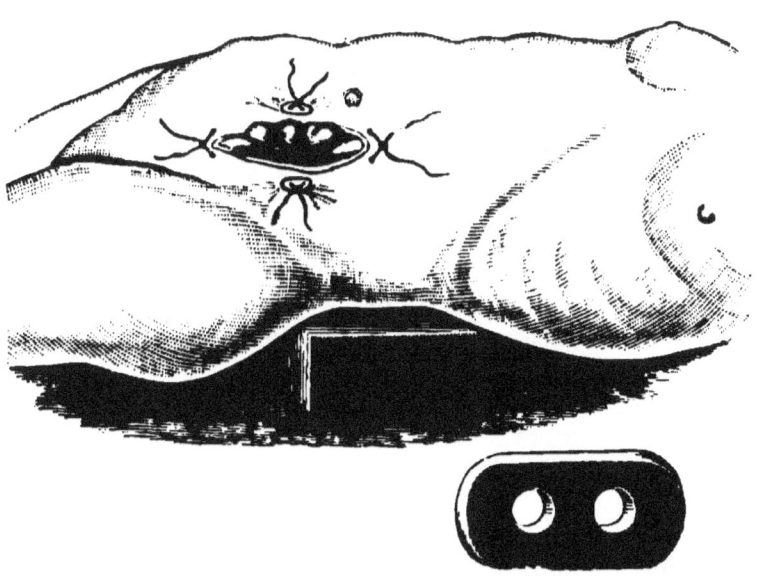

Fig. 58.—Author's Method of performing Laparo-colotomy in the left Linea Semilunaris with fixation of the Gut by leaded Button Suture through the Meso-colon.

incision (1½ to 2 inches) along the outer border of the left rectus abdominis muscle, so that the lower termination is above a line joining the middle of Poupart's ligament with the umbilicus; divide the tendinous structures of the linea semilunaris and peritonæum the entire length of this wound, introduce a small aseptic sponge into the peritonæal cavity, and stitch with fine catgut the peritonæum to the skin by continuous suture all round the wound. If the colon does not directly appear in the incision introduce a finger and feel for the sigmoid flexure as it passes over the brim of the pelvis. The colon can be recognised readily by the longitudinal bands and appendices epiploicæ. Pull out now a good loop of the colon; if there is difficulty in doing this from too short a mesocolon, replace the loop first caught and take a piece higher up the sigmoid flexure where the meso-colon is longer. With a curved needle pass a strong double-sterilised silk suture through the entire thickness of the abdominal wall about one inch from the margin of the wound, through the centre of the meso-colon, and again through the abdominal wall, bringing it out at a corresponding point at the opposite side of the wound. Remove now the sponge from the interior, and tie the double suture on each side tolerably firmly over lead buttons. (*See* Fig. 58.) A single silk stitch is now passed through either angle of the wound, including the longitudinal muscular band, but not the mucous membrane of the gut. The wound is now covered with boric acid, taking care to fill well the angle between the gut and abdominal wall. The operation can now be completed at once by cutting off the protruding portion and suturing the mucous membrane evenly to the skin all round, or the opening of the gut may be left for a few days until adhesions have taken place. I have never experienced any ill effects from opening at once, and where obstruction is very marked or if there is much meteorism, it has obvious advantages. If, however, there has not been urgent obstruction the removal of the protruding intestine may be delayed for a few days, and it is very remarkable the absolute insensibility of the gut, the patient being completely free from pain while it is being cut off with a pair of scissors. Absolutely reliable antiseptics are, of course, indicated at all stages of this operation. The after-treatment of laparo-colotomy is similar to that of lumbar colotomy.

Some surgeons have, in the endeavour still to further simplify this operation, used less perfect methods of suturing, or even dispensed with sutures altogether, relying on a pin or probe passed through the meso-colon and resting on the abdominal wall on either side of the incision to retain the loops of gut. Such a method is highly dangerous, and has been attended with fatal consequences,* hernia of the small intestine taking place beside the colon, while, on the other hand, when not sufficiently secured by suture, the colon has slipped back into the abdominal cavity. If the directions above given are adhered to neither of these accidents could possibly occur, provided the wound is not longer than two inches. If from extreme thickness of the abdominal wall it is necessary to make the incision longer than two inches, it would be wise as an additional precaution to pass a few points of interrupted suture through the skin and muscular coat of the bowel.

Prognosis.—In estimating the probable result of colotomy, it is necessary to classify the cases carefully. Where the operation has been performed for malignant disease, the duration of life can only be prolonged to a certain extent, the disease progressing, and ultimately proving fatal. According to Batt, in 32 per cent. of the cases operated on, death was apparently directly due to the operation when performed for malignant disease; and of 105 that recovered, and whose subsequent history was traced, it was found that six died within two months of the operation, seven died between three and six months after operation, fifteen died between six months and one year, ten died between one and two years, eight

* Kelsey, *New York Med. Journal*, Feb. 18, 1893.

died between two and three years, and one died four and a half years after operation. Of thirty-two patients recorded as being well after the operation, only one had survived two years; and one, one year. It is of course obviously impossible to form any estimate of how long these patients could have lived if colotomy had not been performed. In my own practice I have performed colotomy eighteen times for rectal cancer. In three cases the lumbar operation was selected: one of these patients lived six months, another three years, and one two years; and in fifteen in which laparo-colotomy was selected one lived three years, two two years, three died during first year, while three died within three months after operation from extension of disease, the wound being quite healed. One died on the fourth day. Five are at present alive.*

It has, however, been suggested that this operation has a direct influence in checking the growth of disease. There is in the Hunterian Museum a very beautiful specimen (Fig. 59), No. 2591, taken from a case of colotomy of thirty years' standing, in which the mucosa of the rectum below the opening has undergone atrophy and become villous, these changes being apparently the result of disuse. It is claimed that, as the form of cancer usually found in the rectum is so completely formed of glandular tissue, the abrogation of function of the rectum induced by the colotomy will be followed by atrophy of the morbid growth as well as of the normal structures. I do not think, however, it is safe to draw conclusions as to the result of a certain procedure on a pathological formation from the changes induced in normal structure, as the conditions of growth in the two

* On the next page there will be found in tabulated form particulars of several cases of laparo-colotomy that may be studied with advantage.

CASES OF LAPARO-COLOTOMY.

	Name.	Date.	Sex.	Age.	Hospital.	Private.	Pathology.	Opening of Bowel.	Result.	Observations.
1.	A. T.	Nov. 10, 1884.	F	60	H	—	Cancer of rectum and obstruction.	Immediate.	R.	Lived in tolerable comfort for 2 years.
2.	T. Y.	July 6, 1885.	M	57	—	P.	do.	do.	R.	Lived 18 months.
3.	R. B.	Oct. 26, 1886.	F	49	—	P.	do.	do.	R.	Lived 3 years.
4.	A. N.	Nov. 24, 1886.	F	50	H	—	do.	do.	R.	Rapid increase. Death a month afterwards.
5.	Y. K.	Mar. 3, 1888.	M	27	H	—	Sarcoma filling pelvis. Complete obstruction.	Immediate in sinea alba.	R.	
6.	Mrs. S.	June 4, 1888.	F	45	—	P.	Cancer of rectum and obstruction.	Delayed.	R.	Lived 2 years.
7.	Mrs. W.	Dec. 18, 1889.	F	58	—	P.	Cancer. Recto-vesical fistula.	Delayed.	R.	Lived 18 weeks.
8.	P. B.	Nov. 14, 1889.	M	45	H	—	Cancer of rectum and obstruction.	Delayed.	R.	
9.	Mr. B.	Feb. 24, 1890.	M	55	—	P.	Obstruction due to recurrence of rectal cancer (resected).	Delayed.	R.	Only lived 4 weeks.
10.	Dr. B.	April 24, 1890.	M	60	—	P.	Recto-vesical fistula.	Delayed.	R.	Large calculus with fæcal nucleus removed from bladder 12 months afterwards.
11.	Miss B.	May 28, 1892.	F	58	—	P.	Cancer of rectum and obstruction.	Delayed.	R.	Died of cardiac disease a month afterwards.
12.	M. F. H.	Feb. 27, 1892.	F	45	—	P.	do.	Delayed.	R.	
13.	Mrs. B.	Nov. 19, 1892.	F	70	H	—	Complete malignant obstruction.	Immediate.	R.	Great meteorism and stercoral vomiting at time of operation.
14.	Mrs. S.	Oct. 30, 1893.	F	52	H	—	Complete obstruction.	Immediate.	D	Universal peritoneal adhesions; great difficulty in isolating colon. Death from peritonitis fourth day.
15.	Mrs. H.	Nov. 16, 1893.	F	65	H	—	Obstruction considerable.	Delayed.	R.	

circumstances are so essentially dissimilar; and the dreary record of early death after operation shows conclusively that the progress of disease in these cases is not to any great extent arrested.

That life can be prolonged when death is threatened by obstruction is of course certain, and probably the diversion of fæces from the surface of the malignant growth, and absolute rest ensured for the part, tend somewhat to retard the growth; and any one who has witnessed the relief afforded to a person, suffering from obstruction, by this operation will at once admit the complete justification of the procedure. The immediate result is much influenced by the stage of the disease at which the operation is performed; most of the fatal results being due to what Mr. Bryant has truly called "too late" cases.

Fig. 59.—Villous condition of Mucous Membrane from a case of Colotomy of thirty years' duration.

I believe that the most important evidence that the case is too late for operation is the presence of extensive meteorism. Where the bowel has been hyper-distended it has become in such an atonic condition that it may be unable to recover itself after an outlet has been provided, and if such should prove to be the case, a rapidly fatal result will probably follow. Extensive meteorism is in itself also a very serious complication to the manipulative details of laparo-colotomy. I recently witnessed an operation under such conditions. The moment the abdominal incision was made a very extensive prolapse of small intestine took place, which could not be restrained. The recommendation of Mr. Greig Smith was followed and an incision made in the small intestine, which was then as far as possible evacuated; the incision was sutured, and the intestine returned to the abdominal cavity; the colon was now drawn out, and the operation completed; but the patient never recovered power over the intestine, and died in two days. By making a small incision and immediately plugging this with a sponge while the further stages of the operation are being completed, prolapse of small intestine can be largely prevented, but in very extreme cases it is well to consider the advisability of substituting the lumbar operation.

Of colotomy for imperforate anus, according to Batt's statistics, 47·1 per cent. of the cases were successful. Of course when once the patient has recovered from the operation there is practically no limit to the duration of life, as there is in the case of malignant disease already discussed, but the children are frequently ill-developed, and die early from other causes, comparatively few having reached adult life. In the twenty cases operated on for fistula 18 (90 per cent.) recovered. This would appear to point out clearly that the dangers of the operation itself are comparatively trivial, and that the greater degree of mortality of the other classes is mainly due to the damage done by intestinal obstruction. Of the four cases in which the operation was performed for the relief of ulceration three recovered.

A A—23

In these cases the operation should be so performed that when the ulceration healed the artificial anus could be closed. In the few cases in which it was attempted to close an artificial anus the result of colotomy, considerable difficulty was experienced, and although "Dupuytren's spur" is not so marked as in the artificial anus following hernia, it is sufficient to give some trouble. Mr. Barker* has suggested an ingenious addition to the means at our disposal for the cure of artificial anus. He introduces into the bowel a piece of flexible rubber sheeting one and a half inches long by five-eighths of an inch broad; this is secured on the internal aspect of the orifice by means of two wire sutures, one at each end; the anus is then closed by paring the edges, and inserting sutures in the usual way, the object of the rubber being to protect the wound from fæces. As soon as the wound is closed the wire sutures can be removed, so allowing the rubber to pass away with the fæcal contents of the intestine. Although in the case given by Mr. Barker this plan did not completely answer the purpose intended, it appears to be well worthy of more extended trial.

CHAPTER XXVI.

PRURITUS ANI,

OR, as it has been not inaptly termed, painful itching of the anus, is a most distressing complaint when met with in an aggravated form; patients frequently stating that it is much more difficult to bear than acute pain, and that their lives are rendered absolutely miserable by it. As it may arise from a multiplicity

* *Lancet*, Dec. 18, 1880.

of causes, it may tax considerably the powers of the surgeon to cure. It will, therefore, assist in the consideration of the subject, if we discuss in detail the various diseases of which pruritus ani is a symptom.

Eczema, as one of the not unfrequent causes of pruritus ani, occurring in the neighbourhood of the anus, may be of two forms, viz. the moist, and the dry. In the first, the skin surrounding the anus is red, and exudes a rather copious gummy discharge. When severe, there may be some subcutaneous œdema, and it is attended with smarting pain, in addition to the severe itching. This form demands treatment by soothing applications, such as the linimentum calcis, unguentum zinci, etc.; or, in some cases, it will be found more comfortable to apply boric acid in fine powder dry, which, mixing with the discharge, forms an antiseptic crust, protecting the raw and sensitive surface. In the dry form of eczema the skin round the anal margin is dry and cracked; the surface is covered with dry scales, which, if removed, disclose a red and sensitive surface. This is the form met with in connection with the lithic acid diathesis, and the itching produced by it is very severe. It must, however, be distinguished from a somewhat similar appearance produced by scratching to relieve pruritus arising from other causes. This form of eczema requires an essentially different plan of treatment. Here some of the tar preparations will be found more suitable, such as the compound soap liniment, which consists of equal parts of soft soap, oil of cade, and rectified spirit; or the part may be bathed with a solution of simple tar water. Van Buren recommends the wearing of a pad of prepared oakum in contact with the affected integument. This substance keeps well in place by its adhesive quality, and moulds itself to the parts with which it is in contact; it prevents the morbidly altered surfaces from touching each other (a most essential point), and at the same time keeps them constantly moistened by the tarry exudation it affords. If these means fail, recourse may be had to painting with a solution of nitrate of silver or other powerful irritant, which, by making the eczema more acute, renders the cure more probable. It is well to remember that sometimes, when the skin around the anus is not much affected, there may be an excoriation of the muco-cutaneous margin similar to what is sometimes seen on the lips and eyelids.

Eczema marginatum.—Under this head Hebra has described a disease, which is now proved to be due to the same parasite (trichophyton), which produces ringworm in other parts of the body, the somewhat distinctive characters being only due to the locality of growth. Commencing usually where the scrotum and thigh touch, it spreads along the inguinal fold and backwards over the perinæum, thus involving the anus secondarily. The spreading margin is raised like a wall, and sometimes small vesicles are to be seen; but they are not so frequent as in other parts of the body, owing to the moisture of the invaded parts. In consequence of the intense itching, the appearances are soon modified by scratching, so that the case comes to resemble one of ordinary chronic eczema. It is stated to be much more frequently met with in tropical countries; and that the disease is rare in women. Although seldom originating at the anus, it spreads to it, and owing to the great sensitiveness of the part, produces there an extreme amount of pruritus. The diagnosis can be readily made by observing the raised, spreading margin, and confirmed by making a microscopical examination. If a few of the scales are scraped from the spreading edge, and moistened with glycerine on a slide, the characteristic mycelia and conidia will easily be recognised. The treatment consists in the application of various parasiticides, such as solution of sulphurous acid, perchloride of mercury, or iodine, or ointment of pyrogallic acid (5 per cent.), will sometimes prove effectual. The disease is exceedingly hard to cure, and liable to relapse.

Oxyuris vermicularis.—The presence of threadworms is universally admitted as a cause of pruritus in children; and in adults they likewise give rise to considerable irritation, and should always be looked for in cases of pruritus ani. Contrary to what is usually taught, Heller has shown * that these worms live principally in the cæcum and lower part of the ileum, and that it is principally the females, when about to deposit eggs, that descend into the rectum. The diagnosis can be made sometimes by seeing the worms in the motion just passed, or in the anal folds of skin, and if an injection of cold water is given while the itching is present, a number will probably be expelled. Heller's observation has most important bearing on our treatment in these cases of a very common and a very obstinate disease; as, if the cæcum is the principal habitat of the parasite, it is manifestly useless to treat the rectum by small injections of lime-water, infusion of quassia, etc. The great difficulty in treatment is humorously alluded to by Bremser : † "Just as these parasites are, on the one hand, to be counted among the most troublesome of all those that live at the expense of our bodies, so on the other do they, at the same time, belong to those which are the very hardest to exterminate. Their number is legion. And if, after we have slaughtered thousands, we lay our weapons aside for one moment, imagining ourselves safe from a fresh attack, new cohorts again advance with increased reinforcements. The fæces and intestinal mucus

* Ziemssen's Cyclopædia, vol. vii. p. 752.
† Quoted by Heller, *loc. cit.*

contained in the large intestine behind which they hide themselves, serve them for a breastwork and parapet. If one attacks them from the front with anthelmintics, these become so weakened by the long march through the small intestine that the worms only laugh at them. If we attack them in the rear with heavy artillery the foreposts stationed in the rectum must certainly succumb, but the heaviest enema bombardment cannot reach those encamped in the cæcum, and so long as ever so few remain behind in some hiding place, they, from the amazing rapidity with which they are reproduced, soon again become a large army."

According to Heller no direct reproduction of these worms takes place in the intestine, but the eggs which pass out with the fæces must pass through the stomach, the action of the gastric juice being necessary to allow the worm to escape from the egg by softening the outer coat, and he supposes that the eggs are conveyed to the mouth by the fingers after scratching, having frequently found the eggs under the nails of persons suffering from oxyurides. This shows what an important item in both treatment and prophylaxis absolute cleanliness is. He considers free purgation the best form of treatment, and in aggravated cases the washing out of the entire large intestine by means of Hegar's monster clysters.

Pediculi and scabies may possibly be the cause of pruritus ani, and if so their presence should be easily determined.

Similarly other diseases of the rectum, such as internal piles, fistula, or fissure, may be the only apparent local cause to which the itching can be assigned. A certain number of cases, mostly very inveterate ones, remain, in explanation of which it is impossible to find any local cause, and which must be, in the want of accurate pathological knowledge, described as **neuroses** of the rectum. Usually found in elderly men, but not by any means confined to the male sex, they appear to attack the plethoric and the spare, the rich and the poor alike. The patient is not much disturbed during the day, but when he goes to bed his misery begins. The itching is so intense that it is impossible to avoid scratching, which, instead of giving relief, only adds to the trouble. Sleep at first is impossible, but when at last it comes it is frequently but of short duration, the patient being awakened by the intolerable itching. If an examination be made, the skin around the anus will usually be found devoid of its normal elasticity, and parchment-like. Immediately in the neighbourhood of the anus the normal pigmentation will be absent in patches, the skin here being of a dead white. This is considered by Mollière to be a pathognomonic symptom of the more severe forms. I have seen it very characteristically exhibited in several cases.

From the multitude of remedies which have from time to time been recommended, the difficulty of cure may be inferred. Local measures which will relieve one case will be found absolutely futile in another. In the first place, oleate of mercury with morphia may be tried, and has often proved successful in my hands; bathing with very hot or very cold water, either plain or containing carbonate of soda; the application of tincture of iodine; painting with solution of nitrate of silver, or even the light application of the actual cautery have been recommended. Van Buren says that he thinks most relief is given by the constant application of chloroform ointment, and he recommends patients to go to bed with a large bottle of this beside them, and when the itching comes on to smear the parts well with the ointment instead of scratching. Allingham recommends the wearing of a small pessary in the bowel at night. He says this was suggested to him by the fact that some patients are able to go asleep when they keep the tip of the finger in the anus, but not otherwise. In one case I recently forcibly dilated the sphincter in a very inveterate case of pruritus for the purpose of making a more complete diagnosis, and the operation was followed by complete relief of the itching, although nothing abnormal could be found in the rectum. In a second case, however, there was no marked improvement. Painting the part with a 4 per cent. solution of cocain hydrochlorate has, in some cases, given marked relief, but its action appears to be rather uncertain.

While attending to the local treatment, the constitutional must not be overlooked. Constipation is frequently more or less present, and must be appropriately dealt with. If the patient is very plethoric, a course of saline aperients should be ordered; and if the reverse, iron, quinine, and cod liver oil may be indicated. If nervous and excitable, a combination of chloral and bromide of potassium will usually prove preferable to the internal use of opium or morphia in any form. These are cases in which patience is required both in the surgeon and patient, treatment at times being anything but satisfactory.

The regulation of the diet should be attended to, cases having been known in which special articles of food, such as salmon or shell-fish, have brought on an attack of pruritus.

CHAPTER XXVII.

ATONY OF THE RECTUM.

Rectal constipation. — During health the lower portion of the rectum is empty, except immediately before the act of defæcation, and the impulse to expel the fæces is caused by the descent of a mass of excrement into the rectal pouch. This fact, which was pointed out by O'Beirne[*] more than sixty years ago, although frequently doubted, is substantially true.

If the call to empty the bowel is not responded to at the time, the desire to a certain extent passes off, and when next the bowels move it will be found that what first passes is a hard mass, the more fluid parts having been absorbed by the rectum. Many of the cases of habitual constipation met with in practice owe their origin to a habit of neglecting the calls of nature. In this way the rectal pouch soon becomes tolerant of the presence of fæces, the intervals between defæcation become longer and longer, and the habit of costiveness becomes more and more established. In adults who are capable of taking a sufficient amount of exercise, and those in whom the general health is good, the fæcal mass is, sooner or later, expelled, but habits of irregularity in this respect once started are difficult to overcome. In some people the bowels are evacuated more than once daily, while in others there is only one stool in two or three days; and yet neither of these can be said to deviate from a condition of normal health. It is the irregularity in responding to the call that is a principal factor in the production of this form of constipation in

[*] "New Views on the Process of Defæcation," 1833.

elderly and debilitated people, and occasionally in children there may be habitually incomplete defæcation, resulting in a gradually increasing accumulation in the rectal pouch, the true condition not being suspected, as the bowels occasionally act. After a time a catarrhal flux is produced by the irritation of the mass, and the patient is supposed frequently to be suffering from diarrhœa. This catarrhal discharge, which is not inaptly compared by Cruveilhier to the overflow of an atonic bladder, may be sufficient to soften and break down the mass, thus permitting its expulsion; but at other times symptoms resembling acute obstruction of the bowel supervene; active peristaltic motions take place, often started by taking a purgative internally, which are unable to expel the accumulation of fæcal matter in the rectum; violent colic is often induced, with frequent straining and exhausting efforts to procure an evacuation; profuse sweats break out, the patient becomes cold; the pulse is small and weak; and not unfrequently vomiting occurs; while in many instances retention of urine results from the pressure of the mass upon the urethra; in fact, the degree of collapse produced may be considerable. If an attempt is made to give an injection, it will be found that there is resistance to the introduction of the tube, and that it is almost impossible to force any fluid in. A digital examination will now reveal the fact that the bowel is blocked by a hardened mass of clay-like fæces, which must be broken up mechanically before the rectum can be emptied. For this purpose a lithotomy scoop, or the handle of a spoon, may be used to aid the finger; and as soon as a tube can be introduced copious injections of hot soap-and-water may be with advantage used to assist in the dislodgment. This operation is frequently a tedious, and at all times an unpleasant proceeding, but the relief given is great.

Besides the symptoms detailed, rectal constipation may give rise to other serious complications. Ulceration of the bowel, leading to a future perforation, may result at any part of the large intestine.*

Again, extensive sloughing of the rectum may supervene, as in a very remarkable case that I saw in consultation with Dr. Wright, of Dalkey, in 1884. A widow, aged sixty-five, had what appeared to be an attack of inflamed piles, from which she had suffered before. She stated positively that the bowels had been quite regularly moved. She suffered from retention of urine, and swelling of the vulva came on, followed by diarrhœa. The skin of the buttocks became red, glazed, and erysipelatous-looking, and felt very tense, the inflammation being much more marked on the right side. The skin gave way finally, permitting the escape of a considerable quantity of fæces. Rectal examination now revealed the fact that the entire lower bowel was blocked with fæces. By careful breaking up of the mass, and copious enemas, Dr. Wright succeeded in evacuating what the nurse described as a "bucketful" of fæces. When I saw the patient, at the right side of the bowel, about one and a half inches from the anus, the wall of the rectum had sloughed, leaving an opening as large as a half-crown piece. There was a considerable quantity of fæces in the areolar space external to the bowel, and pressure with the finger extruded lumps of hard fæces as big as walnuts through the large fistulous opening in the right buttock, through which, indeed, almost all the fæces appeared to pass. Notwithstanding the age of the patient, and the gravity of the local manifestations, the constitutional symptoms were slight, and she made a good recovery. I have been since informed by Dr.

* See Lecture by Dr. Bristowe on the consequences of long-continued constipation.—*British Medical Journal*, vol. i., 1885; p. 1085.

Wright that the fistulous track completely closed without any active surgical interference; the bowels act normally every day; and she has no stricture or other inconvenience. This case shows very forcibly how insidious may be the onset of a grave rectal constipation.

Except during the paroxysmal efforts at expulsion, pain is not a prominent symptom, and when present is due rather to the pressure on the branches of the sacral plexus than to the nerves of the rectum itself; thus we sometimes meet with cases of sciatica due solely to the pressure of an over-distended rectum.

Having once got rid of the accumulation, care must be taken to prevent its recurrence. This is best done by alterations in diet, by the use of purgatives, and enemata, and care to defæcate at once when fæces are felt to go down into the lower bowel. Variety in the way of food is of special importance, as if there is too great uniformity diminished sensibility of the intestinal canal will be induced. The habitual use of coarse vegetable food, such as too much bread, oatmeal, and potatoes, induces chronic constipation; while these articles of diet will relieve the constipation which comes on in persons who habitually use too much animal food. Similarly, milk used as a principal article of diet may produce constipation, while with others it acts as a purgative. In the same way aperients must be used with the greatest discretion: if one form ceases to act it is better to change it for another than to increase the dose. But the best treatment of all, as it more nearly simulates the normal stimulus, is the daily use of a small enema. It has been frequently stated that the constant use of enemata leads to the production of rectal atony, but this objection, I feel sure, is chimerical. Probably the best way of all is constantly to change the kind of stimulus applied to the intestinal mucous membrane. Dr. Lee has suggested,* that when a patient is trying to get rid of a rectal accumulation much assistance may be derived from firmly pressing up with the fingers into each ischio-rectal fossa, by this means making the presenting mass more wedge-like and easy to pass.

There is a form of rectal atony met with in hysterical females and hypochondriacal men, whose attention is morbidly fixed upon the evacuation of their bowels, and whose sole interest appears to be directed to the subject of defæcation. In illustration of this subject I cannot do better than quote from the graphic pen of Dr. Weir Mitchell :† "If it happens to you in an evil hour to have one of these cases to treat, with the additional need to treat also the difficulties with which some tender mother surrounds such a case, you are much to be pitied. I recall such an example which I saw in consultation some years ago. It began with a spot of abdominal tenderness over the spleen. Pressure on this caused nausea and vertigo. Then we had convulsions, hysterics, coma, enormous polyuria, and at last constipation. The physician in charge gave this list of the drugs given in four days : night and morning on each day an ounce of castor oil, at midday and bed-time one drop of croton oil ; three drops had been used in one day. The more drugs she took the more she demanded, and yet it was impossible to see that it gave her pain. Meanwhile for the nurse and mother the arrangement for each evacuation was the event of the day. A long stomach tube was carried six or seven inches up the bowel, and half a pint of olive oil injected, then followed one quart to three of flaxseed tea. During the use of the enema one person was occupied compressing the

* *British Medical Journal*, Feb. 10, 1883.
† "Diseases of the Nervous System." London, 1881.

anal opening, so as to prevent the escape of fluid. This help was made necessary on account of the great relaxation of the sphincter, into which a thumb could be passed without any resistance which could be felt to arise from a muscular act. Meanwhile the patient, while insisting on the use of more water, was shrieking with pain. The whole affair took two to four hours, and the patient was, I thought, the least exhausted of those concerned. Sometimes their efforts gave rise to a stool, sometimes there was none for a week, and sometimes under the wild entreaties of the patient these trying scenes were repeated in the night, nurse and mother being aroused to assist. I endeavoured to get this girl out of the control of the family, but I did not succeed, and I believe that her hysteria is now firmly established."

There is a form of atony of the sphincter which gives rise sometimes to a good deal of annoyance, in which it is impossible to expel just the last little portion of fæces, so that there is a difficulty in cleansing the parts after defæcation, and a tendency to have a slight escape of fæces during the day. This condition is not unfrequently associated with piles, but is sometimes independent of them. It is to be treated by copious ablutions with cold water after defæcation.

The paralysis of the rectum following spinal or cerebral diseases and diphtheria presents no points of special interest for consideration here.

CHAPTER XXVIII.

IRRITABLE RECTUM AND NEURALGIA.

IN addition to pruritus and atony of the rectum, there remain some neuroses which require brief notice. Curling has divided these more purely nervous diseases of the rectum into three classes, and the distinctive characters of each appear sufficient to justify the classification: (1) Irritability of the rectum, with frequent desire to expel the contents. (2) Neuralgia of the lower bowel, without any tenderness. (3) Morbid sensibility, where no gross local pathological change is discoverable. Frequently the last two are associated in the same individual, and may exist with a considerable amount of spasm of the sphincter.

(1) **Irritability of the rectum** is sometimes due to a catarrhal condition of the mucous membrane, while at others no such definite cause can be assigned, frequent desire to go to stool, usually at inconvenient times, being noticed. Of this condition many familiar examples will suggest themselves, indicating that when the mind is intensely occupied, chiefly by anxiety, the rectal reflexes may be abnormally high. It is common to hear of students going in for an examination having diarrhœa in the morning before it; and of persons having a desire to defæcate before a train starts, or when it stops at station; and I know a surgeon who has always to retire to the water-closet immediately before undertaking any severe operation. This minor degree of irritable rectum is common enough, but in rare cases it may be so marked that it may materially affect a person's enjoyment of life, or even interfere with business, as in the case of a

clergyman, recorded by Curling, in whom the desire usually came on just before commencing divine service. Cases of this kind are best treated by urging the patient to try to overcome the inclination, in which, with the exercise of a little moral control, they will usually prove successful.

(2) **Neuralgia.**—As in other parts of the body, pain of a more or less severe character may exist at the lower end of the bowel, without any discoverable cause. It is usually found in rather weak and hypochondriacal males, and hysterical females. At other times pain referred to the anus and interior of the rectum may have its origin in an injury to the coccyx, constituting one of the symptoms of the so-called "coccygodynia." These cases are to be treated in the same way that vague neuralgias of other parts of the body are, but it must be admitted that for the most part they are exceedingly unsatisfactory cases to deal with.

(3) **Morbid sensibility.**—The third class is when there is a true hyperæsthesia of some particular spot, which frequently is associated with more or less spasm of the sphincters and levatores ani. These symptoms are usually all much aggravated by passing a motion, the pain afterwards being very severe. In the great majority of patients suffering from these symptoms, a definite cause will be found in the presence of a small irritable ulcer or painful fissure; but in a few, no such obvious pathological condition is to be seen. If, however, the symptoms are really severe, the best treatment to adopt is forcible dilatation.

CHAPTER XXIX.

INJURIES OF RECTUM AND ANUS.

WOUNDS of the rectum may be the result of foreign bodies being forcibly thrust into or through the bowel; of injury inflicted during parturition; of perforation by fragments of bone or other hard substance which have passed through the alimentary tract; of gunshot injury; and of penetration by a fragment of broken pelvis; all of which would be classed under the head of lacerated and contused wounds. Cleanly incised wounds in this region are extremely unlikely to occur, except as the result of surgical operation, either intentionally done, as in the operation of fistula and other similar procedures; or unintentionally, as an accident of the operation of lateral lithotomy.

Direct injury.—A number of cases have been recorded in which an enema tube has been forced through the healthy rectum, and the injection discharged either into the peritonæal cavity, or into the loose areolar tissue surrounding the bowel. So slight is the sensibility of the rectum a little way from the anus, that this accident has happened while the patient was giving himself an injection. Where the perforation occurs into the peritonæal cavity death has in a number of cases resulted; and where the areolar spaces of the pelvis are penetrated, diffuse suppuration, which is likely to prove fatal, is a probable result. Similarly, foreign bodies are sometimes forcibly thrust through the anus, and perforate the rectal wall above. There was some time ago, under the care of my colleague, Professor Bennett, in Sir Patrick Dun's Hospital a man with a recto-urethral fistula,

B B—23

which resulted from his being impaled by a pitchfork. While sliding down a hayrick, one prong passed through the anus injuring the prostatic urethra while the other passed up between his legs without doing any injury. An almost identical case is recorded by the late Mr. Tufnell, in the proceedings of the Pathological Society of Dublin,* in which one prong of a pitchfork penetrated the rectum, dividing the prostate gland in two. The patient died on the fifth day. Mr. Prescott Hewett† published the particulars of a perforating wound of the rectum, from a person falling upon the leg of a chair which penetrated the anus. The result proved fatal. A similar case is related as occurring from a person falling on a stake fixed in the ground, in which the posterior wall of the bladder was much lacerated through the rectum. Recovery followed, urine being evacuated by the urethra after two months.‡ Sir William Stokes mentions a case § in which a lad fell on a long ironworker's tongs, one handle of which penetrated the anus, entered the bladder, and again perforating this viscus opened into the peritonæal cavity. He was brought to the Richmond Hospital in a state of profound collapse, and died shortly afterwards. These cases are interesting as showing how serious injuries might be inflicted upon the abdominal viscera without any cutaneous wound existing; and in the absence of any accurate history, considerable difficulty might be found in making a diagnosis. A case is recorded by Birkett, || in which a patient suffering from rectal ulcer, himself passed an enema syringe through the floor of the ulcer, opening up the loose areolar tissue of the pelvis. This was followed by emphysema of the

* Vol. iv. part 2.
† Trans. Path. Soc., London, vol. i. p. 152.
‡ Holmes' "System of Surgery," vol. ii. p. 722. Second edition.
§ Trans. Acad. Med., Ireland, vol. i. p. 88.
|| Holmes' "System of Surgery," loc. cit., p. 753.

entire lower part of the abdomen ; and as this symptom has occurred in other cases, such as the apparently trivial operation of tapping the bladder through the rectum,* it might prove useful in establishing a diagnosis in obscure cases.

There was a remarkable case of injury to the rectum in Sir Patrick Dun's Hospital under Dr. Bennett's care. The patient was caught in the perinæum by a rapidly moving chain, which inflicted a severe lacerated wound in the region of the anus. When admitted it was found that the lower end of the rectum was separated from its attachments round the greater part of its circumference, and projected at the bottom of a ragged cavity, the case exactly resembling the dissection made for the purpose of amputating the rectum.

During **parturition** injury to the rectum sometimes occurs, the recto-vaginal septum being torn through more or less, in addition to complete rupture of the perinæum ; or even the septum may give way without tear of the perinæum, the child being born through the anus.†

Injuries by substances swallowed.—Owing to the sudden way in which the rectal pouch narrows at the anus, substances, which have thus far safely made the tour of the alimentary tract, may become arrested, and possibly puncture the coats of the bowel and give rise to abscess and fistula. I removed the prong of a horn hair-comb which was transfixing the lower end of the rectum of a young man, and fish bones, pins, and such like substances are not unfrequently found in a like condition. They give rise to a considerable amount of discomfort and pain during defæcation, are easily recognised by digital examination and readily removed.

* *Lancet*, p. 89 ; 1860.
† Bushe, *loc. cit.*, p. 80.

Gunshot wounds.—Dr. J. Marion Sims has pointed out* that bullet wounds of the pelvis, even when passing through rectum, bladder, and peritonæum, are by no means absolutely fatal injuries; and he makes a very forcible contrast between these and perforating wounds of the abdominal cavity. Of seven of the former occurring within his knowledge at Sedan, all recovered, while seven of the latter all proved fatal. Of gunshot wounds of the rectum during the American War, Otis† collected 103 cases with a fatality of 42·7 per cent.; in 34 of these the bladder also was wounded, with a mortality of 41·17 per cent. Dr. Sims attributes the more favourable result in pelvic injuries to the fact of free drainage. When the abdominal cavity is perforated fluids collect in the pelvis, become septic, and lead to a fatal result; but where a man is shot through the pelvis efficient drainage is more likely to result.

Of the incised wounds resulting directly from operation it is not necessary to speak.

Accidental wound of the rectum during the performance of lithotomy does not generally prove troublesome, owing to the free drainage from the wound. In a few cases, however, recto-urethral fistulæ have been established.

A consideration of the histories of rectal injuries shows that where the wound is extensive, and consequently drainage free, the prognosis is tolerably good; whereas punctured wounds in which the conditions of-free drainage are not found are apt to be followed by putrid emphysema, extravasation of fæces, diffuse inflammation, and other more serious septic complications. Mr. Bryant records a case‡ in which a boy, aged twelve years, was impaled on some

---

* *British Medical Journal*, February 18, 1882.
† Ashhurst, "Encyclopædia of Surgery," vol. ii. p. 199.
‡ *Medical Times and Gazette*, May 25, 1878.

area railings which he was mounting when his foot slipped, the point of one of the rails entering immediately internal to the tuber ischii, and two inches from the anus it perforated the rectum twice, and injured the bladder. Mr. Bryant performed cystotomy in order to establish free drainage, and the patient made a good recovery.

I feel convinced that the proper treatment to adopt in cases of punctured wound of the rectum through the anus, especially if low down, would be to divide the sphincter and rectal wall freely up to the point of puncture. With a free vent thus established, neither fæces nor flatus can be forced into the areolar tissue or peritonæum; and thorough drainage is established. In other respects, wounds of the rectum present no peculiar features for treatment, except, perhaps, the restoration of the recto-vaginal septum, for the details of which the reader must be referred to the special gynæcological works.

CHAPTER XXX.

FOREIGN BODIES IN THE RECTUM.

Foreign bodies in the rectum.—Very various substances may reach the lower bowel, either by descending from other parts of the alimentary tract, or by being introduced through the anus.

Any concretions once started in the rectum can go on gradually increasing *in situ*. Foreign bodies, which pass through the entire alimentary tract, are necessarily somewhat limited in size, yet it is astonishing what large and irregularly shaped articles can pass the ileo-cæcal valve. In a case recorded by Pilcher,* a woman suffering from melancholia swallowed a number of nails, pebbles, etc., in the hope of committing suicide, and, subsequently, in the space of six weeks, passed by the rectum three hundred grammes weight of various substances, including nineteen large pointed nails, a screw seven centimetres long, numerous pieces of earthenware and glass, a piece of needle, two knitting needles, a piece of whalebone, etc.

Concretions found in the rectum may consist entirely of fæcal matter, sometimes formed round a small foreign body as a nucleus, such as a chicken bone; or the concretion may be formed of some unusual and indigestible substances swallowed, as the oat-hair concretions met with sometimes in Scotland; or as in a case recorded by Gross,† in which an enormous rectal concretion was composed largely of calcined magnesia.

Foreign bodies introduced through anus. —According to Captain Hamilton,‡ at Balasore on the Bay of Bengal, it was the custom of the people there every morning after defæcation to put into the rectum a dried clay plug, which remained in position till the next morning, when it was removed to allow the bowels to move, a fresh one being subsequently employed. The plugs were found scattered about the ground in great numbers, and proved a puzzle to visitors till the custom was explained to them.

In some countries the introduction of foreign bodies is still adopted as a means of punishment.

The rectum has occasionally been made a hiding place for jewels, etc.; and, according to Poulet,§

* *Lancet*, vol. i. p. 23; 1866.
† Vol. ii. p. 570.
‡ "A New Account of the East Indies," London, 1708
§ "Foreign Bodies in Surgery," vol. i. p. 222

convicts have been known to secrete instruments in the rectum for the purpose of assisting them in escaping from prison. He quotes a case reported by Closmadeuc, in which a young convict died of peritonitis, and at the post-mortem there was found in the transverse colon what is known as a *nécessaire*, which consists of a cylindrical box of wood, closed with a little wax; the pieces which it contains can be fitted one to another so as to serve as a file, saw, etc., the box serving for a handle. In the case alluded to the box was cylindroconical, contained thirty different pieces, and was covered with a piece of sheep's omentum: it was fourteen centimetres long, and weighed six hundred and fifty grammes. It had been habitually introduced by the largest part, so as to be expelled during defæcation. An interesting feature in this case is the way in which the foreign body found its way up to the transverse colon. This was probably due to the conical shape.

As a result of accident, foreign bodies have sometimes slipped through the anus, and they have also occasionally been introduced for the purpose of checking a diarrhœa. This latter cause is not unfrequently assigned by persons who are ashamed to admit the real cause, namely, some perverted sexual impulse. It is astonishing the variety of articles which have been found in the lower bowel. It would of course be useless to recount them all, but the following are a few of those collected by Poulet:* pieces of wood, snails, glass-bottles, crockery, stones, tin-covers, preserve jars, pieces of soap, fork, handle of shovel, pepper-box, pig's tail, upholsterer's nails, ale glass, bottle of mushrooms, etc. The length of some of the foreign bodies found here is considerable. Velpeau† removed an eau-de-cologne bottle over

* *Loc. cit.*
† *Arch. Gén. de Méd.*, 4th série, tome xxi.

eleven inches long; while Laure* records a case in which a wooden club was withdrawn from the rectum of a man aged sixty years, which measured twelve and a half inches; the upper end caused a projection in the right hypochondrium; the patient recovered. It is hard to imagine how such an article could be introduced so far without producing fatal mischief.

Foreign bodies introduced into the rectum may be expelled by natural effort, or they may be more or less tolerated for some time in the bowel, but usually such grave symptoms are soon developed that the surgeon's aid has to be called in. A case is reported by Dr. Weigand,† in which a piece of wood five inches long, the top of a pole which broke off in the rectum, was retained for thirty-one days, and then expelled by natural effort.

As a result of residence in the rectum a foreign body may become incrusted with earthy salts; as in a case recorded by Dr. Dahlenkampf,‡ in which a piece of oak was removed from the rectum which had been there for ten years. It was covered over half its surface with a saline incrustation of a brilliant silvery appearance, which analysis showed to be phosphate of lime. When the foreign body has perforated the bladder, and remains *in situ* for any length of time, the deposit of calculous matter is likely to take place.§

The removal of foreign bodies from the rectum may tax considerably the ingenuity of the surgeon. If the body is large and friable great care must be exercised. In the first place it is well to dilate the anus very thoroughly, when possibly the substance

---

* *Gazette Médicale de Lyon*, 1868.
† Schmidt's "Annalen," vol. iv. p. 95; 1862.
‡ *Heidelberg klin. Annalen*, 1829.
§ Camper, quoted by Poulet, *loc. cit.*, p. 220.

may be expelled naturally ; if not, its removal must be attempted by the aid of forceps, etc.

If the foreign body is a bottle, and the neck presents, a strong forceps may be passed through the neck, widely divaricated, and used as a means of traction. If, however, as is more likely to be the case, the bottom of the bottle presents, greater difficulty will be experienced. In one case* a champagne bottle was removed by perforating the bottom, and getting a wire hook in, a month after the foreign body had been introduced. Possibly the dental engine, armed with a corundum bit, might prove of service in similar cases.

Where other means fail the hand may be introduced into the rectum, and the substance removed by manipulation, the anus being enlarged, if necessary, by an incision back to the coccyx. In other cases, where also other means fail, the body may be broken up by a cephalotribe, or other similar instrument; such treatment is, however, exceedingly dangerous, the laceration of the gut having been, in some instances, extreme. Cripps† mentions a case from the practice of Burnett, in which a man introduced a jam-pot into his rectum ; finding that it obstructed a motion he knocked the bottom, which was uppermost, out with a poker ; this was followed by considerable hæmorrhage, and prolapse of the bowel through the opening ; the rest of the jam-pot was removed with very great difficulty by means of a craniotomy forceps six days later, and the patient eventually recovered.

In some cases the ingenious method of Lefort‡ may be had recourse to ; *i.e.* of injecting soft plaster of Paris into a hollow article, such as an ale glass, and allowing it to harden before attempting removal.

* Pollock, *Med. Press*, 1869.
† *Loc. cit.*, p. 267.
‡ Poulet, *loc. cit.*, p. 255.

Care must of course be taken, if an upper opening exists, that the plaster does not escape above, so forming a concretion.

Where the foreign body is wood, or substance of similar density, the use of a gimlet or screw to fix the object, and assist in its withdrawal, may prove of service.

Where the body has prongs, like a fork or a hook, portion of which has perforated the bowel, the perforating part is to be carefully withdrawn from the mucous membrane, and protected with a forceps, a tube, or piece of cork during extraction. The famous case in which a pig's tail was extracted by Marchettis,[*] illustrates the use of a little ingenuity in this way. A pig's tail, with the bristles cut short, had been introduced into the rectum of a woman. As soon as any attempt was made to pull it out the bristles caught in the mucous membrane. Marchettis selected a hollow reed, and, first tying a string to the end of the tail, he slipped the tube up, and thus extracted the foreign body without difficulty.

Where the foreign body is made of metal wire, it may be divided by means of a cutting pliers, and so rendered more easy of extraction. An electro-magnet has been used for the removal of iron nails.

In cases where the foreign body has got up out of reach into the sigmoid flexure, it may be only possible to remove it by abdominal section. In this way Réalli[†] removed a piece of wood, which was firmly impacted high up, and which resisted all attempts to extract it by the anus. Having opened the abdomen, he attempted to force it down from its position, but failing in this, incised the wall of the bowel, and successfully removed the foreign body. The patient recovered perfectly. In the "History of the American

---

[*] Poulet, *loc. cit.*, p. 260.
[†] *Gaz. Méd.*, July, 1851.

War,"* allusion is made to a case in which a stone introduced into the rectum perforated the peritonæal cavity, and was removed therefrom by abdominal section. A third case is detailed by Studsgaard, † in which a mushroom bottle, six and a half inches long, was introduced into the bowel. The introduction of the hand, after posterior division of the anus, failed in removal. Laparotomy was subsequently performed, and the bottle removed by incision of the great intestine, which was closed by Lembert's suture. The patient made a good recovery.

CHAPTER XXXI.

DIVERTICULA OF THE RECTUM.

Diverticula of the intestine have been divided by Rokitansky into the true and false: the former being congenital and consisting of the entire thickness of the intestinal wall; while the latter are probably acquired, and consist of hernia-like protrusions of the mucous membrane between the bundles of muscular fibres, in this respect resembling strongly the sacculation of the urinary bladder frequently found in obstructive disease of the urethra.

The **true diverticulum**, or that known as "Meckel's diverticulum," is found with tolerable frequency, and it unquestionably has its origin in connection with the development of the umbilical vesicle, and usually is attached to the ileum at from 18 to 24 inches from the ileo-cæcal valve; rarely it is attached much higher, even as far up as the jejunum.‡ The

* Vol. ii. p. 322.
† Soc. de Chirurgie, p. 662; 1878.
‡ Lancet, vol. i.; 1840.

only instance that I have heard of a true diverticulum being found in connection with the rectum is that recorded by Mr. J. W. Hulke, F.R.S.,* which he described as follows: "These parts were taken from the body of a woman, aged twenty-five, who died in Middlesex Hospital, of diffuse puriform œdema of the cellular layer between the abdominal muscles followed by peritonitis after colotomy. The lower end of the rectum is much contracted, indurated, and thickened. The natural structures are replaced by a dense fibroid tissue which has extended through the cellular layer between the vagina and rectum, and also replaces the loose fatty tissues of the ischio-rectal fossa. The skin around the anus is hard and fixed, and it is riddled with many fistulæ which above open into the gut at and above the narrowed part. This part of the rectum is crossed obliquely by a tough fibrous band which separates two channels: one the continuation of the rectum upwards, the other a lateral diverticulum large enough easily to admit the finger running parallel with the rectum, having a mucous lining and muscular coat. The portion preserved is several inches long, but its original length cannot be ascertained, as the free end was cut off inadvertently, the existence of the diverticulum not being known till afterwards. It is plainly no mere mechanical involution of the rectum, such as might be imagined to have occurred from the use of bougies, because she said when her case was taken that they had never been used; and when she came under my own care her sufferings were too great to allow even a digital examination without anæsthesia, so that the question of treatment by gradual dilatation was not even entertained." Mr. Hulke in expressing his belief that this was a true diverticulum, further says that the only other possible solution, that it is a portion

* Transactions of the Pathological Society of London, 1873, p. 87.

of the small intestine, which, after having become adherent to the rectum, has by ulceration opened into it, wholly fails to account for the relations of the true and adventitious tubes. I have reproduced this case nearly in full, as, notwithstanding the imperfect postmortem, there can be no doubt that it was an instance of true diverticulum of the rectum, and as such is of extreme interest from an embryological point of view. The existence of a true diverticulum of the intestine has only been noticed before in the common form attached to the ileum, where during development a canal actually existed. Whether this case could have been in any way associated with the presence of a post-anal gut* can, of course, in the absence of a complete post-mortem examination, only be surmised.

False diverticula are thus described by Rokitansky.† They consist solely of mucous membrane and peritonæum. They occur throughout the entire course of the small and large intestine. They are found in considerable numbers; they occur from the size of a pea to that of a walnut in the shape of round, baggy pouches of the mucous membrane. They form, more especially in the colon, nipple-shaped appendages, which occasionally are grouped together in bunches. When occurring in the small intestine they are commonly developed on its concave side, and are, therefore, placed between the layers of peritonæum. When in the colon the fæces are retained in them, and sometimes dry up into hard concretions. The specimen represented by Fig. 60 was taken from the body of a woman who died of strangulated hernia. The entire colon was closely studded with these little protrusions of mucous membrane, and they descended down into the rectum to

* *See* page 39.
† "Pathological Anatomy," Sydenham Society, vol. i. and ii. p. 48.

within two inches of the anal margin. Each of them contained a little hardened pellet of fæces. In none of them was there the slightest appearance of ulceration

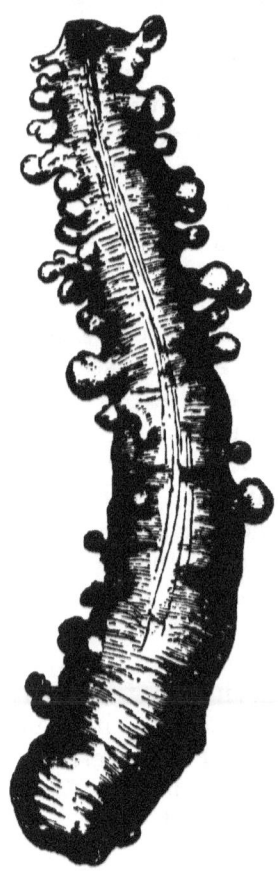

Fig. 60.—False Diverticula of Rectum.

or other trouble; and they were in no way associated with the hernia which had proved fatal. I have since seen two similar cases exhibited at the Dublin Biological Club, one by Dr. Lentaigne and the other by Dr. Bewley. As far as one can learn they are seldom productive of symptoms during life, although from their appearance ulceration and perforation similar to what takes place in the vermiform appendix would seem a probable result. Cases are, however, recorded by Mr. Sydney Jones and Mr. C. Hawkins, in which false diverticula of the sigmoid flexure became attached to, and finally opened into, the bladder.* Dr. Platt records a case † of a little girl, aged nine years, who suffered from symptoms of intestinal obstruction. Upon making a digital examination, a soft elastic swelling was felt in the anterior aspect of the rectum, and leading into this was an orifice which felt like the os uteri. The long tube could be passed up the rectum without difficulty. The child died, and it was found at the post-mortem

* Lond. Path. Soc. Trans., vol. x. pp. 131 and 208.
† *Lancet*, vol. i. p. 42; 1873.

that the small intestine was strictured ; and the swelling in the rectum is described as a hernia of the bowel beneath the peritonæum. It is probable that it was a case of false diverticulum.

The disease described by Physick under the name of "encysted rectum," and by Gross under the name of "sacciform disease of the anus," consists in an enlargement of the sinuses of Morgagni which are situated between the columns of Glisson at the margin of the anus, and protected by the anal valves. These, normally, are extremely small ; but as age advances they occasionally become considerably dilated, so as to admit the tip of the finger. They may ulcerate in consequence of their becoming receptacles for little masses of fæces. (*See* page 119.) This may be followed by suppuration, and may be attended with very severe pain. They can be diagnosed by being hooked up with a strabismus hook, or bent probe, and the treatment consists in snipping away the little anal valve in front of them, which will entirely obviate any danger of further accumulation. If they result in the formation of fistula or painful fissure, suitable treatment for these conditions will have to be undertaken.

INDEX.

Abraham, Dr. P. S., sarcoma, 342
Abscess, Ischio-rectal, 64
——, Marginal, 66
——, Stercoral, 61
——, Tubercular, 65
Adams, Dr. J. E., colotomy and excision, 373
Adelman, rupture of rectum, 208
Adeno-carcinoma, 333
Adenoma of rectum, 297
Ainsworth, double vaginal anus, 45
Albuminuria, in connection with rectal disease, 121
Allingham, colotomy, 380
——, introduction of hand, 19
——, speculum, 10
——, treatment of abscess, 64
Alveolar sarcoma, 343
Amussat, colotomy, 380
——, congenital malformation, 34
Amyloid degeneration of rectum, 197
Anal valves, 118, 136
Angioma, 318
Annular carcinoma, 331
Anus, Forcible dilatation of, 10
—— preternaturalis in ano, 107
Ashton, multiple anus, 41
Atony of rectum, 409
Atresia ani, 26
—— ——, operations, 29
—— —— urethralis, 42
—— —— vaginalis, 42
—— —— vesicalis, 42
Aveling, operation for vaginal anus, 47

Bacillus tuberculosis, 109
Bacterium coli commune, 289
Balfour, development of rectum, 21
Bamberger, follicular ulceration, 128
Banks, Mitchell, use of enemata, 264
Bardenheuer, excision of rectum, 375

Barduzzi, gummata, 193
Barker, Mr. A., angioma, 318
——, closure of colotomy wound, 402
Barnes, cure of polypus, 310
Bartels, intestinal ulceration, 121
Barton, operation for vaginal anus, 47
Bassereau, condylomata, 191
Batt, colotomy, 388
Bell, Sir C., reduction of prolapse, 209
Bellelli, bilharzia hæmatobia, 302
Benham, pile clamp, 279
Benign growths in rectum, 296
Bennett, Prof. E. H., recto-urethral fistula, 106, 417
Benton, fiddle-shaped bag, 291
Bewley, false diverticula, 430
Bilharzia, a cause of polypi, 302
Billroth, alveolar sarcoma, 343
——, excision of prolapse, 225
——, excision of rectum, 363
——, multiple polypi, 298
Birkett, injury of rectum, 418
Bodenhamer, classification of malformations, 22
——, colonoscope, 16
——, phleboliths, 243
Bonière, gonorrhœa of rectum, 54
Bose, lipoma, 315
Bougies, Conical, 183
——, Use of, 14
——, —— in stricture, 175
Bowlby, fibrous polypus. 300
Boyer, irritable ulcer, 142
Bremser, oxyurides, 405
Bristowe, Dr. J. S., spasmodic stricture, 148
Brodie, Sir B., fistula, 72
——, probe-pointed director, 87
Bryant, colotomy, 380
——, injuries of rectum, 420
Bumstead and Taylor, chancroid, 190

C C—23

Bursa fabricii, 22
Bushe, cause of ulceration, 115
———, congenital malformations, 34

Calcification of piles, 243
Callisen, colotomy, 380
Cancer of anus, 350
——— of rectum, 321
——— ———, cases of excision, 379
——— ———, complications of excision, 370
——— ———, Excision of, 360
——— ———, excision when high up, 373
——— ———, Extravasation of fæces in, 354
——— ——— in early life, 325
——— ———, linear proctotomy, 359
——— ———, medical treatment, 358
——— ———, method of suture in excision, 367
——— ———, modes of recurrence, 366
——— ———, Œdema of leg in, 354
——— ———, Pathology of, 325
——— ———, perinæal excision, 367
——— ———, removal by scoop, 359
——— ———, Statistics of, 321
——— ———, symptoms, 350
Cancerous obstruction, 353
Carbolic acid injection in piles, 271
Carcinoma myxomatodes, 338
Cardiac disease, Relation of, to piles, 251
Carrington, lympho-sarcoma, 348
Catarrhal proctitis, 50
Chancre of rectum, 191
Chancroid of rectum, 189
Chassaignac, excision of rectum, 363
Child-birth a cause of rectal ulceration, 115
Chronic proctitis, 52
Churchill, Dr. F., 45
Clover's crutch, 9
Coates, Dr., excision of piles, 280
Coccygodinia, 416
Coccyx, Removal of, 31
Cohnheim, bleeding in inflammation, 51
———, embryonic origin of tumours, 335
———, intestinal hæmorrhage, 250
Colectomy, 360
Colles, piles symptomatic of stricture, 174
Colley, Mr. Davies, delayed opening of colon, 392
Collins, statistics of colotomy, 22
Colloid cancer, 337

Colonoscope, Bodenhamer, 16
Colotomy, 380
———, Accidents during, 384
———, Amussat's method, 382
———, author's method, 394
———, Cases requiring, 381
——— combined with excision, 372
———, delayed opening of bowel, 392
———, Hernia at site of, 387
———, Littré's method, 387
———, Lumbar, 382
———, statistics, 387
Columnar piles, 247
Complete prolapse, 203
Compound external pile, 241
Condylomata, 191
Congenital malformations, 20
——— ———, statistics, 22
——— ———, Varieties of, 23
Constipation, Rectal, 409
Copeland, division of sphincter, 143
Cripps, Mr. Harrison, multiple polypi, 298
———, spasmodic stricture, 148
———, varieties of cancer, 331
Cruveilhier, colloid cancer, 337
Curling, excision of rectum, 363
———, colotomy, 380
———, congenital malformations, 34
Curschmann, irritable ulcer, 140
Cutaneous pile, 241
Cylindroma, 330
Cystoma of rectum, 317
Czerny, excision of cancer, 363

Dahlenkampf, foreign body in rectum, 424
Danzel, case of dermoid cyst, 314
Deep fistula, 91
De la Faye, sacral anus, 39
Dermoids of rectum, 312
Development of rectum, 20
Diagnosis of rectal disease, 1
Dickinson, intestinal ulceration, 121
Dieffenbach, artificial anus, 40
———, excision of rectum, 364
———, procto-plastic operation, 188
———, treatment of prolapse, 219
Digital examination, 4
Diphtheria of anus, 55
Discharge, a symptom of rectal disease, 3
Dittel, elastic ligature, 96
———, operation for recto-urethral fistula, 106
Diverticula of rectum, 427
Dolbeau, enchondroma, 317

INDEX. 435

Downes, Mr. E., pile clamp, 279
Dumarquay, division of sphincter, 143
Duncan, Dr., lupus of anus, 125
Dupuytren, operation for prolapse, 221
Dysenteric proctitis, 53
—— ulcer, 129
Dysentery a cause of stricture, 154

Ectropion recti, 51
Eczema marginatum, 404
—— of anus, 403
Elastic ligature, 96
Electric light for examination, 9
Electrolysis of piles, 274
Embryology of rectum, 20
Enchondroma, 317
Encysted rectum, 119
Epithelioma of anus, 330
Ergotin in prolapse, 213
Esmarch, case of lipoma, 315
——, classification of malformations, 22
——, large adenoma, 307
Esthiomène, 125
Ewer-shaped anus, 174
Examination of rectum, methods, 3
Excision of rectum, 360
——, combined with colotomy, 372
External piles. (*See* Piles.)
—— rectal sinus, 75
Exner, pathology of cicatrix, 293

Faget, excision of cancer, 360
Fenn, ergot in piles, 271
Ferrand, injection for prolapse, 214
Ferrous sulphate in piles, 266
Fibroma of rectum, 308
Fine, colotomy, 380
Finny, Prof. J. M., gangrenous abscess, 44
Fissure, 132
——, Ætiology of, 132
——, Modern pathology of, 136
——, relation to nerves, 133
——, Symptoms of, 138
——, Treatment, 142
Fistula bimucosa, 103
—— of rectum, 69
—— ——, Blind, 74
—— ——, Causes of, 70
—— ——, Complex, 76
—— ——, Diagnosis of, 78
—— ——, Division of, 86
—— ——, Histology of, 77

Fistula of rectum, Incomplete, 74
—— ——, —— treatment, 90
—— ——, Incontinence of fæces after, 95
—— ——, Paquelin cautery in, 100
—— ——, position of internal opening, 72
—— ——, recto-vesical, 103
—— ——, statistics, 69
—— ——, suture and immediate union, 100
—— ——, treatment, 82
—— ——, —— by ligature, 96
Follicular ulceration, 121
Forcible dilatation of anus, 10
Foreign bodies in rectum, 421
Fournier, chancroid, 189
Frankland, phleboliths, 243
Fruit stones, a cause of obstruction, 167

Gangrene of prolapse, 203
Gangrenous inflammation, 67
Gersung, sphincterplasty, 365
Givard, injection of piles, 271
Glairy mucus, Discharge of, 52
Glisson, Columns of, 114
Glycerine, Use of in piles, 265
Godard, Congenital malformations, 84
Godin, cancer in early life, 325
Goeschler, congenital malformations, 34
Gonorrhœa of rectum, 53
Goodsall, Dr., bougies, 15
Gosselin, ætiology of stricture, 153
Gowlland, cancer in early life, 325
Granular papilloma, 310
Greig Smith, treatment of distended gut, 401
Gross, sacciform rectum, 431
——, trichiasis recti, 95
Gummata, 193
Gunshot wounds of rectum, 420
Gymnastic exercise for piles, 267

Hæmorrhage from external piles, 242
—— from internal piles, 248
Hæmorrhoids. (*See* Piles.)
Hæmorrhoidal ulcer, 116
Hamilton, Captain, rectal plugs, 422
Hamilton, Dr. E., classification of piles, 247
Hartmann and Lieffring, bacteria in piles, 239
Haussmann, nature of cancer, 329

Hawkins, Mr. C., diverticula, 430
Hebra, eczema marginatum, 404
Hegar's retractor, 9
Heller, rectal parasites, 405
Heuck, excision of rectum, 363
Hewett, Mr. P., wounds of rectum, 418
Hilton, nerves of anus, 133
———, white line, 132
Hodgkins' disease, 343
Horner, muscle of mucosa, 199
Houston, nitric acid for piles, 268
———, strawberry pile, 248
Humphrey, Sir G., proctotomy, 185
Huguier, l'esthiomène, 125
Hulke, diverticulum of rectum, 428

Immediate union of fistula wound, 100
Impaction of fæces, 409
Imperforate anus, 24
Incontinence of fæces after fistula operations, 95
——— after excision of rectum, 365
Injuries of rectum, 386
Internal piles. (*See* Piles.)
——— rectal sinus, 75
Intussusception of rectum, 227
Invagination of rectum, 226
Irritable rectum, 415
——— ulcer, 132
Ischio-rectal abscess, 64
Itching of anus, 402

Jones, Mr. T., closure of colotomy wound, 386
———, Dr. Sydney, false diverticula, 430

Kelsey, bougies, 15
———, carbolic acid injection in piles, 271
———, classification of malformations, 22
———, injection for prolapse, 213
Kleberg, excision of prolapse, 222
Klebs, sputum a source of rectal tuberculosis, 123
Koch, tuberculosis, 109
Kolliker, post-anal gut, 39
Kraske, excision of rectum, 373

La Coste, sacral anus, 39
Laminar carcinoma, 331
Laparo-colotomy, 391

Laparotomy for foreign body, 427
Lee, Dr., clamp for piles, 251
———, rectal constipation, 413
Lebert, gonorrhœa of rectum, 54
Lefort, foreign body in rectum, 425
Legg, Dr. Wickham, obstruction by cherry stones, 170
Leisol, gummata, 193
Lentaigne, Dr., false diverticula, 430
Linear cauterisation, 218
——— proctotomy, 184
——— ——— in cancer, 359
Lipoma, 314
Lisfranc, excision of cancer, 360
Lister, Sir J., curetting of cancer, 360
Little, Dr. James, case of piles, 252
Littré, colotomy, 380
Luke, ligature of fistula, 96
Lumbar colotomy, 382
Lund's insufflator, 383
Lupoid ulcer, 125
Luschka, multiple polypi, 298
Lymphoma, 321
Lympho-sarcoma, 343

Macan, Dr., case of myoma, 320
McDonald, Dr. A., lupus, 125
McDonnell, Dr. R., pessary, 212
Madelung, operation for colotomy, 43
Maisonneuve, forcible dilatation, 10
Malassez, histology of stricture, 162
Malformations, Congenital, 20
Malignant neoplasms, 321
Manual examination of rectum, 16
Marchettis, foreign body in rectum, 426
Marginal abscess, 65
Marsh, Mr. H., angioma, 319
Marshall, colectomy, 360
Martin, operation for vesical anus, 43
Mason, ætiology of stricture, 153
Massage cadencé, 143
Melanotic sarcoma, 344
Mesenteron, 20
Metastatic abscess, 58
Mickulicz, excision of prolapse, 225
Mollière, classification of malformations, 22
———, fistula, varieties of, 71
———, gummata, 193
———, partial prolapse, 199
———, pruritus ani, 407

INDEX. 437

Morgagni, sinuses of, 118
Morgan, congenital malformations, 25
Mott, operation for prolapse, 221
Multiple anus, 41
—— polypi, 298
Myoma of rectum, 320

Nævoid pile, 248
Naphthaline in colotomy, 384
Nepven, melanotic sarcoma, 348
Neumann, chancroid, 190
Neurenteric canal, 39
Neuroses, Rectal, 406
Nicoladoni, excision of prolapse, 225
Nitric acid for piles, 268
Nussbaum, excision of rectum, 372
Nux vomica for prolapse, 213

O'Beirne, rectum empty in health, 366
Œdema of leg, a symptom of rectal disease, 2
Ointment introducer, 117
Ossifying cancer, 348
Oxyuris vermicularis, 405

Paget, Sir James, syphilitic growths, 192
Pain, a symptom of rectal diease, 2
Papendorf, classification of malformations, 22
Papilloma, 310
Paquelin, thermo-cautery, 215
Parasites of rectum, 405
Pars-caudalis intestini, 39
Partial prolapse, 198
Parturition, Injury of rectum in, 419
Patteson, case of polypus, 336
Péan and Malassez, chancroid, 189
Pediculi about anus, 406
Pelvic cellulitis, 58
Pelvi rectal space, 71
Periproctitis, 57
——, Circumscribed, 62
——, Septic, 57
——, ——, treatment, 60
Peritonitis, a sequence to stricture, 171
Pessary for prolapse, 212
Phleboliths, 243
Phthisis, Relation of, to fistula, 108
Phyma, 63
Physick, encysted rectum, 119
Pilcher foreign body in rectum, 422
Piles, 228
——, Ætiology of, 229

Piles, Cutaneons, 241
——, Excision of, 280
——, External, 239
——, ——, Hæmorrhage from, 242
——, ——, Inflammation of, 240
——, ——, Suppuration of, 242
——, ——, Treatment of, 244
——, Histology of, 237
——, Internal, 247
——, ——, Actual cautery in, 286
——, ——, Bleeding from, 248
——, ——, chemical caustics, 268
——, ——, Complications of, 251
——, ——, Crushing, 278
——, ——, Electrolysis of, 274
——, ——, Fissure a complication of, 260
——, ——, Forcible dilatation for, 273
——, ——, Injection of, 260
——, ——, Ligature of, 275
——, ——, mucous discharge, 250
——, ——, operative treatment, 268
——, ——, palliative treatment, 261
——, ——, polypus, a complication of, 260
——, ——, relief of congestion by operation, 295
——, ——, strangulation and gangrene, 256
——, ——, symptoms, 248
——, ——, use of purgatives, 263
——, pathology, 230
Pillore, colotomy, 380
Pollock, Mr. G., crushing piles, 279
Pollosson, colotomy and excision, 372
Polypus of rectum, 296
—— ——, Connection of, with cancer, 304
—— ——, Spontaneous separation of, 301
Port, case of dermoid, 312
Post-anal gut, 39
Pott, phyma, 63
Poulet, foreign bodies in rectum, 422
Pozzi, adenoma of rectum, 498
Proctitis, 49
——, Bloody discharge in, 51
——, Catarrhal, 50
——, Dysenteric, 53
——, Gonorrhœal, 53
——, Treatment of, 55
Proctodæum, 21
Procto-plastic operations, 188
Proctotomy, 184
Prolapsed hernia, 206
Prolapsus recti, 198

Prolapsus recti, complications, 206
—— ——, excision, 222
—— ——, linear cauterisation, 218
—— ——, Rupture of, 207
—— ——, treatment, 209
Protrusion, a symptom of rectal disease, 2
Pruritus ani, 402
—— ——, treatment, 407
Purser, secondary malignant growths, 331

Quain, arteries of rectum, 114
——, cancer in early life, 325
——, case of Talma, 164
Quénu, rupture of rectum, 207

Récamier, forcible dilatation of anus, 10
——, massage cadencé, 143
Rectal sinus, 75
Rectorraphy, Circular, 377
Recto-vesical fistula, 103
Reeves, ignipuncture, 290
——, operation for fistula, 100
Reflex pains in fissure, 135
Rhagades, 192
Ribes, fistula, position of, internal opening, 72
——, sinus of Morgagni, 118
Richardson, B. W., case of fistula, 355
Richet, hæmorrhoïdes blanches, 251
——, multiple polypi, 298
——, pelvi-rectal space, 71
Ricord, chancre of rectum, 191
Rizzoli, operation for vaginal anus, 47
Robert, case of lipoma, 317
——, operation for prolapse, 221
Rodent ulcer, 152
Roe, case of gangrenous piles, 259
Rokitansky, false diverticula, 429
Rollet, gonorrhœa of rectum, 54
Roser, ectropion recti, 51, 199
Rupture of rectum, Spontaneous, 207

Sacral anus, 38
—— excision of cancer, 373
Salmon, method of ligaturing piles, 276
——, treatment of fistula, 89
Saugalle, lipoma, 315
Sarcoma of rectum, 341
Scabies of anus, 406
Schœning, cancer in early life, 325

Scott, Prof. J. A., photographs, 304
Sentinel pile, 242
Sepsis, Prophylaxis of, 61
Serremone, irritable ulcer, 140
Simms, Dr. J. Marion, gunshot injuries, 420
Simon, introduction of hand into rectum, 16
Simpson, hernia at colotomy wound, 387
Smith, Mr. H., clamp for piles, 286
——, —— for prolapse, 216
——, Mr. Stephens, suture of fistula, 100
Spasmodic stricture, 148
Speculum, Rectal, 8
Sphincter, Forcible dilatation of, 10
Sphincterplasty, 365
Stenosis of anus, 23
Stercoraceous abscess, 61
Stokes, Sir William, injury of rectum, 418
Storer, Dr. H., method of examination, 13
Strawberry pile, Houston, 248
Stricture of anus, 188
—— of rectum, 148
—— ——, Annular, 159
—— ——, Colotomy in, 184
—— ——, Complications of, 163
—— ——, differential diagnosis, 179
—— ——, Electrolysis of, 187
—— ——, Excision of, 187
—— ——, Histology of, 162
—— ——, Proctotomy in, 184
—— ——, Symptoms of, 165
—— ——, Traumatic, 156
—— ——, treatment, 180
—— ——, Valvular, 159
Subcutaneous injection for prolapse, 213
Syme, fistula, 73
Symington, anal canal, 136
Symptoms of rectal disease, 1
Syphilis, Congenital, 194
—— of rectum and anus, 189
Syphilitic stricture, 153
—— ulcer, 189
Syphiloma, Ano-rectal, 194

Tape-like fæces, 167
Teratoma of rectum, 312
Thermo-cautery, 215
Thomson, Mr. W., colotomy, 384
Thrombosis of veins after rectal operations, 293
Traumatic stricture, 156
Trelat, syphilitic fistulæ, 195

INDEX.

Treves, melanotic sarcoma, 344
——, nature of cancer, 329
Trichiasis recti, 95
Todd's dilator, 184
Torments at anus, 136
Tubercular abscess, 65
—— ulceration, 123
Tuberculosis, Rectal, 109
Tuberous carcinoma, 331
Tufnell, recto-vesical fistulæ, 418

Ulceration of rectum, 113
—— ——, Classification of, 116
—— ——, Dysenteric, 129
—— ——, Follicular, 121
—— ——, Hæmorrhoidal, 116
—— ——, Irritable, 132
—— ——, Lupoid, 125
—— ——, Syphilitic, 189
—— ——, Traumatic, 115
Ureters, Involvement of in cancer, 355
—— opening into rectum, 49

Vaginal anus, 42
Van Buren, forcible dilatation, 16
——, chancroid, 190
——, linear cauterisation, 218
Vance, ulceration of rectum, 120
Velpeau, excision of rectum, 363
——, fistula, 72
Venous pile, 247
Vermiform appendix, Connection of to rectum, 108
Verneuil, gummata, 193
——, operation for malformation, 81
——, proctotomy, 185

Verneuil, veins of rectum, 230
Vicq d'Azyr, operation for vaginal anus, 46
Vidal, injection for prolapse, 213
Vienna paste, Use of, for piles, 269
Villous tumour of rectum, 310
Virchow, lipoma, 315
——, melanosis, 344
——, rectal sepsis, 59
Voillemier, congenital malformations, 34
Volkmann, excision of rectum, 362

Wagstaff, cancerous obstruction, 353
——, case of stricture, 165
——, ossifying cancer, 348
Walsham, introduction of hand into rectum, 17
Ward's paste, 266
Weigund, foreign body in rectum, 421
Weir-Mitchell, rectal constipation, 409
Whitehead, excision of piles, 281
Willems, sphincterplasty, 365
Williams, statistics of cancer, 323
Woillemier, linear cauterisation of piles, 289
Wounds of rectum, 417
Wright, Dr., case of fæcal extravasation, 411

Zappula gummata, 193
Ziegler, colloid cancer, 338
——, ulceration, 130
Zohrer, statistics of malformation, 22

PRINTED BY CASSELL & COMPANY, LIMITED, LA BELLE SAUVAGE, LONDON, E.C.

www.ingramcontent.com/pod-product-compliance
Lightning Source LLC
Chambersburg PA
CBHW022112300426
44117CB00007B/685